D1429892

SURPLUS - 1
LIBRARY OF CONGRESS
DUPLICATE

River Pollution: An Ecological Perspective

River Pollution:
An Ecological Perspective

River Pollution:
An Ecological Perspective

S.M. Haslam, M.A., Sc.D.
Botany School, University of Cambridge

Illustrated by Y. Bower

LIBRARY
LINCOLN MEMORIAL UNIVERSITY
Harrogate, Tennessee

220925

Belhaven Press
(a division of Pinter Publishers)
London, and New York

QH541.5
.S7
H37
1990

© Sylvia Haslam 1990

First published in Great Britain in 1990 by
Belhaven Press (a division of Pinter Publishers),
25 Floral Street, London WC2E 9DS and PO Box 157, Irvington, New York

Paperback edition reprinted in 1992

All rights reserved. No part of this publication may be
reproduced, stored in a retrieval system, or transmitted by any
other means without the prior permission of the copyright holder.
Please direct all enquiries to the publishers.

British Library Cataloguing in Publication Data
A CIP catalogue record for this book is available from the British Library

ISBN 1-85293-073-X
ISBN 1-85293-218-X (Pbk)

Library of Congress Cataloging in Publication Data
Haslam, S.M. (Sylvia Mary), 1934–
    River pollution: an ecological perspective / S.M. Haslam :
illustrated by Y. Bower. – 1st ed.
    p.   cm.
Includes bibliographical references.
ISBN 1-85293-073-X
1. Stream ecology. 2. Water – Pollution – Environmental aspects.
I. Title
QH541.5.S7H37   1990   90-326
574.5'26323-dc20   CIP

Filmset by Mayhew Typesetting, Bristol, England
Printed and bound in Great Britain by SRP Ltd, Exeter

To
Mrs Y. Bower and Mrs P.A. Wolseley

in appreciation of their expert help
and illustrations over the years.

We can command Nature only by obeying her.
Francis Bacon

Be not deceived: God is not mocked: for whatsoever a man soweth, that also shall he reap.
Galatians 6: 5

If you drive Nature out with a pitchfork, she will soon find a way back.
Horace

Contents

LIBRARY OF CONGRESS APR 1 2 1994 COPY COPYRIGHT OFFICE W

Preface

During this century, river pollution has greatly increased: increased in the numbers of rivers affected, though often lessened in the worst-affected rivers. The awareness of its damaging effects, to the rivers and the life they contain, to the seas to which the rivers flow, and on the people drinking the water, is spreading. It is to be hoped that this awareness will lead to the abatement of the nuisance.

This book is intended both for river specialists, and for undergraduates and those with a general interest in pollution. The general reader will, the writer hopes, find a readable text and decorative illustrations. The Tables, and the later parts of Chapter 4 and 5 can be omitted. The specialist will find much more detail in the Figures and Tables, which should be read in addition to the text. Most Figures need more attention than an equivalent area of text.

The text is biased towards vegetation. That is the speciality of the writer. More importantly, vegetation is the most obvious and the most easily interpretable of the living components of the river ecosystem. The pollution in most European lowland rivers can be assessed using this book.

Literature

References, including some not cited in the book, are listed at the end. However, special notice can be given to the following:

1. Zoology
Hynes, H.B.N. (1960). *The Biology of Polluted Waters*. Liverpool University Press.
   The classic book (for English readers) of animal behaviour in relation to pollution.
Hellawell, J.M. (1986). *Biological indicators of freshwater pollution and environmental management*. London: Elsevier.
   A most comprehensive source-book of the literature on the effects of pollutants. Diagnostic methods, and the effects of habitat disturbance.

Many tables in this text come from here, by kind permission of author and publisher. These are, though, a tiny selection of the numerous tables in that book. References for the tables used are in the original, but are omitted here. The reference list is very full.
Warren, C.E. (1971). *Biology and water pollution control*. Philadelphia: W.B. Saunders Co.
   An interesting book, mainly on animals, about the philosophy and effects of pollution, from an American standpoint.

2. Botany
Books by the writer. To keep the present book readable, repeated references to the writer's earlier work have been omitted. The evidence for the principles of plant behaviour not obvious from this book is given and discussed in these other books.
Haslam, S.M. (1978). *River Plants*. Cambridge: Cambridge University Press.
   Plant life in the river, vegetation in different river types, pollution. (Middle-level.)
Haslam, S.M. & Wolseley, P.A. (1981). *River vegetation: its identification, assessment and management*. Cambridge: Cambridge University Press.
   As indicated in the title – i.e. pollution assessment. For Britain. (Low to Middle-level.)
Haslam, S.M. (1987a). *River Plants of Western Europe*. Cambridge: Cambridge University Press.
   Comprehensive research on many aspects of river vegetation, including pollution. Full reference list. (High-level.)
Haslam, S.M. (1982b). *Vegetation in British Rivers*. London: Nature Conservancy Council.
   Hundreds of River maps, with brief text. Mainly 1970s. Provides a data base from which later trends and changes, as well as the then pollution, can be assessed.
Haslam, S.M., Sinker, C.A. & Wolseley, P.A. (1982). *British Water Plants*. Taunton: Field Studies Council, Nettlecombe, Williton, Taunton, Somerset. An extract from this, to identify common river species is in Appendix 7.

Identification for the general reader, and using vegetative parts. Nomenclature here follows this, except for *Scirpus* (formerly *Schoenoplectus*) *lacustris*.

Methods for the examination of waters and associated materials. *Method for the use of macrophytes for assessing water quality in rivers and lakes, (1986).* Her Majesty's Stationery Office, London.

One of the British Standard Methods Booklets.

3. General

Netherlands Central Bureau of Statistics. Environmental Statistics of The Netherlands. The Hague.

A marvellous compendium of pollution and ecological status. A condensed version in English is published alternate years.

Methods for the examination of waters and associated materials. London: Her Majesty's Stationery Office.

A series of specialist booklets describing Standard Methods for the assessment of anything from cadmium to chlorophyll. The biological methods are listed in Appendix 3.

Directives of the Commission of the European Communities.

These set legal standards for the Community. They are obtainable from the Commission's offices in each country. Some relevant titles are in Appendix 4.

Water Research Centre (1977). Water Purification in the EEC. Oxford: Pergamon Press, Commission of the European Communities.

Contents as its title suggests. Rather out-of-date, but not replaced.

The vegetation data and analyses in this book are from the writer's data-base of *c.*33000 field sites, unless cited from elsewhere.

The landscape strips at the top of some pages may be viewed as decoration, or studied for detail: including over-farmed, exhausted land; pollution; innumerable causes of river damage; the amount of care given to rivers in different situations, etc.

References in the text to Germany are to the Federal Republic.

Acknowledgements

This research was partly funded by the Commission of the European Communities (under Contracts 079-71-1 ENV UK, 105-76 ENV UK and ENV-710-UK(H)), the Department of the Environment, and other bodies, whose support is gratefully acknowledged. The help given by the University of Cambridge Botany School and Department of Applied Biology has been much appreciated.

The new illustrations were drawn by Mrs Y. Bower, and those reproduced from the writer's earlier books were drawn by Mrs Y. Bower and Mrs P.A. Wolseley, the latter endorsed by name. Most grateful thanks are due for these!

I am much indebted to Mr D.F. Westlake, Mr H.A. Hawkes and Mrs D. Phipps, who gave of their valuable time to review the botany, zoology and history respectively; sincere thanks. (They are not responsible for the writer's opinions or material).

The following most kindly helped in various ways: Mrs T. Bone (typing), Mrs R.A. Collyer (HMSO information), Mrs M.P. Everitt (research), Mrs S.M. Hornsey (inking).

Permission to use copyright material was very kindly given by Blackwells, Cambridge University Press, The Master and Fellows of St Catharine's College, Cambridge, University of Durham Botany Department, Elsevier, Professor A. Kohler, Liverpool University Press, Nature Conservancy Council, Severn-Trent Water, Sønderjyllands Amstkommune and the Field Studies Council.

July, 1989
S.M.H.

1
What is pollution?

A sudden little river crossed my path
  As unexpected as a serpent comes.
  No sluggish tide congenial to the glooms;
This, as it frothed by, might have been a bath
For the fiend's glowing hoof — to see the wrath
  Of its black eddy bespate with flakes and spumes.

So petty yet so spiteful! All along,
  Low scrubby alders kneeled down over it;
  Drenched willows flung them headlong in a fit
Of mute despair, a suicidal throng:
The river which had done them all the wrong,
  Whate'er that was, rolled by, deterred no whit.

Which, while I forded, — good saints, how I feared
  To set my foot upon a dead man's cheek,
  Each step, or feel the spear I thrust to seek
For hollows, tangled in his hair or beard!
  It may have been a water-rat I speared,
  But, ugh! it sounded like a baby's shriek.

Robert Browning
'Childe Roland to the Dark Tower Came'

In spite of the vivid description, this river is not too bad: the willows and alders are not being killed by the polluted water which drenches them, and water rats may be present. In the most grossly polluted rivers, no large plant or animal can survive. This book is concerned with the myriad streams which, like Browning's, can support some life, and which are small enough to be able to have large (macrophytic) vegetation. What happens to the unfortunate life they support? Where does the pollution come from? And how else does Man harm the rivers?

What are rivers?

Rivers and streams are courses of water flowing along a bed on the earth towards the sea. They are passages of aquatic life that cross the land, passages whose water connects in one direction

a)

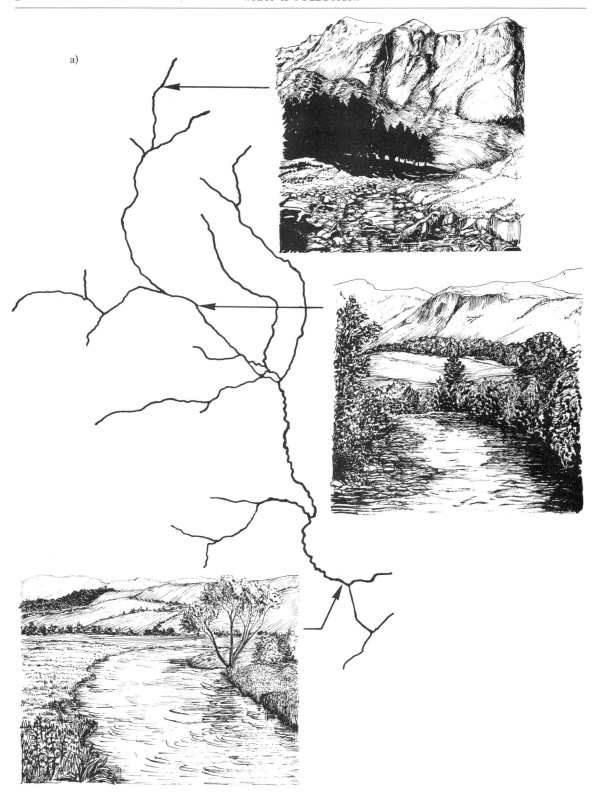

Figure 1.1 Mountain and lowland rivers. a) Mountain river on Resistant rock (hard: not susceptible to erosion or solution, e.g. granite, gneiss, slate, schist). b) Lowland river on clay

Resistant   Limestone   Sandstone   Clay

Lowland

Upland

Mountain

Alpine

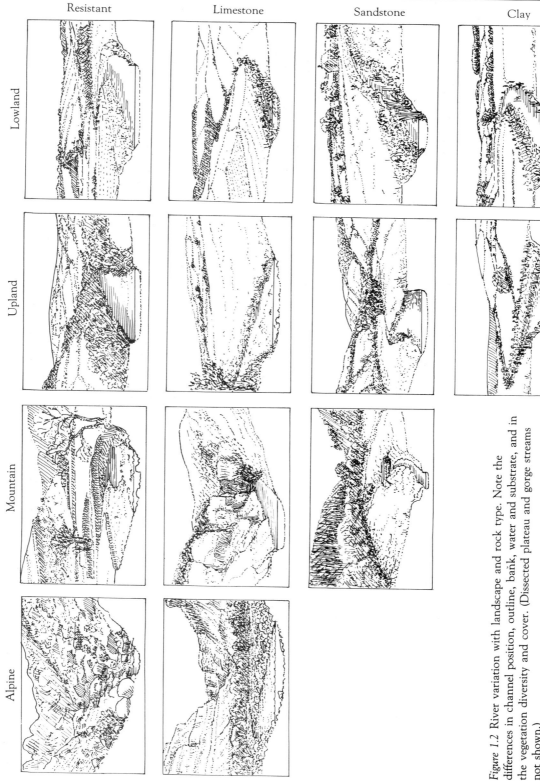

Figure 1.2 River variation with landscape and rock type. Note the differences in channel position, outline, bank, water and substrate, and in the vegetation diversity and cover. (Dissected plateau and gorge streams not shown.)

Source: P.A. Wolseley in Haslam, 1987a

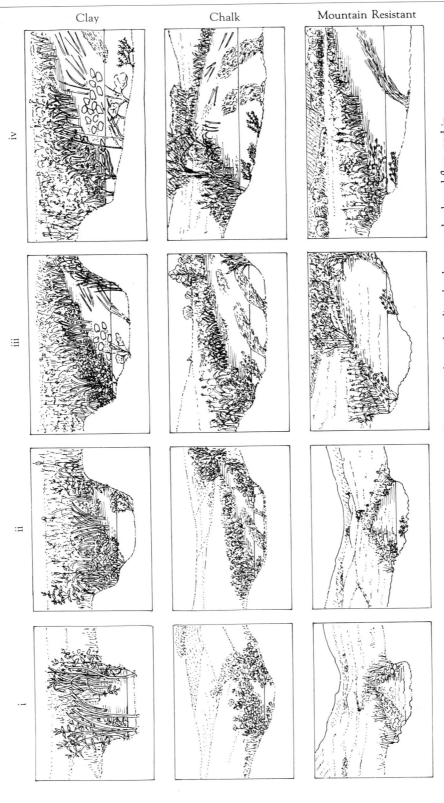

Figure 1.3 River variation with stream size and rock type. Note the differences in channel outline, bank, water depth and flow, and in the vegetation type, diversity and cover. Size i, up to 3 m wide, no water-supported species. Size ii, up to 3 m wide, at least one water-supported species. Size iii, 4–8 m wide. Size iv, 10+ m wide (in this book, mostly 10–25 m: very large rivers rarely included)

Source: P.A. Wolseley in Haslam, 1987a

only, downstream. They vary from source to mouth, from stream to stream, from country to country, from field to wood. Man, with his insatiable wish to classify, imposes arbitrary patterns on this natural continuum of water. For example, streams 4–8 m wide are labelled size iii, wider watercourses size iv. Call this a mountain stream, and that a lowland river.

Understanding has to start with small, discrete parts: but it is the total continuum which is the river. River classifications are numerous and are summarised in, for example, Haslam, 1987a. Most are based on animals (such as Persoone 1978; Huet, 1954; Jones & Peters, 1976) or topography and hydrology (Gessner, 1975; Haggett and Charley, 1972; Holmes, 1978; 1983), though a few are based on other characters, such as diatoms (for example, Descy, 1979). The (British) Freshwater Biological Association classification may prove useful for monitoring. Figures 1.1–1.3 illustrate stream variation and the classification used in this book (from Haslam, 1987 and Haslam and Wolseley, 1981). This classification uses the three following criteria:

1. **Water force** is the most important factor affecting river vegetation. It depends on the combined effect of landscape and precipitation regime through the year, and the interaction of precipitation and rock type. It is complex, but within one rock type and climatic region it is related to landscape, so rivers may be classed in terms of water force through landscape criteria (Figures 1.2 and 1.3).
2. **Rock type**, though of lesser importance, can be defined most easily and so, in Figure 1.2 is used as the first classificatory character. Rock type acts, most crucially, through river chemistry, but also through water force, in that hill shape and porosity influence water behaviour; and through substrate texture, in that erosion, solution, particle size and stability vary with rock type.
3. **Upstream-downstream** variation. Flowing downstream, a river generally becomes larger — though not in the even steps shown in Figure 1.3. Curiously, although, the categories shown in Figure 1.3 apply to vegetation patterns throughout west and central Europe, they do not in the far south.

What does Man do to rivers?

Man's actions are determined by his expediency. If it makes Man's life more convenient, less expensive or pleasanter, the river and its aquatic life will be sacrificed. Actions to benefit the river come only when its state displeases man: when it carries cholera or cadmium, when its ugliness offends, or when species or habitats he now thinks important are being lost. These are the basics of river damage and pollution. Verneaux (1976) says: 'The ultimate objective of human activity in the field of conservation management and utilization of aquatic resources is the preservation of as complete a biological structure as possible in terms of abundance and diversity'. Would that it were!

Man's activities are superimposed on the determining natural factors: rock type, water force (landscape) and upstream–downstream position (size). Man pollutes, but not only pollute. His overall effects are rarely negligible and may be severe enough to alter or, indeed, to override the effect of the natural determinands.

Man alters the size, shape, texture and position of the stream channel. He alters the drainage to the stream, the ground water level around the stream, and the quality, speed and turbulence of the flow within the stream. Such changes may be massive. Changes are made for flood protection and prevention; drainage and irrigation of farmland; using water supply for settlement and industry, including mines; waste disposal; navigation, commercial and recreational; power generation, formerly water mills, now electricity (including cooling water); recreation and amenity; fishing; and other less destructive pursuits. Man also alters the use of the land and so changes both the chemical and physical characteristics of the streams draining that land. Man changes the original vegetation, typically woodland and marsh, to grassland and crops. Forest removal may decrease precipitation and increase the speed of run-off (flash floods) and erosion, thus altering flow and increasing sedimentation in the streams. Lack of trees means lack of log-dams, which alter flow and sedimentation, loss of the tree leaf fall which alters river chemistry and animal food supply, and absence of shade, which encourages water plants. Reafforestation provides shade, leaf fall (sometimes toxic, see below), etc. Ploughing alters run-off and erosion; fertilisers and biocides are added to the run-off to the streams. The effect of Man's alterations is even greater when the vegetation is removed altogether and replaced by concrete! No rain can penetrate the ground: all runs swiftly to the stream; no soil chemicals or soil can be carried, only urban litter.

What is pollution?

The *Oxford English Dictionary* gives two definitions, of which the earlier is 'to render ceremonially or morally impure' and the later 'to make physically impure, foul or filthy'. For many years river pollution has meant only the latter. Very recently, however, overtones of the morally impure have appeared among those speaking of environmental pollution. This is excellent, since it could produce a climate in which pollution becomes unacceptable.

So what is pollution today? Definitions multiply! What Man wants from the river will determine how he defines pollution. Tendron and Ravera (1976) list its components as mechanical, organic, chemical, thermal and radioactive. This is unusually wide, and in this book mechanical effects are termed structural damage, and not pollution. Surveying definitions, Hellawell (1986), notes ones as diverse as 'something present in the wrong place, at the wrong time, and in the wrong quantity' (Holdgate, 1971) and 'that which is due to adding something which changes the natural quality of the water so the riparian owner does not receive the natural quality of the stream' (Wisdom, 1976). As Tendron and Ravera (1976) point out, definitions based on for example, what fish find — apart from what anglers think — are conspicuous by their rarity. Pollution here is entirely or partly chemical. It begins a series of changes moving a habitat further from what it would be without Man. The effects of pollution always give rise to changes, varying according to the intensity of the pollution, in the structure or nature of the plant or animal communities (from Descy, 1976e). Equivalently, pollution is shown by the effects of substances added or removed by Man (after Haslam, 1978a).

Substances entering streams enter a near-closed environment. Pollutants may move downstream with the water, or be deposited on the bed, accumulating particularly in silt, and only later being washed down. They may be chemically altered. Most substances leave the habitat (i.e. water and sediment) only by being washed to sea, dredged, or leaked from the bed, though a few may diffuse into the atmosphere. Toxins from towns, factories, farms, fields and mines thus accumulate and become concentrated in streams, polluting them and any larger rivers into which they flow. Within the river chemical (e.g. oxygenation) and biological (e.g. microbial) agents may render the pollutants harmless, but some substances are resistant to such processes even when in very low concentrations. Other substances are decomposed slowly so downstream purification is ineffective if concentrations are high or if further impurities are constantly added downstream, for example from a series of sewage works.

Polluted rivers come in more colours than there are colours of the rainbow. Most commonly waters occur in shades of grey and beige, but some may be quite clear: many toxins are neither coloured nor particulate. The writer has recorded black (N. England), white (Scotland, Italy), bright red (The Netherlands), china blue (Belgium), orange (Wales), pink (from sediment, Madeira), as well as innumerable dull and browny tones.

Is there natural pollution?

On the definition used here, no. There are, however, many habitats unsuitable for most aquatic life for reasons unconnected with Man. That which is natural is not necessarily wholesome. Radon gas, from the breakdown of uranium, seeps through the ground to springs. It occurs much more widely than was previously thought, and the radioactivity can be high and hazardous. Radioactive materials such as uranium are washed down from the land into rivers. Life of some kind occurs in hot springs, but only up to 94°C (chemobacteria). Animals rarely occur over 40°C, though a wide range of bacteria and blue-green algae grow over 45°C (Castenholz and Wickstrom, 1975). Sulphur springs have a very restricted ecology (of e.g. *Thiobacillus*), and likewise alkaline ones (of e.g. *Chloroflexus*). Acid springs may bear *Cyanidium calderium*, whose optimal pH is about 2.3 (Castenholz and Wickstrom, 1975). To say that life, mainly microbial life, occurs in such extreme conditions is not to say that ordinary river life can do so. It merely means that where equivalent extreme habitats are created by Man, there those same micro-organisms can potentially grow well.

Organic decomposition from natural causes can be damaging. There is no difference — except in size — between the effects on a stream of a dead sheep, a dead buffalo or a — small — uncontrolled slaughter house effluent. Oxygen deficiency may be caused by excessive leaf-fall as well as by sewage or weed cutting: all may kill fish, and to the fish it is irrelevant which.

What grows in rivers?

All the main groups of organisms and micro-organisms are represented in rivers (see also Chapters 3–5). In temperate rivers, mammals are sparse, but include beaver, otter, water-rat, water vole and coypu. Birds are more common and various, and include the waterfowl (e.g. mallard, coot) and the waterside birds (e.g. kingfisher, heron). Freshwater fish, of course, abound: both salmonids like salmon and trout, and coarse fish such as bream and roach. Reptiles and amphibia are fewer, but the next group down, the larger (macro-)invertebrates, are numerous, as are the protozoa, bacteria and other micro-organisms. The macrophytes, the larger plants, are mostly flowering plants (angiosperms) and, especially in mountains, mosses; but ferns, horsetails and liverworts do occur, and some larger algae are also included (stonewort and filamentous growths of, for example, *Enteromorpha* and *Cladophora*). Smaller algae are numerous: periphyton (around macrophytes) may be abundant, and this and the benthic (bottom-living) algae and the phytoplankton (free-floating algae) contain many diatoms. The other freshwater algal groups are also represented: green, blue-green, yellow-green and red algae.

Rivers ordinarily have a wide range of plants and animals. Each has its own habitat range of each factor — water force, rock type and upstream–downstream variation. Hence the presence of one species shows these specific conditions are present. Within the possible range of each critical component, each species has a band in which it can grow well, peripheral bands of both less and more of the factor in which development is poor, and outside these, bands in which it cannot occur at all. A group of species

therefore indicates much more than a single one, since the habitat must be suitable for each of these species, narrowing the possible ranges of, say, rock type or oxygen status.

When human interference alters river chemistry or temperature enough to alter the biota (the river life), pollution is present, whether or not the altered biota can occur naturally somewhere else. This is obvious in extremes. If the temperature of, say, the Seine at Paris were raised to 80°C, no one would say that because such temperatures occur in Iceland's springs the Seine had not been polluted. However, there is a school of thought which says that if, for example, a salmon river is altered to a coarse fish river, this is not pollution because coarse fish rivers occur widely. This reasoning is false. The river in question, if unpolluted, is a salmon river.

Rivers differ in their ability to absorb change. Adding a given amount of a plant nutrient may affect plants in a stream with a low natural level of that nutrient, therefore polluting the river, while having no effect on a stream already rich in that substance.

What causes pollution?

Any dissolved or particulate matter which Man can alter! (See Figures 1.4 and 1.5 for some of the ways pollution is caused.) The range of possible substances is immense, and classification is needed. No classification is comprehensive since the same pollutant may arise from different sources, and intermediates occur between categories of sources and of effects. When considering the possible range, it must be emphasized that substances for which no suitable tests are available are not therefore either absent or benign. Deaths occurred from heavy metals long before instrumentation permitted their detection. Mériaux (1982c) interestingly separates toxic substances which destroy biological structure and lead to its loss (for example, biocides, heavy metals, detergents) from excess nutrients, which upset ecological structure and lead to its simplification. Pollutions can also be divided on their intensity (effects gross to negligible), persistence (susceptibility to downstream purification), and continuity (sporadic, chronic).

Perhaps the easiest separation is by source. Here again it must be remembered that

Figure 1.4 Polluting objects. How many occur in your home and place of work?

Figure 1.5 Polluting processes

any chemical found in nature can theoretically be introduced to a river, by Man, as a pollutant. Man has also introduced new substances, of course. Consequently, when sodium chloride (common salt) enters a stream because it drains from salt-pans, or there is a high tide, and affects the ecology, the stream is not polluted. Yet the same effects from de-icing salt washing off roads, from worked salt pans or from sewage works, are pollution.

Domestic effluents

These contain sewage, varying from untreated to fully treated. Formerly this was almost all. Now they also include a wide range of chemicals, many highly toxic (what, after all, are disinfectants used for?). What happens to the medicine and food additives – including preservatives – we eat? More go to the river than the grave! Europe-wide, domestic effluents are the most damaging form of pollution. Tables 1.1 and 1.2 summarise the sewerage and treatment position of the then EEC countries in the mid-1970s. Sewerage, the provision of channels or pipes to remove dirt, has somewhat increased since. Farm effluents may also be organic and similar considerations apply.

Primary treatment is the settling and screening out of grit and other solids. Secondary treatment is by activated sludge or biological filters, breaking down a good deal of the organic material. The best biological treatment can produce water as good as that from tertiary treatment (90 per cent removal of organic matter is considered satisfactory — see Hynes, 1960). Tertiary treatment may be with micro-strainers, sand filters, or chemical or biological treatment in lagoons, etc. Treatment ultimately separates solids from the waters, which pass to watercourses. The Netherlands has the declared aim of producing such waste waters as will have little or no impact on the receiving waters. The solids, the sludge, have also to be disposed of, usually by incineration, as landfill, or (decreasingly) dumped at sea. None of these ways is ecologically satisfactory, and it is high time recovery techniques were used to recycle used materials. Water reused for domestic supply — or indeed needing treatment before being first used for domestic supply — may be treated by any or several of: storage (precaution against temporary gross pollution of surface waters), coagulation (of unwanted substances), sedimentation, or rapid

Table 1.1 Pollution due to sewage in the different EEC countries

	% of population with sewerage[a]	Effluent discharge mainly to
Belgium	30	Rivers
Britain	94[b]	Rivers
Denmark	84	Lakes, rivers
France	44	Rivers
Germany	79	Rivers
Ireland	61	Sea (rivers)
Italy	58	Rivers (sea)
Luxembourg	70	Rivers
The Netherlands	92	Drains, etc.

[a] Untreated sewage is more toxic than treated material.
[b] The quality of British effluent is lower than the high percentage of sewerage quoted would seem to indicate.

Source: Water Research Centre (1977, Table 1.1)

gravity filtration, usually through sand filters (to remove solids, ammonia and nitrate), absorption on carbon (usually to remove taste and odour, occasionally as emergency treatment), disinfection, usually with chlorine, and aeration, usually by cascade or spray (Water Research Centre, 1977).

Belgium and Italy have the worst pollution of those countries covered in Table 1.2, with the least secondary treatment and doubtful law enforcement. In Denmark and Ireland, the least sewage passes to rivers (it goes to the sea or lakes instead). France has a large rural population, low sewerage and particularly bad pollution of small streams. Britain has exceptionally high sewerage and secondary treatment, but is no more success-ful than is Germany, with — at that time — much less secondary treatment. In Germany, in contrast to France, severe pollution tends to be concentrated in large rivers. Britain, with equiva-lent total pollution, has its dirtiest rivers spread more evenly between stream sizes.

Noticeable pollution comes from using streams for waste disposal, increased by treatment plant inadequate for the density of population: inade-quate by original design, or by overloading a works which is too small. Secondary causes are untreated storm water, contamination of the source water, or abstraction of water from the river so decreasing the dilution of the waste water. In fact, where sewage is well treated,

Table 1.2 Comparison of sewage treatment in different countries.

	B	Br	D	F	G	Ir	It	L	N
% population with sewerage	30	94	84	44	79	61	58	70	92
% population with primary treatment only	1	<10	20	?	20	28	8	20	6
% population with primary and secondary treatment	6	>80	23	<40	44	4	7	70	50
Effective law enforcement	?	+	+	+	+	−	−	+	+
Population (millions)	9.5	55.5	5	51	61	3	55	0.3	13.5
Area (100 km^2)	30	242	43	550	250	69	335	2	33
Population: area (%)	32	23	12	9	32	4	16	15	41
Industry	+++	+++	+	+	+++	−	++	+	++

Abbreviations for countries: B, Belgium; Br, Britain and Northern Ireland; D, Denmark; F, France (including Corsica); G, Germany; Ir, Eire; It, Italy; L, Luxembourg; N, Netherlands.

Source: Water Resource Centre (1977)

untreated urban run-off may be the principal source of pollution. Treated sewage effluent is very different chemically from untreated but diluted effluent, and the effects on streams differ also (see Chapter 10).

It may seem incredible, but there are no reliable analyses of the total chemicals in any natural waters, let alone in the (more complex) effluents and few that account for the majority of the dissolved and suspended solids. If typical water analyses are examined, the tested substances (ammonia, chloride, etc. and perhaps a few heavy metals etc.) usually add up to 20–30 parts per million (ppm). The total dissolved substances, however, if measured, amount to perhaps 200–300 ppm, leaving the vast bulk of them unaccounted for — and then there is the composition of the suspended solids to consider. Humic (plant breakdown) substances are usually ignored and considered inert, but — as was predictable — it is now known they do influence processes in rivers — including toxicity and purification.

Humic acids are the most abundant macromolecular substances on earth, comprising 60–80 per cent of soil/organic matter, and thus constantly washing into rivers (Hayes, 1987). (They are various and large, with molecular weights from about 2000 to 1.5 million. Fulvic acids, another plant breakdown product, are also present and bond to them.)

The Water Research Centre (1984) lists several thousand polluting organic substances which have been found in rivers (including from industrial and farming sources, see below), in the following groups:

Organometallics
Mercaptans and
  miscellaneous sulphic
  compounds
Phenols
Quinones
Heterocyclics
Surfactants
Optical brighteners
Ethers
Aldehydes
Ketones
Aliphatic acids
Aromatic acids
Esters
Alcohols
Arylalkanes
Alkanes
Alkenes
Aminoacids and proteins

Carbohydrates
Steroids
Pigments, enzymes, vitamins,
  nucleosides and
  miscellaneous compounds
Alkyl-substituted polynuclear
  aromatic hydrocarbons
Aliphatic amines and
  derivatives
Cyanides and azo-
  compounds
Nitro- and nitroso-
  compounds
Organophosphorous
  compounds
Pesticides and herbicides
Aliphatic organohalogens
  excluding pesticides and
  herbicides.

(Far too many of these occur in drinking water — see for example, Fielding et al., 1981.)

To this list must be added all the inorganic solutes, nutrients, metals, salts of many kinds, etc. Analyses of a few river soils by capillary chromatograms with a little gas chromatography or mass spectrometry showed too many organic compounds to resolve, though the size of the major peaks was related to the vegetation-assessed pollution.

Since detergents are not natural river

chemicals, their occurrence can be used to indicate sewage content (unlike nutrients or simple organic compounds, whose origin is various). In Britain, a highly-sewered country (Table 1.1) a survey of 500 small streams showed detergents in nearly all lowland and many hill streams.

Physical, chemical and biological interactions in streams can alter the organics present in the environment, precipitating them, making them water-soluble, binding them to solids or even making them inert. Streams may act as vehicles for the transport and mobility of pollutants like biocides and oils (Choudry, 1983).

Faecal bacteria indicate organic effluent, domestic or farm. Ordinary biological treatment of sewage does not remove pathogens. *Escherischia coli* lives in water for longer, and is commoner, than most such bacteria, and its presence indicates that one or the other faecal organisms may also be present. These could include typhoid, cholera, bacillary and amoebic dysentery, polio and various parasitic worms (Hynes, 1960).

In addition to domestic effluent, there is also household rubbish. In remote parts of Europe rivers are often still the main means of disposal, but usually this is used as landfill or incinerated, etc. creating pollution likely, in due course, to reach the river (Figure 1.6).

Figure 1.6 The river as rubbish dump (Madeira)

Industrial effluents

The effluents are highly diversified and have little in common. Organic effluents will come from, for example, sugar beet, dairy and jam factories, and from oil and other petrochemical works. Inorganic effluents will result from, for example, steelworks, car, and other heavy industry; particles, dust and metals from mines and quarries; wash from gravel extractions. These are but a few of the many.

The industrial effluents passing through sewage treatment works are, of course, treated with the domestic effluent, which increases the total pollution from the works, as described above. Factories are, of course, distributed differently from houses. Nowadays, many are discharged direct to sewers, but some are still discharged direct to rivers, hopefully after local treatment.

Road run-off should perhaps be included here. Rubber, bitumen and other tyre derivatives, heavy metals, petrochemicals and other hydrocarbons from exhaust fumes, petrol and oil, glass,

aggregate, tarmac derivatives and particles, derivatives from shoes, de-icing salt in winter, dead leaves, dead animals and spills from any type of load may all be present. Road and urban run-off is far from clean! As Hynes (1960) points out, a gallon of oil can spread to cover four acres of water, and alter the surface tension and gas exchange of this, and hence damage the fauna.

Factory emissions to the air may also eventually reach, and influence, streams (see acidification section). Radioactivity may be emitted by nuclear power plants.

Excess — waste — heat is produced by industrial processes, particularly by power plants. A third of British power is generated along the River Trent, whose temperature is raised by 5°C in consequence, and the evaporation in the cooling towers could, at most, appreciably reduce flow (Lewis, 1981). The Sambre below Charleroi is raised by 9.3°C (Descy and Empain, 1984). Heat pollution also occurs when ground water which has previously fed a stream is abstracted and 'replaced' by run-off and sewage works effluent.

The ground water temperature is stabilised, that of the surface water is not. Warmer water alters the habitat response of organisms.

Farming

Even primitive farming by removing trees alters the physical and chemical states of rivers (see above and Chapter 11), so how much more does modern intensive farming! Ploughing, especially deep and first-time ploughing, alters the soil structure, releasing vast quantities of soluble nutrients (nitrates, etc.) into the streams. Additionally, fertilisers are added (for example, per hectare, 190 kg of nitrogen, 54 kg of phosphate (P_2O_5) and 52 kg of potash (K_2O)). It took several decades, but wash-out of these has now contaminated both ground water and river water. Biocides are many and various (examples include MCPA, a hormone weedkiller, and Isoprotoron, a residual weedkiller). Although, of course, they are applied in much lower quantities, they are active at much lower concentrations, and are now present in drinking water (see, for example, Fielding et al., 1981). Aquatic herbicides and, frequently, containers of land biocides are put straight into watercourses, and tractors and aeroplanes may spray into the water.

Sediment has been described as the greatest pollutant in the USA. The more the soil is bared (ploughing, recreation) and disturbed (deep ploughing, digging, etc.), the more soil is washed out. Cropping, tillage, burning, drainage, conifer afforestation, ditching, deforestation, recreation and industry all increase sediment yield. The sediment delivery to British streams has not been 'natural' for millennia. For example, the four-day turnip harvest yielded an extra 35 tons of sediment to the River Almond, Scotland. Extra sediment modifies channels and flow regime, substrate chemistry and texture, deposition, erosion and stability (Lewis 1981).

Organic effluents from farms, slurry, silage, etc, can cause serious pollution (see Chapter 11, and the discussion of domestic effluents above).

Eutrophication and acidification

These are natural processes which may be increased, modified or reversed by pollution.

Minerals and anions leached from the rocks and soils accumulate in lakes and lower reaches of rivers, raising the nutrient, or trophic, status. This eutrophication also comes from fertiliser, nutrient-rich effluents such as sewage, (mineral) sedimentation, and the removal of acid layers on the land (such as bog; see above and Chapter 11). It may also come from industrial emissions.

The reverse of eutrophication is acidification: the lowering of pH and the loss of nutrients. Two processes, often linked, but which may occur separately. Rain causes leaching and wash-out, so aids nutrient loss. Acid bogs, Calluna heaths, conifer forests, etc, and places where impeded drainage encourages nutrient-poor environments, give rise to acid and nutrient-low wash-out into rivers. (Heavy rain may cause flushes from bogs which may be acid enough to kill fish.) Nutrient-poor rocks and soils with low buffering capacity to neutralise acids favour acidification, as do any processes removing basic ions, including the loss of broad-leaf forest. Downstream eutrophication, as described above, necessarily means upstream loss of nutrients.

In more recent times, factory emissions have, in some parts, greatly accelerated the process. In Scandinavia, fisheries loss was noted first in the 1920s and became severe in the 1950s (Muniz, 1984). In North-West England, however, pH and alkalinity has remained stable for half and perhaps all, a century, perhaps longer (Sutcliffe and Carrick, 1986); it may be that the maximum emissions reached this area from the nearby industrialization in the early nineteenth century. Sulphur and nitrogen compounds are the main components (the form of the nitrogen compound, of course, determines the nature of its effect). In addition, hydrocarbons, ozone, hydroxides and peroxides may be present, and, near factories with other toxic fumes, those other compounds also. The effect of 'acid rain' is, consequently, more damaging than would occur solely from a change in pH (Edwards et al., 1987; Ineson, 1986). As much may be deposited as solids and mist as in rain. (That deposited on the land may be filtered and decreased before reaching the rivers.)

Discussion

It is rare indeed for a river to receive a single pollutant. The plants and animals are normally affected by a mix of substances, making the water different in both content and balance to its natural state. Pollution research tends to divide into two types. Laboratory work can be done where organisms are grown in artificial surroundings and where there is a known concentration of, usually, a single pollutant, or where organisms from the rivers are analysed for one (or a few) pollutants. Such studies give clear numbers, but are often doubtfully related to field conditions. In contrast, river research looks at the effect of a mix of pollutants arising from, say, two settlements with septic tanks draining to the river, four with sewage treatment works of varying efficiency, one containing an electrical goods factory, two factories discharging direct to the river, one petrochemical, one jam, five pig units, three cow units, and the usual fertilised and biocide-treated fields. These studies are necessarily inexact as to chemicals present. Acrimony is apt to develop between the groups involved in each type of research, one insisting that known numbers alone are accurate, the other that the important thing is that harm is being done. Pollutants may interact both with natural river substances, (humic acids, see above), and with each other. In addition, their effect on organisms may be altered by the presence of other toxins. Hellawell (1986) comments that the effect of mixtures of pesticides may be more poisonous than predicted from the known effects of each. Levels of nutrients below sewage works which are individually harmless are, combined, toxic to vegetation (Haslam, 1987a). For separate, or combined pollutants, toxicity may vary with temperature, with pH, with oxygenation and other characters of the waters. Heavy metals, for instance, are less toxic in harder waters.

After a pollutant enters a river, it becomes diluted downstream. This does not remove the substance! However, poisonous substances may be rendered less harmful, or even harmless. They may be oxidised (particularly organics), bound (adsorption, chelation) or precipitated (often metals). Suspended solids settle out and so, within the water, decrease downstream. Solids build up near their entry, but are gradually washed downstream (Hynes, 1960). As sewage is the principal river pollutant, the organic recovery pattern is one of the most important (Figure 4.15).

2
Pollution, ancient and modern

No real judgement should be passed on anti-pollution measures until they have been at work about a century.

(Hynes, 1960).

Discharging of all sorts of highly obnoxious matter into our waters is an affront to civilised values and damaging to our culture.

(Warren, 1971).

Few sights are more displeasing than the black and greasy waters and the scum-rimmed stagnant banks, with an occasional dead fish floating dully on the surface, of a polluted river, near the outskirts of a crowded town. The sun and the wind only intensify the stench, the houses nearby become unhealthy and unpopular, and there is a general air of depreca-tion and dejection. Even the cattle suffer from the impurity of their drinking water.

(Rodgers, 1947-8).

Since pollution is due to Man, it has been on earth as long as Man. It has, however, increased, gathering potentially devastating momentum over the past two centuries: increased, that is, in its total impact on the planet — it has always been gross in some of the places where men have gathered together. It is the scale which has now changed.

In quantity and quality, Man's waste products are in proportion to his numbers and his technology. The environmental impact depends also on his distribution: the waste from one can be absorbed where that from a million cannot. The population of Britain is estimated to have been about a million in Saxon times, the same again in Norman times, and, despite the loss of a third in the Black Death of the fourteenth century, to have risen to 5 million in Tudor times. Since then the rise has been rapid: 11 million in 1801, 16.5 million in 1831, 40 million before the mid-twentieth century, and 60 million only forty years later. (Between 1600 and 1934 the population of Europe is estimated to have increased from 100 million to 525 million.)

Technology proceeds erratically, however, in contrast to this consistent increase in population. The comforts of life found in Roman civilization were not equalled, let alone exceeded, until the twentieth century.

Domestic pollution

In the Classical world, cities could be large and fair. In Antioch in the first century AD, for instance, running water reached all houses, so that public fountains were for ornament only (so wrote Libianus). They had washing machines, automatic dispensers of temple holy water, water sprinklers to reduce fires, and, of course, pleasure gardens and swimming pools (Morton, 1936). The Roman world in general was most competent at bringing fresh water to cities. Where, as in many cities in Asia Minor, long aqueducts were essential, the cities died under barbarian invasion. Where the aqueducts were merely desirable, as in Lyon, the city survived.

Though we associate water supply and waste disposal, this is a recent development. Although Eshnuro, in 2500 BC, had brick-lined sewage pits in kitchens etc., with outlets to covered street drains (Warren, 1971), the foul drainage of cities like Rome was foul in both senses: used for waste, and disgusting. Such sewers were also inadequate. Indeed, sewerage as now understood is a nineteenth- and twentieth-century practice, and an incomplete one at that (Table 1.1). Older sewers were designed for storm and drainage water, and it was only when population, and hence waste, became excessive that practice changed. Human waste was forbidden to be discharged to sewers until 1815 in London, 1823 in Boston, and 1880 in Paris. In 1847 the cesspools of London houses were allowed to be connected to sewers. In the 1850s, two American towns had a water carriage system, bringing fresh, and disposing of waste, water (Warren, 1971). It took even longer for industrial discharge to rivers to become legal. The 1847 (British) Gas Works Act prohibited effluent discharge, and it was not until 1937 that British industry had the right, under certain conditions, to discharge to sewers (Hynes, 1960).

Until very recently those dealing with nuisance from waste were primarily concerned with solid waste. The collectors of night soil have been important in cities for most of history. Country practices varied. Manure is used in farming and before household chemicals became abundant human manure was more acceptable. Indeed, China had long-term waste disposal on land (Warren, 1971).

Small settlements could be hygienically designed. St. Bernard of Clairvaux designed monasteries to have incoming water first for power (for flour milling, with its negligible pollution), drinking supply and fish ponds, the stream then dividing to form the cloister general supply and drainage and a larger stream used for any dirty industrial processes, such as tanning. A sixteenth-century German village is shown in Figure 2.1. The drinking water is clean, from a well. In the stream, pollution (sheepskins) is downstream of fishing (poles for nets). A little sediment pollution will come from the farming. Welsh farmhouses may be on streams, with intake above, privy below, and a waterwheel for power. These simple principles did not apply in larger communities. Privies, let alone earth closets and cesspits, could be deplorably inadequate. The acceptance of unhygienic practices seems incredible, which highlights the contribution to Britain by Beau Brummel (personal cleanliness), Florence Nightingale (personal, house, town, army and hospital cleanliness), Lister (antiseptic cleanliness), and the countless thousands who cleaned the nation in the face of vilification from those objecting to change, particularly to costly change.

The medieval street in Figure 2.2 has an open sewer: a state of affairs lasting all too long, with refuse poured on to the street from all dwellings and workshops. Cleaner water for drinking and washing is obtained from upstream. Perchance these towns, with central flowing gutters, were the lucky ones. Others, like Edinburgh, could have refuse piling ever higher on the streets.

Rules and regulations were numerous, and ineffective in the long term, if not in the short. For instance, in 1372 a writ to the mayor and corporation of the City of London forbade the casting of dirty rushes, dung and refuse into the Thames, and in 1553 Henry VIII again prohibited the practice (Rodgers, 1947–8). There was a general opinion also that, if it cannot be seen, it is not there, i.e. that only solids caused pollution. A village stream had complex rules

Figure 2.1 Sixteenth-century village, Germany (C. Pencz). Drinking water comes from the well. In the river, the fish nets are upstream of pollution from the hides. Note stream type, banks, and the proximity of ploughing to the river.

Figure 2.2 Medieval city with stream, open sewer, along the high street

permitting discharge of cesspits and gutters by
night only (8 p.m. to 4 a.m.) (Parker, 1975) which
assumes that the water was clean again by 5 a.m.
By 1846 a traveller to Italy could report that the
medical people attributed the black hollow cheeks,
sunken eyes and general ill health of the Floren-
tines to the water being impregnated with lead.

Cambridge (Greenhaugh, 1980) may be cited to
show the historic pattern. In 1215, the King's Ditch
(named for King John) was an open gutter-channel,
no doubt the latest modern improvement of the
day. It reached the Cam downstream of the city
centre – but passing Slaughterhouse Lane early in
its course, at least by the sixteenth-century. Later
that century the Grey Friars built a new conduit to
bring clean water, selling the old one to the town.
In 1614 Hobson's Conduit brought (again) clean
drinking water to the town centre, and runnels
cleansing the main streets. From 1610 to c.1800 the
sewers, the foul drainage channels, were gradually
covered over, bringing fresher smells. Bureaucratic
delays in improving the environment are not confined to the twentieth century! The first sewage
works was opened in 1895, though storm and
surface water drained directly to the river until the
1960s. The fine University of Cambridge has for
centuries boasted of its fair river, and organised
boat races — downstream of the town. Considering
the repeated nineteenth-century complaints about
the intolerable nuisance and foul state of the river

— the undergraduates no doubt repeatedly falling into this sewage — the boasts seem surprising. Dredging (as occurred, for example, in 1869) would decrease the nuisance for perhaps two decades. Here sewage, solid and liquid, refuse and all other garbage reached the river. The earlier system of division has now partly returned: rubbish is collected separately, sewage separated into solid and liquid, with only the latter entering the Cam. In the nineteenth-century all rotted slowly, unpleasantly and dirtily in the water. As late as 1887, river keepers could be employed full-time removing dead animals from the river. It was not that 1887 had innumerably more dead animals than in 1987, it was the attitude: I don't want it, throw it in the river.

It was not just animal corpses which reached rivers. Dickens, in *Our Mutual Friend*, written in 1865, could describe those who made their living from human corpses in the Thames, and George Eliot's books record drowning as a normal way of death. Suicide, murder and accident are still with us. It is just the cultural attitude to the river which has changed.

'Drowned puppies, stinking sprats, all drenched in mud
Dead cats, and turnip tops, come tumbling down the flood.'

Dean Swift

Cambridge is a small, non-industrial town, not poor, with well-meaning local authorities. What, therefore, happened to towns with none of these advantages?

Industrial as well as domestic filth were at their worst during the peak years of the Industrial Revolution (earlier, therefore, in England than elsewhere) when town populations exploded, new industries sprang up, and when local bureaucracy had neither the understanding nor the will to cope (see Figures 2.3 and 2.4). The Mersey of these times was described as follows:

'Dank and foul, dank and foul
By the smoky town with its murky cowl
Foul and dank, foul and dank
By wharf and river and slimy bank.
Danker and darker the further I go
Baser and baser the richer I grow.'

Charles Kingsley

Wolverhampton and Sheffield were described as:

'Grim Wolverhampton lights her smouldering fires
And Sheffield, smoke involved: dim where she stands
Circled by lofty mountains, which condense
Her dark and spiral wreaths to drizzling rains
Frequent and sullied.'

Anna Seward

How and why did these appalling conditions occur? Having sewers to remove dirt is praiseworthy. This aid to town cleanliness is almost new in European towns. Sewers, however, do not clean. Septic tanks, cesspits and sewage treatment works clean to varying degrees. One reason for the recent worsening of EEC pollution is the provision of sewerage without enough accompanying increases in purification. The river suffers.

The inertia of local authorities is seldom more evident than in matters concerning cleanliness and water. The squalor and dirt in Figure 2.4, (Victorian) is even worse than in Figure 2.2 (medieval). While everyone, including the rich, lived in the mess of Figure 2.2, in Figure 2.4 it is but the poor. Except that these slums would often be beside the dwellings of the Great, and not only smells can pass round corners. Diseases do, too. Epidemics of water-borne diseases abounded in the nineteenth-century. In the 1832 cholera outbreak in Ely, the upper social strata, including the upper working class, were largely immune: their water came from well pumps. The unfortunate rest obtained their water from the Great Ouse (Holmes, 1974): even at a later date, this river was considered to carry the equivalent of 10 per cent of London sewage (Miller and Skertchley, 1878). Public health is cleanliness for all.

Dirt came through the inevitable — and perpetual — difficulty of disposing of waste, and plain dirtiness. Brooms have always been available!

Panic is the most efficient spur to action. With the gradual and unwilling acceptance that dirt means disease, and that disease in the poor, who do not pay taxes, spreads to the rich, who do, improvements spread. The main decrease in mortality and increase in health in the last two hundred years is marginally due to advances in medical research, but mainly to improvements in cleanliness of water, home and town. Life expectancy at birth, in the worst industrial areas of Nottingham, was barely thirty in 1820; in Britain as a whole it was fifty (average for men and

Figure 2.3 Waterfront from early in the Industrial Revolution

women) in 1900, rising to seventy-four in 1988. Those who rail at the inexcusable delays in removing nitrates and biocides from our drinking water should remember it took two hundred years to cover the sewers of Cambridge. The time-scale is, and was, long, Europe-wide. The blindness of our ancestors to contamination by dirt is inexplicable — as inexplicable as our blindness to the current dangerous pollutants will seem to our descendants.

The Royal Commission on Sewage Disposal, set up in 1898, and producing ten Reports, began the modern attitude to river pollution in Britain. This set a standard — copied intermittently in many countries — the Royal Commission 20/30 standard. That an effluent diluted at least eight times in a river should have a Biological Oxygen Demand (BOD), dissolved oxygen absorbed in five days, not over 20 parts per million, and suspended solids, not over 30 parts per million.

The five river classes were very clean, clean, fairly clean, doubtful and bad. Other standards are numerous, often more appropriate to later effluents of more complex chemistry than just sewage, but so far, more transient.

Oxygen demand is not confined to sewage. Table 2.1 illustrates the range, and Table 2.2 some of the 'germs' occurring with sewage.

The electrical generation must remember that until recently all households had open fires. Papers and burnable refuse cannot be put into the oil central heating system or the microwave cooker. More refuse has to leave the house — and much more now enters in the form of packaging, mail, newspapers, etc.

Changing times mean changing waste. Treated sewage has less organic solids, and more semi-degradable substances, such as nitrates. Instead of dead dogs there are detergents, bleach and

Figure 2.4 Nineteenth-century overcrowding. Dirtier than Figure 2.2

petrochemicals. Bridges, once built, cause no pollution. The numerous fords they replaced had slight, but definite, pollution. For a century liquor from coal gas works was a major concern: now gas is piped to Britain from the North Sea. Obviously the effects on ecology differ. Pollutants vary in their biodegradability. Sewage degrades easily, for instance: the trouble comes from its quantity, not its quality. In contrast, fossil fuels are very resistant to bacterial decomposition and also coat substrates and render them unsuitable for many organisms (see, for example, Warren, 1971). Some past influences can be elucidated, since, for instance, septic tanks are still common. Large quantities of cabbage leaves and dead pigs, however, are not. Even so, different regions of different countries show different ecological effects in their rivers (see Chapter 10).

Farming

The deliberate husbandry of plants and animals for food, clothes and other purposes is, of course, a very ancient practice, necessarily influencing most European rivers, often to an unrecognisable extent (see also Chapters 1 and 11). Farming patterns have varied with period, with culture, with topography, rock type, soil type, and climate, and with technology. These stamped the landscape of Europe, making German fields different to French, Danish to Italian and indeed, Kent to Sussex (adjoining English counties), Jutland to Sjælland (Denmark) and Mallorca to Minorca (Balearics). The character of the rural land is its farming, past — unless recently obliterated — and present. Thus the character of

Table 2.1 Approximate biochemical oxygen demand (BOD) of typical effluents and freshwaters

	Range of BOD (mg l^{-1})
Mountain streams	0.5–2.0
Lowland brooks	2.0–5.0
Large lowland rivers	3.0–7.0
Sewage effluents, crude sewage	200–800
treated	3–50
Farm wastes, pig	27 000–33 000
poultry	24 000–67 000
Silage liquor	60 000
Abbatoir	650–2 200
Meat packaging and processing	200–3 000
Fruit canning	635–2 100
Vegetable processing	480–4 400
Sugar beet	3 800–4 200
Sugar refining	210–1 700
Dairies, milk	300–2 000
cheese	1 800–2 000
Breweries	500–1 300
Distilleries	over 5 000
Tannery	250–5 000
Textile waste	50–1 000
Paper making	100–400
Petrochemicals	200–8 000

Source: adapted from Hellawell (1986)

the streams draining that land is determined by its management.

Farming involves forest clearance, marsh drainage, grassland establishment, conversion to arable land, and drainage and irrigation as required. The agricultural revolution of the late Dark Ages had a profound effect on farming, hence also on run-off to rivers. This was the invention of the hard collar and the deeper plough, allowing farming on clay and marshland. In Britain, under-drainage of fields spread from the mid-nineteenth century. In parts, as on much of the French chalk, even the larger tributaries are now underground. Such alterations change the chemical as well as physical characters of the water — and when a stream is no longer there, its flora and fauna are lost.

By altering run-off of rain, the wash-off of sediment to the river is altered also. Farming practices tend to speed run-off and remove the protective cover on the land. In America, where population is low and farming lavish, pollution by sediment has, indeed, been cited as the greatest pollution

Table 2.2 Pathogenic organisms known to occur in sewage effluents or associated with sewage treatment

Organism	Disease or condition	Comments
Viruses		
Polio virus	Poliomyelitis	Found in effluents, but not proven to be water-borne transmission
Infectious hepatitis virus	Infectious hepatitis	Only virus for which water route has been proven, epidemiologically
Bacteria		
Salmonella typhi	Typhoid fever ⎫	Common in sewage and effluents
Salmonella paratyphi	Paratyphoid fever ⎬	in epidemics
Shigella spp.	Bacterial dysentery	Source of infection, mainly polluted water
Bacillus anthracis	Anthrax	Spores resistant
Brucellosis spp.	Contagious abortion in livestock, undulant or Malta fever in man	Infection normally from contact or infected milk but sewage suspected also
Mycobacterium tuberculosis	Tuberculosis	Isolated from sewage? Possible mode of transmission
Vibrio cholerae	Cholera	Transmission by polluted water
Leptospira icterohaemorrhagiae	Leptospirosis, Leptospiral jaundice (Weil's Disease)	Carried by rats in sewers
Protozoa		
Entamoeba hystolytica	Amoebic dysentery	Contaminated water, tropical countries
Metazoa		
Schistosoma spp.	Bilharzia ⎫	
Taenia spp.	Tape worms ⎬	Spread by application of sludge as
Ascaris spp.	Nematode worms ⎭	agricultural fertiliser

Source: Hellawell (1986)

(Lewis, 1981). Excess sediment chokes, smothers, and creates instability as well as altering chemistry.

In the nineteenth century, along with so many other changes, came the main drainage of ordinary British farmland. Ditches, of course, had been used as boundaries and storm drains from earliest times, collecting and diverting water down pathways ordained by Man. Drainage and flood prevention were much pursued in Roman days, and increased again — intermittently in both time and space but often to a high level — from the Middle Ages (Haslam, 1987a). In contrast, over much of Europe, uniform patterns have now been at work for the past century or so, culminating in the post-1945 explosion of straightening channels, deepening channels, removing channels, and speeding up the movement of water. Both timing and intensity vary nationally. For instance, Britain, Germany and Denmark have straightened most, but Denmark has managed to retain better quality in the streams. These changes cause habitat destruction, chemical change, and also the imposition of uniformity on diversity. This latter imposition is also seen in other major changes of the twentieth century.

Livestock traditionally feed on grass in summer, on hay in winter, and manure the fields they graze. More recently other crops have been grown for feed, like turnips, and, very recently, feed barley, with increasing effects on field and therefore on river chemistry, both from cultivation, and by the removal of stock from the fields — where, for conservation of habitat, animal density was often strictly controlled — to close quarters in barns. Organic waste used to be necessary and valuable as fertilizer. Now fertilizer in the bag is cheap. Too often it is: Why bother with the muck? To the river with it! Slurry from livestock (used on the land) and silage liquor from plants, (both highly poisonous products) are not supposed to be discharged untreated, but all too often they are, whether or not by accident. Just think of the waste. Instead of complete recycling within the farm, the wanted nutrients are imported and the unwanted pollute (even treated material is still polluting).

As farming intensifies, the land, and hence the run-off to the river, becomes more nutrient-rich. From the marsh to the poor grassland to the improved grassland and — at present finally — the much-fertilised and perhaps previously-ploughed grass field. Grassland often occurs by rivers, storm flows having brought silt to fertilise. These, then, were valuable and cared for, whether as water meadows or other penned water systems, or just as flood meadows. (Many still remain, near-dry and with the water system lost.) Nutrients were removed from the stream to help the farmer. Now it is the reverse. Also, ley (short-term) grass has more nutrient run-off than permanent grassland because of the ploughing and consequent loss of both soil and soluble nutrients.

Ploughing is, of course, necessary for arable farming. The deeper the ploughing — as has happened in recent years — the more the wash-out. Increasing field size may likewise increase wash-out of top soil. It was early realised that arable fields, from which nutrients were constantly removed by crops, required fertilising to restore the nutrients. Lime, livestock dung, pigeon dung (pigeons were kept to eat and to fertilise), shrimps, rabbits' feet, and any other available nutrient-rich material was used. The Norfolk four-course rotation, with its legume crop to fix and add nitrogen to the soil (turnip for winterfeed, grain for bread, grass for manuring) was, perhaps, the culmination of the traditional system: management for maximum yield with minimum outside materials. Crop failure from disease and deficiency was, of course, still rife. Enrichers were sought more widely in the nineteenth century: guano from South America (nitrogen), bone fertilizer (phosphorus); and then, the mineral fertilizers. The period after 1945 marked the explosion of the powders. Powder for this, powder for that. Add it, and the crop will grow fine. With a delay of a few decades, the powder — very predictably — reached the rivers, both by direct leaching and via the ground waters, which are now far from fine over much of Europe. It is curious to reflect that the obtaining of nutrients for arable crops has been a major European preoccupation for millennia. It has taken only decades to reverse the position and make surplus nutrients the problem.

To channelling, the plough and the fertiliser, a fourth factor must be added, one whose use accelerated after 1945: the biocide — for the wart and for the worm, for the moth and for the mould, for the blight and for the mite. Though some of the chemical remains in the plant and is

220925

harvested away from the field, the rest may reach the river, either directly or via decomposing and degrading remains in the soil. Concentrations in the river and in drinking water are lower than those of fertilisers, and public concern is, so far, less. There seems a happy view that intent controls results. The biocide was put on for the mite. Therefore it is impossible it should harm Man or plants. It is also true that other river pollutants may be so damaging that biocide damage is undetectable (see Chapters 9, 10, 11).

Having stressed the damage done by modern farming to rivers, it should also be said that Britain, with one of the densest populations and a large part of the country mountainous and infertile, is now self-sufficient in food (though not in individual types of food).

Mining, quarrying and gravel extraction

Metals for tools and appliances, stone for building, have been wanted by Man for millennia. Some sites persist, some are transient due to exhaustion of supply or changing technology. The tin mines of Cornwall, which exported to the Phoenicians, were noted as polluting the River Dart (with 'tinny sands') in Spenser's *Faerie Queene* in the sixteenth century, and are still active today. Goldmines are frequently soon exhausted, stone quarrying varies much with cultural patterns. Streams are often wanted in mines, for washing ore, and very complex channel systems may be developed — for example, a seven-mile leat across the Welsh hills in Roman times to create the head to wash gold ore (Limbrey, 1983), and the sixteenth-century 'hushes' (large earth dams) to impound water for the Pennine lead mines. This lead mining was probably pre-Roman, reached its peak near 1800, and ceased in the 1930s (though other Pennine ores showed different patterns). Hundreds of kilometres of underground tunnels were created, many with flowing water (Johnson, 1981). Mine water may be unfit and poisonous. It will contain the metal mined, which will often be toxic, like lead. If used for washing, the sediment washed will be deposited and, of course, choke, smother and render unstable the river it enters.

Where the ore is treated on site, streams may also be wanted for power — for example, the 1610 *Britannia* describes the English Weald as full of iron mines. In order to make and refine the ore, brooks were diverted, turning meadows into pools, so developing the head of power needed to drive the hammer mills. The Forest of Dean and Derbyshire also had both iron deposits and wood for working the iron. Not until the Industrial Revolution was iron worked by coal and steam, and the industrial areas shifted to south Wales, the Pennines, and other areas.

Coal mines have led to some of the dirtiest rivers, both directly because of the wash and the quantity of coal required, and indirectly through the development of heavy industry on coalfields. Twentieth-century coal mines are less polluting (see Haslam, 1978a, for effects on vegetation).

Gravel extraction from river beds is one of the most important sources of pollution in Italy. Vast amounts of sediment, in suspension and deposit, are released downstream (see Haslam, 1987a, and Chapter 10).

Other industrial pollution

Water mills, harnessing the power of flowing water, were available in the late first century BC, and spread across Europe and the Mediterranean with the speed characterising changes perceived as good, becoming widespread by the end of that century. These were undershot wheels, requiring large volumes of water. Overshot wheels, turning on tiny streams (if with a good head) appeared in the Mediterranean by the fifth century AD. Up to the tenth century these mills were mainly corn mills, from which pollution was minor (the corn not being in the water). Then their uses multiplied: water-driven trip-hammers for fulling and wool (whose names are perpetuated in such towns as Shepton Mallet, sheep farm, in Somerset, England); and power-driven forges for metal work. By the fourteenth century, textiles, fulling, dying, tanning, sawing, forging, grinding anything from olives to ore, polishing anything from weapons to gems, all could be done by the water mill (Whyle, 1962), when also there were some 10000 floating mills on the rivers of Europe. Blast furnaces started in the fifteenth-century (Crossley, 1981). Pollution therefore followed (Figure 2.5), and could be as gross as for any later development (for example, in the Quantock Hills, with wool products), except that it was more local, and more confined to the river (compare Figures 2.5 and 2.3). This was the first European

Figure 2.5 Sixteenth-century mill. Pollution from processing hides or wool

Industrial Revolution and was accelerated by the loss of manpower through the Black Death in the mid-fourteenth century.

Although pollution from water mills went straight to the river, other industrial dirt was also likely to pollute. In 1307, the London Fleet was described as: 'by the filth of the tanners and such others, was sore decayed, also by raising of wharfs, but especially by diversion made by them of the new Temple, for their mills. . .and divers other impediments.' The remedial measures were only partly effective (Rodgers, 1947–8). The armoury town of Augsburg, in fifteenth-century Germany, must have produced much waste. Descriptions such as that in Court (1938) of the English Midlands shortly before steam power, suggest rivers becoming unpleasant. 'A countryside in course of becoming industrialised; more and more a strung-out web of iron-working villages, market towns next door to collieries, heaths and waste . . . being covered by the cottages of nailers and other persons carrying on an industrial occupation in rural surroundings.'

River deterioration was gradual and passed largely unnoticed. Izaak Walton in *The Compleat Angler* (1653) described rivers full of fish which, by 1900, could not support this. Defoe, writing a little later, described the water use for the cloth industry near Halifax: 'If the house was above the road, it [the stream] came from it, and cross'd the way to run to another; if the house was below us, it cross'd us from some other distant house above it, and at every considerable house was a manufactory or work-house, and as they could not do their business without water, the little streams were so parted and guided by gutters and pipes, and by turning and dividing the streams, that none of these houses were without a river, if I may call it so, running into and through their houses.'

Doubtless there were regulations, as usual, for the disposal of solid waste. Steam power (coal) spread from the 1770s in England; the new mills, freed from the necessity of river power, spread to the hills, and the density of industry and population this permitted led to the full horrors of the Industrial Revolution, first in England, then (sometimes diminished) elsewhere. Arthur Young described landscapes such as Figure 2.3 as 'sublime': 'That variety of horrors art has spread at the bottom; the noise of the forge, mills etc. with all their vast machinery, the flames bursting from the furnaces with the burning of coal and the smoke of the limekilns are altogether sublime'. Well, in a picture, perhaps. The slum was born (Hoskins, 1953). Water power created no smoke or dirt, did not allow for dense population, and mills had to be well apart from each other to make use of the power available. Steam had no such constraints and the industrial rivers became appalling. Every stream was fouled with chemical fumes and waste (Crossley, 1981), from a much greater variety of industrial processes than ever before.

The records of nineteenth-century deterioration are endless: The Irwell (North-West England) between 1820 and c.1885 (Holland and Harding, 1984); the Sirhowy and the Ebbw, in the Welsh coal valleys (Edwards *et al.*, 1984). Here ironworks grew along the rim of the coalfields at the heads of the valleys in the late 1700s. By 1841 the people were there: no sewerage, no privies, surface drainage poor and direct to the river, and — what a surprise! — cholera. By 1878 there was piped water, and some sewers, which piped direct to the river. Only in 1912 was a sewer laid to the coast, but this still overflowed into the river. It was replaced in 1971, and the steelworks' effluent was treated in 1973. The river improved. But it improved far more from a different cause: history was leaving the Welsh coal streams, industry was closing, and the effluents produced were less. The 1781 description of the Sirhowy, with plenty of trout and banks gay with flowers, soon passed with the development of the ironworks, the 'lawless practices of the forge men' making the river poisonous. By 1861 the few remaining trout were tainted and inedible.

The Birmingham Tame has a similar story. In 1826 it was not injurious, and the upper river was used for water supply. This was stopped in 1871: industrial discharges had killed fish for nearly twenty years. In 1901 it was said: 'I believe after heavy storms the sewage of Birmingham is turned direct into the river without being treated at all, the result being that for many miles the river [is] simply an offensive drain' (Lester, 1975). Some of the later history of the Tame is given in Chapter 4. For all of its awfulness, the population increased during the period at a rate never seen before.

Different countries and different rivers have peak pollution at different times — the Rhine, for instance, was at its worst in the period 1950–70 (Friedrich and Müller, 1984). It depends on culture, population, technology and regulation.

River improvements — sewerage, sewage treatment works, legislation to require factories to clean waste water — developed slowly, all too slowly, during the later nineteenth and early

Table 2.3 Toxic substances present in industrial effluents

Substance	Source
Acids	Chemical industries, battery manufacture, minewaters, iron and copper pickling wastes, brewing, textiles, insecticide manufacture
Alkalis	Kiering of cotton and straw, cotton mercerising, wool scouring, laundries
Ammonia	Gas and coke production, chemical industries
Arsenic	Phosphate and fertiliser manufacture, sheep dipping
Cadmium	Metal plating, phosphate fertilisers
Chlorine (free)	Paper mills, textile bleaching, laundries
Chromium	Metal plating, chrome tanning, anodising, rubber manufacture
Copper	Plating, pickling, textile (rayon) manufacture
Cyanide	Iron and steel manufacture, gas production, plating, case hardening, non-ferrous metal production, metal cleaning
Fluoride	Phosphate fertiliser production, flue gas scrubbing, glass etching
Formaldehyde	Synthetic resin manufacture, antibiotic manufacture
Lead	Paint manufacture, battery manufacture
Nickel	Metal plating, iron and steel manufacture
Oils	Petroleum refining, organic chemical manufacture, rubber manufacture, engineering works, textiles
Phenols	Gas and coke production, synthetic resin manufacture, petroleum refining, tar distillation, chemical industries, textiles, tanning, iron and steel, glass manufacture, fossil fuel electricity generation, rubber processing
Sulphides	Leather tanning and finishing, rubber processing, gas production, rayon manufacture, dyeing
Sulphites	Pulp processing and paper mills, viscose film manufacture
Zinc	Galvanising, plating, rubber processing, rayon manufacture, iron and steel production

Source: Hellawell (1986)

twentieth centuries. Hudson (1984) comments that as long as factories, steam engines, railways and dirt were inseparable, the planners' hands were tied. Electric power and motor transport allowed the development of clean — or at least very much cleaner — industrial estates. This gives too rosy a picture of river pollution, though. Changing technology meant some decrease in visible pollution, in fact much decrease in the worst industrial areas. But the range of chemicals, especially toxic chemicals, increased, and industrial pollution has spread to previously unaffected areas. That 'clean' industrial estate has polluting run-off from its access motorway (see Chapters 1 and 4) and from its concreted ground, pollution from its effluent treatment works and, perhaps, polluted water from the tip where its poisonous waste is dumped. The dirt may have changed, but it is still there (see Table 2.3). Between the world wars it seemed possible that pollution could eventually be stopped. The technology was available for most of the main pollutants. In the 1980s we are, as a century ago, unwilling to direct our energies to develop the means of rendering safe our current pollutants. 'We cannot always anticipate new technological developments, but we can learn the needs of aquatic life, so that when these developments arrive, their probable effects on life can be evaluated. Planning for water quality control must reach beyond the waters, to where and how we live and work. The patterns of land use and industrial development are as important as the regulation of existing uses for [river] quality' (Warren, 1971).

3
Ecosystems and pollution

'Even the most satisfactory effluent is not river water and must, therefore, produce some change in the receiving water.'

(Hynes, 1960).

The study of river life

Water is the medium of life on earth. All living organisms are composed mainly of water, and those of rivers are partly or wholly immersed in water. Life evolved in water, with water, and with the physical and chemical characters of natural water, and this balance is necessary. By its presence any organism changes the river. Macrophytes (larger plants) provide habitat, shelter and indeed food for many animals and smaller plants, being thus desirable or necessary for the good development of these animals and smaller plants. Other changes made by one group may be detrimental to another: swans may destroy macrophytes, blooms of the alga *Prymnesium* produce a toxin which may kill fish. The natural river bears a natural balance of plants and animals, some interdependent, some competing, some independent. Man alone alters the habitat for whim, as well as for survival.

Why should river life be studied, apart from Man's insatiable curiosity? Firstly, for its own sake. We have an obligation to maintain and conserve life forms other than our own. This was clearly perceived as long ago as when the Book of Genesis was told. The command to replenish the earth came before that to subdue it and have dominion over it. More recently the priorities have been reversed. Secondly, it must be studied for our sake. Is the water fit for drinking? Bathing? Cooling in industry? Can the fish from it be safely eaten by Man? Will living beside it make children unhealthy?

How can river quality be measured? It can be examined chemically. For any given substance, in the river the chemist has to make many observations over a long period of time to obtain the average value and the possible range of its concentration. This is time-consuming. It may also be ecologically inaccurate since it may be extremes rather than averages which kill: low dissolved oxygen kills fish, and it is no less lethal if it happens on only one day in four years. Also, isolated pollutions such as one lorry-load of milk may be missed completely, and the resultant damage be inexplicable. As Hynes (1960) points out, the numerical exactitude of individual chemical results gives a spurious confidence to the planners.

Rivers can also be examined biologically. The plants and animals in a stream have, if moving little, received its water constantly during their lifetime, and their state on a given day represents the cumulative effect, on them, of that water — means, extremes and all. A plant growing for three years means that no discharges lethal to that plant have occurred for three years. Moving animals likewise integrate the total effect of their habitat, but it may be more difficult to determine where a heron became poisoned than where a stickleback did.

Chemical analysis may show the general state of water or soil for, say, oxygen, nitrate, cadmium and perhaps up to twenty other substances. Organisms integrate the effects of all the many thousands of substances present. Since they are living, their (biological) data are less precise than those of the chemist. Macrophytes assess and register, for the trained eye, all the factors influencing macrophytes. Likewise, diatoms, for their habitat factors, and eels, invertebrates and rotifers, for theirs. The basic tenet of interpreting is the maxim of Dr A.S. Watt that 'the plant (or animal) is always right'. If theory and chemical measurements show that a plant cannot grow, and the plant is there, theory and chemistry are wrong. (The writer once commented to some young researcher: How odd you found *Berula erecta*, it usually needs high calcium. Subsequently the information arrived: the plant *was* growing by lime-rich rubble. The plant *was* 'always right'!). Our ability to exploit this, however, relies on a proper understanding of the plant's behaviour.

Receiving waters

If exactly the same effluent or other pollution is added to a range of streams, will the ecological effect be the same? No. The effect will vary with both the quantity and the quality of the receiving water; the quantity is important because an effluent diluted a thousand times is much less harmful than one diluted twice. This is well known and accepted — even by planners. The quality is important because streams vary much in water chemistry. They vary with rock type, (natural) downstream changes, type of vegetation on the land, and with the soil developed — hence also with Man's use of that land (see Chapter 11). Chemistry also varies with the other pollutants already present in the stream. Assuming the following criteria can be met in clean streams (i.e. that river life is not reduced by boats or the high water force of Alpine rivers, etc) then, to be even moderately acceptable, pollution should permit:

1. A varied river life, including vegetation covering at least 50 per cent of water under 1 m deep (excluding heavy shade, spates, etc.), and composed of varied species of varying shapes, colours and placing. No toxicity to plants or animals on the bank during floods.
2. A good fishery, with fish safe for human consumption. (A good test! Many people satisfied with a river because they can see the — eatable — fish would recoil in horror at the thought of actually eating them.)
3. Water usable for irrigation on food plants; for drinking by animals whose products will be eaten by Man — and, preferably, usable for drinking by Man.
4. Water not causing skin rashes, diarrhoea, eye infections, etc., to those in or over it. The Thames, whose purification was so praised in the 1970s, now fails this criterion — among others.

The receiving waters can be classed as:

1. **Fertile streams of robust quality.** These are nutrient-rich and organic-rich, for example, middle and lower clay streams, lower clay-mixed streams, and (clean) canals in lowland fertile farmland (e.g. Figures 1.2, 1.3: clay). The water is well buffered, rich in many solutes. A good coarse fishery is supported. A wide variety of substances of low toxicity can be added in moderation without altering the fundamental character of the river. This has its own dangers. Because of the general physical

damage prevalent in such rivers (see Chapter 8), which reduces the flora and fauna from their full, ideal state, mild extra chemical damage may come to light only through food poisoning or skin rashes — unless research on the river has been more intensive than usual.

2. **Moderately robust, moderately fertile streams**. These are, for example, lower limestone and sandstone rivers, upper clay streams, and the most fertile of the Resistant rock streams. They therefore comprise more variety than type 1. They are nutrient-medium and organic-medium. Some are lime-rich, none are lime-poor. Game fisheries (of salmonids) are present in most, but not all, types. All waters need care to prevent toxic substances entering. Greater care, though, is necessary in type 2 than in type 1 to prevent pollutants from altering the character of the river. Several septic tank overflows would be unnoticed in type 1, but would need watching in type 2, where certainly the refuse from a whole village would be damaging (unless much purified or diluted). The typical semi-cleaned effluent from an organic factory (dairy, jam, sugar-beet) will cause damage. It takes less extra chemicals than in type 1 for the river character to alter.

3. **Moderately fragile streams**. These include nutrient-low, organic-low (lime-variable) streams such as upper-middle lime and sandstone ones, farmland streams on Resistant rocks, grassland hill streams, etc. The variety of stream and vegetation types is thus greater than in type 2, in the same way as type 2 variation is greater than type 1. The solute content is less, and the added variations more significant. The water and soil type is characteristic, but poorly buffered and easily altered by incoming discharges or major changes in land use.

4. **Fragile streams**. These are nutrient- and solute-deficient, except that they may be rich in humic derivatives from peat, particularly acid peat or podsols. Included here are, for example, the acid sands streams of the English New Forest; the French Les Landes; the New Jersey Pine Barrens; the most nutrient-poor of the chalk seepages; Resistant rock streams either under the influence of acid peat or leaf fall, or excessively solute-deficient; and the *Potamogeton coloratus* calcareous nutrient-poor

streams of Germany and France. All such are easily upset and altered, not just by effluents but also by minor changes in land use. Their survival depends on the survival of the land use (and effluent) practices under which they developed.

The Pent (Figure 6.14) has fragile chalk seepages, with rather nutrient-poor (oligo-mesotrophic) clean vegetation, some species being discoloured or shortened in the manner characteristic of nutrient deficiency. Just downstream there runs a motorway with, of course, toxic and organic run-off. Even that small part of the muck draining to the upstream side of the motorway destroys not just the nutrient-poor character of the stream, but also its general chalk character: the vegetation is of a species-poor, partly-polluted, nutrient-rich clay stream at best, or of just pollution at worst (Blanket weed). Clay streams, being rich in nutrients, as noted above, can 'carry' more pollution than other stream types. It is probable that this run-off could be added without much effect to a clay brook, and certainly to a large clay river. Added to the upper Pent, however, it renders the stream no longer fit for the river life of this water type. Similarly, Morgan and Philipp (1986) note that under pollution the characteristic New Jersey Pine Barrens vegetation is replaced by a community with many marginal or non-indigenous species. (The alkalinity is low, indicating low buffering capacity, and the nitrate levels rose under pollution.)

Therefore, the greater the buffering and neutralising capacity and total solute content of the stream, the more extra substances can be added without changing the stream character. Also, the better the physical structure (see Chapters 6 and 7), the greater the ability of the communities to withstand an unsatisfactory chemical status (Figure 3.1).

When a stream receives consecutive mild pollutants, each having but little effect on the ecology, the total effect will depend on the degree of downstream purification and the distance between effluents. This may seem obvious, but many an unfortunate downstream effluent producer has been blamed for the total damage caused by several effluents. This arises when the capacity of the stream to absorb pollution without major change has been reached, and an

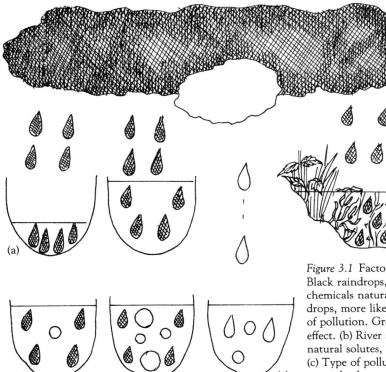

Figure 3.1 Factors influencing the effects of pollution. Black raindrops, pollution of a type very unlike the chemicals naturally present in the river. White raindrops, more like natural river chemistry. (a) Dilution of pollution. Greater dilution decreases ecological effect. (b) River solute concentration. The more the natural solutes, the less the effect of pollution. (c) Type of pollution. The closer the pollutants to the natural solutes, the less the effect on the river. (d) Physical structure of river. The better the habitat in other respects, the greater its resistance to the effects of pollution.

over-dramatic response occurs with the next effluent. If much purification takes place, of course, damage is less, and vice versa.

Recovery in time, and recovery in space (downstream self-purification), show similar changes in community. In both, change is towards the proper community for the stream.

Where the plants and animals are found

Different parts of the ecosystem are affected differently by pollution. Rooted macrophytes absorb chemicals from both soil and water. Generally, pollutants are most concentrated in silt, and least so in coarse particles and water. Therefore macrophytes usually reflect soil more than water status: though, for example, a cyanide spill will kill parts in the water, with subsequent recovery from underground organs. When effluents are improved the water immediately improves, but it takes some years for polluted silt

to be washed out. Dredging helps. The degree of recovery in the Lincolnshire Witham seen over some seven years was rapidly achieved — very temporarily! — by dredging, in the Yorkshire Went. There polluted water washed back from the Don into the mouth of the Went. Before dredging, macrophytes were absent. The polluted soil was removed by dredging early in 1976, and eight species were present later that year, in the good soil and bad water. By 1978 only three pollution-tolerant species remained, as pollutants again accumulated in the soil.

Different horizontal layers of river communities are therefore affected differently by pollutants. Micro-organisms of some sort are found throughout. Macrophyte roots otherwise go down furthest, but burrowing worms, etc., are also soil-dependent. On the ground surface are macro-invertebrates, some of the non-rooted mosses, diatoms, other algae, etc., and of course the lower parts of flowering plants, rooted or, (like *Ceratophyllum*) non-rooted. Here depositing particles affect the habitat, whether coming direct from pollution or being washed down within the stream. Unless these are particularly toxic, thick enough to be smothering (or the organism spends much time within the soil layer) water quality will be the more important. In the main body of the water are most fish, plankton, many macrophyte shoots and the invertebrates, diatoms and other life forms living thereon, large algae, diving fowl and mammals while diving, etc. On the water surface are floating macrophyte and other plants, and various invertebrates such as water skaters, and duck. Finally, above the water surface, yet still part of the river ecosystem, is the vegetation – herbs and trees – of and beside the bank, the invertebrate populations they support — which may spend part of their life cycle in the river — and the resting places, burrows and homes of the otters, dippers, kingfishers, and other mammals and birds. These are the least directly affected by pollution. Some may be killed by poisonous flooding. But everything entering the water (even by roots only) or feeding on anything which is, or has been, in the water, may suffer.

Communities

A community is a definable unit of plants, animals or both. To become a specialist in the behaviour and ecology of any one group of river organisms is unusual, to become one in many has so far been impossible. Community studies therefore focus on the group — the macrophytes, the fish — rather than on the stratum of the water or on the site. Such communities can be defined in as many ways as there are researchers. Since definitions generate as much heat as light, this book keeps to generalities.

Vegetation can be classified and interpreted through rock type, water force (via landscape) and upstream–downstream patterning (Haslam, 1978; 1982a; 1987a; Haslam and Wolseley, 1981). In phytosociology (see, for example Shimwell, 1971), the 'association' is a plant community of definite composition presenting a uniform physiognomy and growing in uniform habitat conditions (Flahault and Schröter, 1910). Recent developments in computer analysis (see, for example, Castella, 1987; Balocca-Castella, 1988, for regression analyses with canonical correction; and Holmes, 1983; and Wolseley et al., 1984 for

TWINSPAN) use the end-groups of the analysis as communities. However, computer work – even in 1990 – depends on the field knowledge of the user, and can never substitute for it. For instance, Blanket weed (trails of filamentous algae, usually *Cladophora*) grows well with pollution, on weirs in rather cleaner streams, with disturbance in lowland streams, and as an ephemeral between storms in near-clean hill streams. Unless the computer is told all this, it will, by amalgamation, make interpretation impossible.

Pollution effects on communities

Pollution basically imposes a uniformity: at worst, polluted rivers are sterile — irrespective of the original, natural stream type, and irrespective of the exact chemicals concerned. This is ecological uniformity at its most extreme. In the next stage, species are few.

Each species has, as described above, its own habitat range. Knowledge of these ranges elucidates habitat conditions. All river species can grow in the absence of Man. Species now associated with pollution would have occurred in leaf-fall-rich habitats (for example, Chironomids, tubifex), in brackish water (for example, *Potamogeton pectinatus*), or in other conditions. Pollution makes it possible for some species, those more tolerant to the particular pollution, to extend their distribution, while much curtailing the occurrence of others. Table 3.1 summarises the impact of pollutions. The effect on vegetation is to:

1. Reduce species diversity. Under chemical stress fewer plants occur of any species, and these are increasingly confined to the best physical conditions for that species e.g. shelter, or water 30 cm deep. Eventually they are absent from some sites, so site diversity drops. The species most sensitive to that pollutant decrease first, the tolerant persisting longer. Therefore the community is increasingly composed of species tolerant to that type and level of pollutant.

However, in habitats of mixed influence — for example chalk/clay, fresh water/slightly brackish water — diversity may be higher than in either separately. Pollution of low toxicity may occasionally have the same effect.

2. Increase pollution-favoured species. Few species are so favoured by pollution that they enter a site because of it, but *Potamogeton*

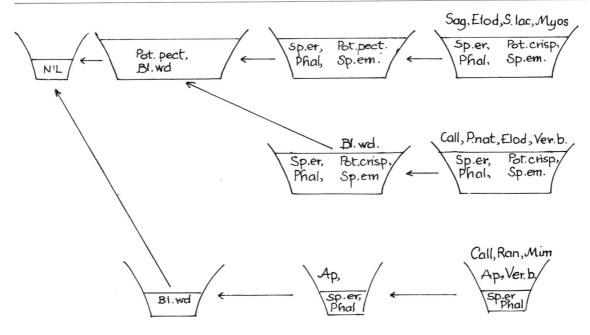

Figure 3.2 Pollution produces uniformity from diversity. Pollution increases from right to left. Gross pollution, on the left, leads to empty channels. This is the same in all stream types. Clean streams have different species assemblage. (See Table 4.1 for species names represented by these abbreviations.)

pectinatus and Blanket weed are, over most of Europe. Others do so under more particular conditions — for example, *Potamogeton natans* (some industry), *Myriophyllum spicatum* (salt). All these occur, of course, when Man has simulated something close to the proper conditions of another place. *Cladophora* is near-ubiquitous. It is important to note therefore that where *Cladophora* grows profusely and forms trailing masses obvious to the naked eye, it is Blanket weed — weed able to blanket the river, if abundant enough.

3. Reduce biomass and cover. As species decrease in abundance so they decrease in biomass, first the more sensitive species then, as dirt worsens, the more tolerant ones. Unlike the observations of species presence, biomass measurements are destructive to the habitat, time-consuming, and often inaccurate as a guide to pollution (biomass depends on storms, incident light and temperature, etc., and is complex to interpret in regard to pollution alone). Cover, fortunately, is less variable and so more indicative.

Although no one group indicates and assesses the total river, macrophytes should be the best single indicator since they are large, stationary and have other groups depending on them.

For invertebrates the same underlying principles apply but the practice is rather different. Assessment is normally destructive to a minor extent, the animals being removed from the river, and killed, for identification and counting: community and posterity both being lost from small areas. In the purest waters invertebrates become sparse (while macrophytes still do well). Consequently, under mild and reasonably natural pollution, such as sewage, invertebrate diversity may increase, only decreasing with worse discharges. Also, as for example Thienieman (1954) points out, as conditions become more extreme (deviate more from normal) and diversity decreases, so the numbers of individuals of the remaining species increase. The number of species of invertebrates — in most habitats — far exceeds those of macrophytes. In practice this is an advantage to the macrophyte worker, identification being easier. In theory,

Table 3.1 Environmental implications of the discharge of sewage and industrial effluents

Factor	Principal environmental effect	Potential ecological consequences	Probable severity	Remedial or ameliorative action	Comments
Degradable organic matter					
1. High biochemical oxygen demand (BOD) caused by bacterial breakdown of organic matter	Reduction in dissolved oxygen concentration	Elimination of sensitive species, increase in some tolerant species; change in community structure	Dependent upon degree of deoxygenation, often very severe	Pretreatment of effluent; ensure adequate dilution	BOD can be reduced by adequate treatment
2. Partial biodegradation of proteins and other nitrogenous material	Elevated ammonia concentrations; increased nitrite and nitrate levels	Elimination of intolerant species, reduction in sensitive species	Variable, locally severe	Improved treatment to ensure complete nitrification; nutrient stripping possible but expensive	Adequate treatment is best solution
3. Release of suspended solid matter	Increased turbidity and reduction of light penetration	Reduced photosynthesis of submerged plants; abrasion of gills or interference with normal feeding behaviour (see inert solids below)	Moderate, usually local	Provide improved settlement, ensure adequate dilution	
4. Deposition of organic sludges in slower water	Release of methane and hydrogen as sulphide matter decomposes anoxically Modification of substratum by blanket of sludge	Elimination of normal benthic community Loss of interstitial species; increase in species able to exploit increased food source	Variable, may be severe	Discharge where velocity adequate to prevent deposition	Tends to be local
Other poisons					
1. Presence of poisonous substances	Change in water quality	Water directly and acutely toxic to some organisms, causing change in community composition; consequential effects on prey–predator relations; sub-lethal effects on some species (changes in behaviour, etc.)	Highly variable, depending upon substance and its concentration	Increase dilution	Difficult to generalise
Inert solids					
1. Particles in suspension	Increased turbidity. Possibly increased abrasion	Reduced photosynthesis of submerged plants. Impaired feeding ability through reduced vision or interference with collecting mechanisms of filter feeders (e.g. reduction in nutritive value of collected material). Possible abrasion	Variable, often moderate	Improve settlement	Inert solids change the character of the substrate and are unstable. They provide no additional nutrition
2. Deposition of material	Blanketing of substratum, filling of interstices and/or substrate instability	Change in benthic community, reduction in diversity (increased number of a few species)	Variable, often severe	Discharge where velocity adequate to ensure dispersion	

The effects of the three major categories of effluents, namely degradable organic matter, toxic substances and inert solid particles, are considered separately. Many effluents are composed of more than one type and the proportions of these vary according to the source. When higher water standards are required, ways to meet them are normally found.

Source: adapted from Hellawell (1986, Table 4.1)

invertebrate interpretation should be more sensitive — but then macrophytes are recorded for the whole site, invertebrates just from one small part where they have to be extracted (and extraction should be uniform, but regrettably is not necessarily so).

Similar patterns apply with fish (salmonids are more sensitive than coarse fish), diatoms, etc. (see Chapters 4, 5, and 7).

The longer a polluted habitat exists, the more likely an adapted community is to develop.

Other assessments and discussion

Effects of pollution are also tested by the amount of toxins found in dead organisms. Here nice chemical analyses give numbers which we all like. The snag is that the organism may be suffering from non-analysed as well as analysed chemicals (as a determination of, say, lead in the river, these analyses are good. If, though, lead levels are high enough to be harmful, for example, and the plants are plainly damaged, there is a strong temptation to say that the plants are damaged by lead — and not continue testing, so failing to discover that the plants are really poisoned by diquat). Finally, pollution can be tested in the laboratory, by growing organisms — often non-native ones like rainbow trout — in water to which known amounts of pollutants have been added. These again give clear results, but their relevance to field conditions may be doubtful. And even the water may not be the same as that in the river. Tap water and well water differ from river water. In addition, the trout do not necessarily remain unstressed in the testing environment. Community studies show what *is*, what is in the river, in a way sometimes difficult to interpret, and always impossible to interpret accurately. Since no one knows the composition of a river (Chapter 1), no one, consequently, can say exactly what concentrations of what chemicals are indicated by what community. The degree of pollution, the total response cumulative to all factors, yes, that is indicated. The two approaches are complementary. The birds of prey would not have been saved without analyses proving that some of their number were killed by such pesticides as dieldrin. On the other hand, to record the absence of a few lethal substances in a plainly polluted river, allowing planners to call

it 'clean', is no service to anyone. Oddly enough, this does happen.

Natural stream communities of all organisms are many and varied. Pollution constricts this variety, and is therefore wrong.

Communities lacking in components have not been widely studied — in fact, are seldom studied. Because of lack of knowledge, sites are not studied from mammals to viruses. It is, however, known that macrophytes and some invertebrates occur in waters with levels of heavy metals too high for fish, and likewise some fish and invertebrates occur in macrophyte-free waters. Different groups thus vary in response to different pollutants, as do different species within those groups. Macrophytes are the most sensitive to — and so the best indicators of — mild organic pollution. They respond first to its toxicity, invertebrates first to the oxygen depletion. Invertebrates are the more sensitive to heavy metals, however, Macrophytes accumulate the metals in their tissues, but are more tolerant. Affected plants are smaller — but in a river it is seldom possible to say that a plant with three shoots would have ten if heavy metals were absent! Incidentally, algal blooms may poison cattle, sheep, pigs, ducks, etc. — the blooms develop because of pollution.

You can throw Nature out with a pitch fork, but she will always come back. What comes back is not necessarily what, or as soon as, one would like, but something does come back. The ecosystem is a whole. Stress it in one part and it will respond — if not in the same place, then in another. Deficiencies, excesses and constrictions will interact.

Purification by macrophytes

To finish this chapter on a happier note, many pollutants lose their toxicity (are fixed or broken down) more rapidly in the presence of plants. Macrophytes produce oxygen (by day), and organic substances from sewage, etc., are broken down by oxidation. Periphyton may be increased by mild pollution, and this again improves stream oxygen.

Much more importantly, macrophytes purify water and sediment. Their epiphytic micro-organisms break down toxins, their root exudates develop microflora which kill pathogenic bacteria and cause mineralisation, and substances are taken up and altered within the plants (see, for example, Kickuth, 1976; Kickuth and Tittizer, 1974a; 1974b; Lawson, 1985).

Some practical examples are the incorporation of phosphates into phytates and denitrification (using *Phragmites communis* see Kickuth, 1976); the (considerable) removal of *Escherischia coli*, *Salmonella* and *Enterococcus* bacteria (using *Mentha aquatica*, *Alisma plantago-aquatica*) (see, for example, Seidel and Kickuth, 1967; Seidel *et al.* 1967); and removal of petrochemicals (using *Phragmites communis*, see de Maesener *et al.*, 1982). Records of reduction of for example 99 per cent of nitrate-nitrogen, 10–98 per cent of metals, and between 96 and nearly 100 per cent of

bacteria (Schmitz *et al.*, 1982) are far from uncommon in studies of natural marshes. Encouraging marsh edges to polluted rivers (particularly *Phragmites communis*), therefore aids purification. Existing fringes should be conserved, as should little-managed land use beside the rivers (Chapter 11).

This purification by plants has, of course, been used to treat sewage and other effluents. The most efficient design is the root zone treatment with *Phragmites communis* (Kickuth, 1984; and see, in English, Lawson, 1985). The beds are larger than those for conventional sewage treatment, but in the right circumstances can produce purer water even from chemical factories than does conventional treatment (also see Haslam, 1987a).

4
The effect of pollution on plants

'The plant is always right.'

A.S. Watt

Macrophytes: distribution

The most-polluted sites will not bear marophytes. When conditions improve so that tolerant species can occur, Europe-wide these are most likely to be *Potamogeton pectinatus* and/or Blanket weed (depending on habitat and pollution). With less pollution, more species grow, first those more tolerant, then those less tolerant to pollution (Figure 4.1). In the worst places it is ecologically irrelevant what the pollutant is: no macrophytes from cyanide is the same 'community' as none from sewage. (Toxic and purification processes differ in the two, of course.) However, where pollution is but moderate, so that a number of species can grow, different pollutants are more

likely to have different effects, for example more *Groenlandia densa* and *Oenanthe fluviatilis* will grow in chalk streams with septic tanks than in those with sewage works. Pollution unifies, and the greater the pollution, the greater the uniformity of the vegetation. The community is both reduced, and skewed, from its natural to its polluted state.

River maps are valuable for demonstrating pollution patterns. River macrophytes cannot get worse than in the lower Neckar (Figure 4.2), with its 'nil', 'nil', 'nil' records. There is much industry, especially in the cities of Stuttgart and Heilbronn, poor sewage works, and intensive farming (Pinter and Backhaus, 1984).

The Skjern (Figure 4.3) near-clean, and the Idle (Figure 4.4) moderately polluted, make an interesting pair. Both are recognisably lowland sandstone rivers: the Neckar could be on any rock type! The Skjern has up to twelve species

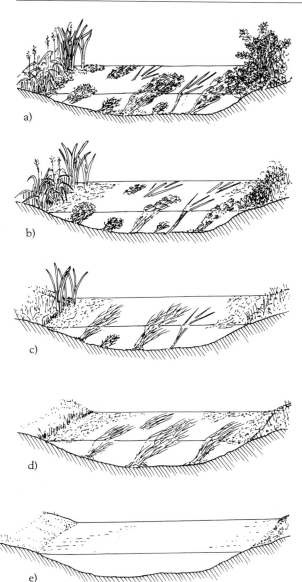

Figure 4.1 Effect of increasing pollution on macrophytes. (a) clean, (b)–(e) increasing pollution. Species present: (a) *Ranunculus* sp. (abundant), *Apium nodiflorum*, *Callitriche* sp., *Myosotis scorpioides*, *Phalaris arundinacea*, *Potamogeton crispus*, *Rorippa nasturtium-aquaticum*, *Sparganium emersum*, *Sp. erectum*, *Veronica beccabunga*. (b) *Callitriche* sp., *Myosotis scorpioides*, *Phalaris arundinacea*, *Potamogeton crispus*, *Ranunculus* sp., *Sparganium emersum*, *Sp. eretcum*, Blanket weed. (c) *Potamogeton crispus*, *P. Pectinatus*, *Sparganium emersum*, *Sp. erectum*, Blanket weed. (d) *Potamogeton pectinatus* (abundant), Blanket weed. (e) No macrophytes. (P.A. Wolseley)

per site, high cover, the expected downstream increase in diversity, and few sites, even upstream, have less than five species. Table 4.2 shows pollution-tolerant species. Many species of the Skjern are sensitive to pollution — they do not occur in Table 4.2. (Tolerant species may of course occur in the cleanest sites. All species have a 'natural' habitat, without Man-made pollution.)

The Idle, in contrast, has but few species per site, and most of these are pollution-tolerant, listed in Table 4.2. Pollution here is a controlling factor:

1. Low site diversities, few species per site (31 of the 40 sites have four or less species).
2. Blanket weed and *Potamogeton pectinatus* are frequent and often abundant.
3. *Lemna minor* agg., *Sparganium erectum*, and *Enteromorpha*, pollution-tolerant species from Table 4.2, are frequent.
4. Among the water-supported species only nine of the 40 sites bear species not found in Table 4.2, that is, bear non-tolerant species. Of these nine, four are in fact out of the main pollution pathway.

The upper site on the Ryton tributary was, in 1976, a shallow depression below a factory estate, with eight species and no water force. Pollution hinders root development, making polluted species easier to wash out (see below). In 1977, with a low but definite flow, there were five species, four effluent-tolerant.

The Alzette (Figure 4.5) is worse. It can be held up as a good example of polluted communities, and a good example of the effects of intermittent extra pollution. Industry and towns are plentiful, especially near the source. Species, on the other hand, are few — and pollution-tolerant! All but six species records, over three years of survey, are from the most pollution-tolerant group, and three of the remainder from the semi-tolerant group. (Incoming propagules can grow for a short time in rivers too polluted to support them properly.) *Potamogeton pectinatus* is abundant. Therefore the vegetation looks all right to the uninitiated: plenty of it, and pleasing to the eye (Figure 4.1). And, indeed, similar vegetation can occur in clean, brackish-water places. But here, woe betide anyone putting their hands in the water. Such rash behaviour brings rashes on the skin.

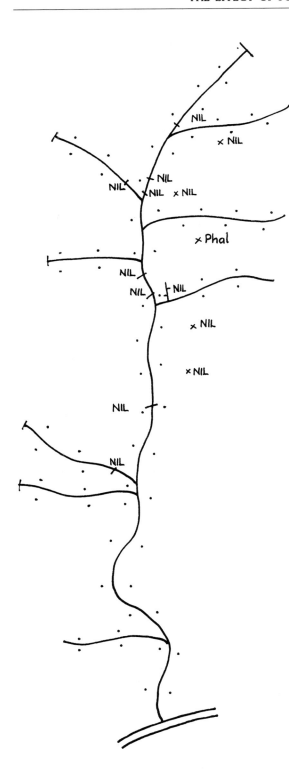

Figure 4.2 River map. Lower Neckar, Germany (from Haslam, 1987a). Gross pollution, devoid of macrophytes. Keuper. Lias etc. lowland, 1978 (L). The River maps in this and other chapters are presented in a stylised format for easy comparison. The flow is from top to bottom (unless indicated otherwise). Where the rivers enter the sea, this is indicated by shading, and where they enter larger rivers, the rivers are drawn.

The survey method is described in Table 7.3. Minor tributaries (crosses) are shown only if they have recorded sites. Rock-type boundaries (solid line across river) are marked only where rivers are crossed (sometimes giving a misleading impression of rock distribution in the catchment) and where this is not referred to in the text. The landscape is indicated by the density of dots along the river (plain, lowland, upland, mountain and alpine). The abbreviations used for the species names and the order in which the species are listed are given in Table 4.1. Maps are drawn to different scales (to illustrate different points). L indicates large, over *c.*50 km long, M medium *c.*25–50 km long, S small, *c.*10–25 km, and T tiny, 0–8 km

Extra effluent entries here may remove even the *Potamogeton pectinatus*, which then returns a little downstream where the effluent is more diluted and purification is perhaps just starting. In the worst parts a 'toxic line' is seen on the banks. This shows where river water reaches and poisons the bank vegetation in summer. It occurs in the upper river and below some sewage works in the lower river. A purification pattern can be seen below sewage works, more strikingly in 1980 than in 1985: a few tolerant species other than *Potamogeton pectinatus* – Blanket weed – *Phalaris arundinacea* actually come in. The general pattern is the same from 1978 to 1985, though with no improvement or deterioration.

River vegetation normally changes all the time. Within a stable environment both plant losses and plant gains come from the same species assemblage, the same vegetation type. That is, incoming plants may not be the same species as outgoing ones, but both are typical of the particular habitat. There is both stability of community, and instability (variation) in the actual species present at each site. Changes come from storms, silt accumulation and erosion, growth, maturity, spread, interactions and death of plants, grazing, and all the varied effects of Man from pollution to dredging or fishing.

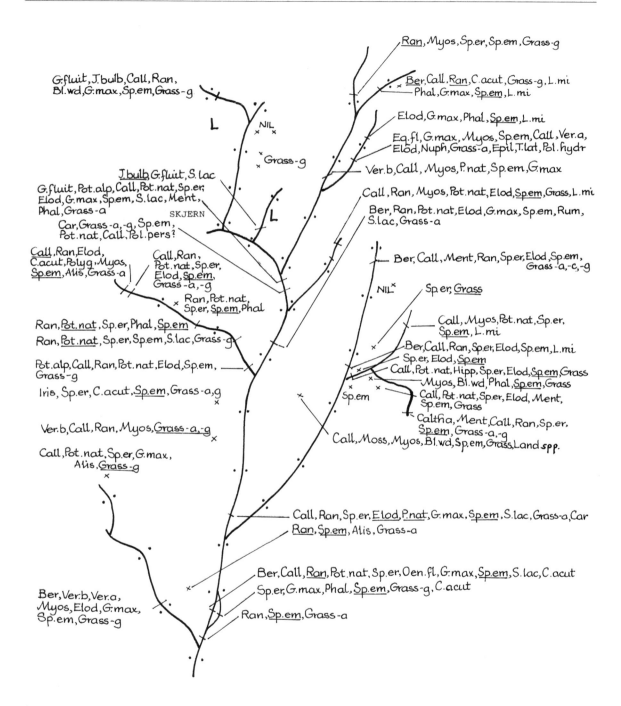

Figure 4.3 River map, Skjern, Denmark (from Haslam, 1987a). Fairly clean. L = Lignite, ochre streams (see Chapter 10). Fluvial sand (mainly), lowland. 1977, 1978 (L). Notes as for Figure 4.2

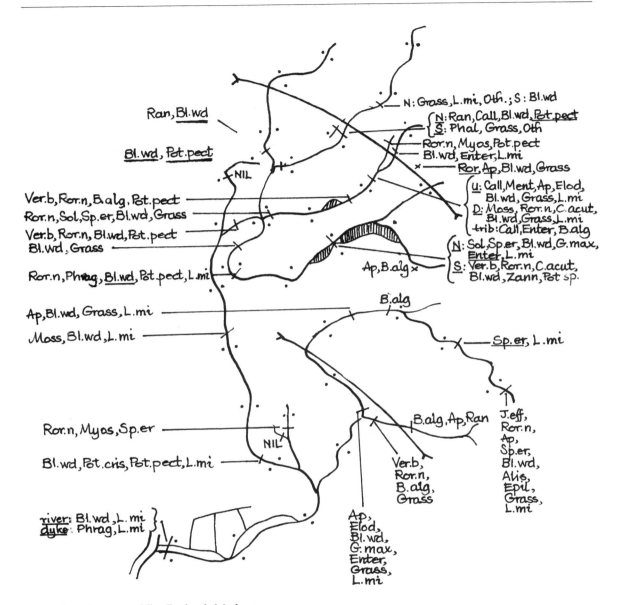

Figure 4.4 River map, Idle, England. Moderate
pollution. Sandstone, lowland, 1976 (M). Notes as for
Figure 4.2

Table 4.1 Macrophyte species names, and order of presentation, on river maps.
The contractions used on the maps are listed in the order in which they are placed on the site lists on the maps. This is roughly in nutrient status order (see Chapter 7). However, as different counties differ, and a single order is desirable for easy comparison in these maps, the order here is similar to, but not the same as, that of any one of the various countries. Additionally, species may behave differently in unusual habitats within a country.

(Brown)		Trapa	*Trapa natans*	S. lac	*Scirpus lacustris*
Sphag	*Sphagnum* spp.	Hydroch	*Hydrocharis morsus-ranae*	Nuph	*Nuphar lutea*
Dros	*Drosera* spp.	Iris	*Iris pseudacorus*		
Erioph	*Eriophorum angustifolium*	Ror.n (Ror)	*Rorippa nasturtium-aquaticum*	**(Unbanded)**	
Litt	*Littorella uniflora*		*(Nasturtium officinale)* agg.	Alder	*Alnus glutinosa*
Meny	*Menyanthes trifoliata*	Sol	*Solanum dulcamara*	Ar,Arun	*Arundo donax*
Pot.poly	*Potamogeton polygonifolius*	Ap	*Apium nodiflorum*	Car.pan	*Carex paniculata*
Car ros	*Carex rostrata*	Myos	*Myosotis scorpioides*	Cup	*Cupularia viscosa*
Ran (flamm)	*Ranunculus flammula*	Pot. nat	*Potamogeton natans*	Cyp	*Cyperus* spp., usually Cyp.b,
		Ver. a	*Veronica anagallis-aquatica*		C. *badius*; Cyp. f, C. *fuscus*
(Orange)			agg.		and Cyp. l, C. *longus*
Ran.hed	*Ranunculus hederaceas*	Sp. er	*Sparganium erectum*	Eq. fl	*Equisetum fluviatilis*
Ran omi	*Ranunculus omiophyllus*			Eq. pal	*Equisetum palustre*
Myr. alt	*Myriophyllum alterniflorum*	**(Mauve/Purple)**		Grass-a	*Agrostis stolonifera*
Call. ham	*Callitriche hamulata*	Elod	*Elodea canadensis* agg.	Grass-g	*Glyceria fluitans/plicata* (with
J. art	*Juncus articulatus*	Pol.hydr, Pol.pers	*Polygonum hydropiper* agg.		short leaves)
El. ac	*Eleocharis acicularis*	C. acut	*Carex acutiformis* agg.	Grass sp.	Small grasses other than the
Sc. fl.	*Elegeton (Scirpus) fluitans*	T. ang.	*Typha angustifolia*		above
J. bulb	*Juncus bulbosus*	T. lat	*Typha latifolia*	J. eff	*Juncus effusus*
G. fluit	*Glyceria fluitans* (long leaves)	Phrag	*Phragmites communis*	J. inf	*Juncus inflexus*
		Pot. perf	*Potamogeton perfoliatus*	J. sp.	*Juncus* spp. other than those
(Yellow)		Polyg. a	*Polygonum amphibium*		listed
Pot. alp	*Potamogeton alpinus*	Bl. wd	Blanket weed (except as above)	L. mi	*Lemna minor* agg.
Phal	*Phalaris arundinacea* (hills	Groenl	*Groenlandia densa*	L. poly	*Lemna polyrhiza*
	and nutrient-poor areas)	Zann	*Zannichellia palustris*	Ment. u	*Mentha aquatica*, special
Moss	Mosses (hills only)	Oen. fl	*Oenanthe fluviatilis*		Corsican form
Bl. wed	Blanket weed (sparse in hills	G. max	*Glyceria maxima*	Myosoton	*Myosoton aquaticum*
	only)	Alis. p	*Alisma plantago-aquatica*	Oen. aq	*Oenanthe aquatilis*
Oen. cr	*Oenanthe crocata*	Pot. luc	*Potamogeton lucens*	Osm	*Osmunda regalis*
Caltha	*Caltha palustris*	M.spic	*Myriophyllum spicatum*	Pot. sp	*Potamogeton* spp. other than
Pet	*Petasites hybridus*	Pot. cris	*Potamogeton crispus*		those listed above
		Phal	*Phalaris arundinacea*	Ran. rep	*Ranunculus repens*
(Turquoise)		Ac	*Acorus calamus*	Sal. e	*Salix eleagnos*
Mim	*Mimulus guttatus*			Sal.sp.Salix	*Salix* spp. other than the
L. tri	*Lemna trisulca*	**(Purple/Red)**			above
Cerat s	*Ceratophyllum submersum*	Epil	*Epilobium hirsutum*	Sc. hol	*Scirpus holoschoenus*
Ver.-b	*Veronica beccabunga*	But	*Butomus umbellatus*	Sc. sp, Scirp	*Scirpus*, other than those
Ber	*Berula erecta*	Cerat	*Ceratophyllum demersum*		listed above
Bid	*Bidens cernua*	Sp. em	*Sparganium emersum*	Tam	*Tamarix gallica*
Call	*Callitriche* spp. other than C.	Enter	*Enteromorpha* sp.	Oth	Other aquatic spp.
	hamulata	Sag	*Sagittaria sagittifolia*	Land	Land sp. or spp. (including
Moss	Mosses (lowlands)	Pot. nod	*Potamogeton nodosus*		*Urtica dioica*, sometimes
Ment	*Mentha aquatica*	Pot.pect	*Potamogeton pectinatus*		entered separately)
Ran	*Ranunculus* spp. (Batrachian)	Rum	*Rumex hydrolapathum*	B. alg	Benthic algae (as obvious
Catab, Grass-c	*Catabrosa aquatica*	Ror.a	*Rorippa amphibia*		green patches)

Table 4.2 Pollution-tolerant species of rivers of different countries
4, tolerant species; 2, semi-tolerant species; 3 and 1 as appropriate; +, species occurring frequently but not tolerant to sewage and town effluent.

	Belgium	Britain	Denmark	France	Germany	Luxem-bourg	Ireland*	Italy (Po plain)	South Norway
Agrostis stolonifera	4	2	4	4	4	4	4	2	2
Glyceria spp., short leaves	2	2	2	2	2	2	4	2	2
Blanket weed	4	4	4	4	4	4	4	4	2
Potamogeton pectinatus	4	4	4	4	4	4	4	4	
Potamogeton crispus	4	4	2	4	2	2	2	2	
Sparganium erectum	2	4	4	2	4		4	2	2
Phalaris arundinacea	2	4	2	4	4	4	2	+	2
Sparganium emersum	2	4	4	+	4	2	4	2	2
Phragmites communis	2	+	2	3	1		2	2	4
Butomus umbellatus		2	2	2		2	2		
Iris pseudacorus	2			2	+	2	2	2	
Nuphar lutea		2	+	4	2		2		
Rorippa amphibia/austriaca	2	2		2	+		2	4	
Scirpus lacustris	2	4		4			4		2
Lemna minor agg.	2	2	4	+	2	+	+	+	+
Potamogeton nodosus	4			4	2	2		+	
Glyceria maxima	4	2	2		2				
Enteromorpha sp.		4			2		4		2
Carex acutiformis agg.	+	+	+	+	2			4	
Myriophyllum spicatum		+		2	+		2		
Land spp., including grasses				2				2	
Acorus calamus									2
Carex spp., other tall spp.	+			+			+	4	+
Elodea canadensis	+	+	+	+	+	+	+	2	
Polygonum amphibium		+	+	2	+				
Polygonum hydropiper agg.	+	+	+	+	1		+	4	
Rumex hydrolapathum		+	+	2	+				
Typha latifolia	2	+	+	+	+		+	+	+
Apium nodiflorum		3		+			+		
Mimulus guttatus		3							
Ranunculus spp.	+	1	+	+	+	+	+	+	

*See also Table 7.10

Source: Haslam (1987a)

Where changes alter habitat, the vegetation also changes. The plants that are lost are made good, at least in part, by species more typical of and better adapted to the new conditions: whether this be, say, increased pollution or deeper water. River changes induce macrophyte changes — slowly. If the river change is rapid, for example because of a toxic spill, plant death is also rapid (see Haslam, 1987a, for a full description).

The Tame (Figure 4.6) was singled out as awful in Chapter 2, and awful it still was, in 1974. A major clean-up led to macrophytes — of a sort — being present in most sites in 1976, with several species often appearing where tributaries dilute the water further. By 1979, between four and six species per site was usual, and no site had only

Potamogeton pectinatus and Blanket weed. The incoming species are appropriate to the habitat: they tolerate pollution. Sensitive species will, of course, be reaching the Tame, but will not develop. The river's current state is still shocking — all should be, at the least, like its cleaner Blyth tributary — but it is greatly improved on 1974. Birmingham has a large population, heavy industry, light industry, and motorways, so the pollution is all that is undesirable. The Cole tributary shows a cyclical pattern, poor in 1974 and 1979, but better in between, the clean-up being overtaken by increased pollution.

The Tame flows into the Trent, also a polluted river. Its upper stretch (Figure 4.7) is worse than the Alzette, covered with vegetation — but practically nothing except *Potamogeton pectinatus* until

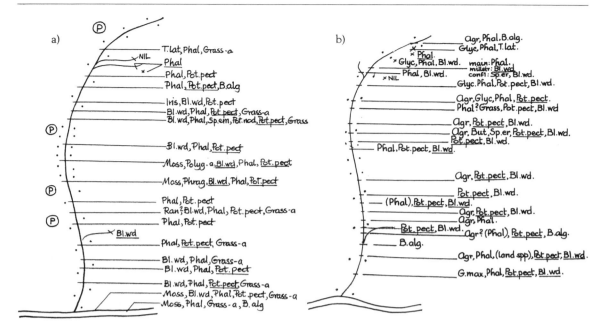

Figure 4.5 River map, Alzette, Luxembourg. Severe pollution, (a) 1980, (b) 1985. P, pollution. Sandstone (etc.), lowland with local gorge (S). Notes as for Figure 4.2

the cleaner Sow tributary provides dilution downstream. Dilution is important! Little change occurs here over the years: pollution has so restricted the species that can occur, that little variation can be expected unless pollution changes (or something drastic like dredging happens).

For another example of intermittent pollution, but here mild pollution, the Suran (Figure 4.8) can be cited. Note that where the pollution occurs the species lists are short, and the limestone assemblage of *Ranunculus – Berula erecta – Callitriche* is blotted out. Further down the whole river community becomes rather distorted, with the loss of characteristic chalk stream species, and the appearance of some tolerant species like *Nuphar lutea*.

The upper Welland (Figure 4.9) illustrates the combined effect of many small pollutions. If clean, this would be better than the Blyth tributary in Figure 4.6 because its general habitat is similar. In fact, it is worse. A town discharges near the source, and over fifty villages discharge into tributaries or rivers along its way. The Welland is worst upstream, with the town pollution. The additional effluents prevent a proper

downstream recovery. Diversity (especially of water-supported species) and cover are too low, Blanket weed too abundant, and pollution-sensitive species are too sparse.

Contrasting pollution effects are seen in the Emmer and the Nethe (Figure 4.10). These are similar in size, rock type and landscape. Both are polluted, but one has low site diversities, much Blanket weed (and good downstream recovery, little downstream pollution). The other is species-rich with much *Potamogeton pectinatus*, over-many tolerant species, and little downstream change. The two patterns reflect different types of effluents. The former probably having higher anions and ammonia (Haslam, 1987a).

Most Sicilian rivers, including the Belise (Figure 4.11), are polluted from source since much settlement is on the limestone hills. This far south, the rivers are wide and deep during the winter rains, but narrow or even dry in summer. The inner channel has water more of the time, so is polluted longer, and is the more polluted the less the water. The outer bed has less pollution, less scour and less flooding, and tends to bear marsh: some of these species are southern, not seen in

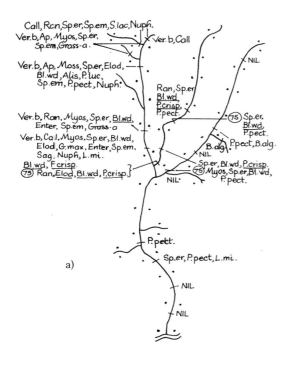

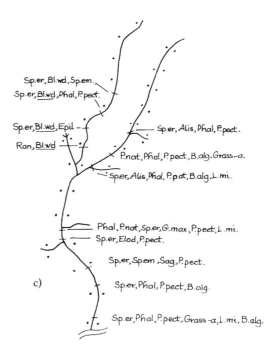

Figure 4.6 River map, Tame, England. Gross to moderate pollution. (a) 1969–72, (b) 1976–7 (c) 1979. Clay, lowland. Notes as for Figure 4.2

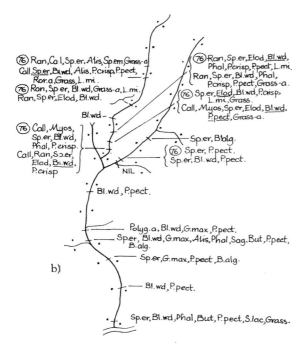

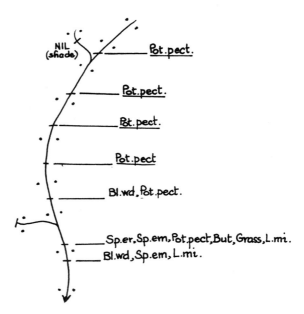

Figure 4.7 River map, Upper Trent, England. Severe pollution. Clay, lowland, 1974 (M)

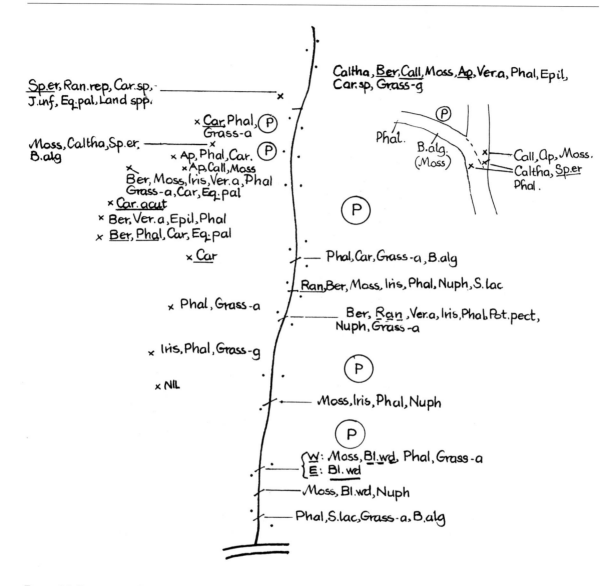

Figure 4.8 River map, Suran, France (from Haslam 1987a). Intermittent pollution, increasing downstream. Limestone, lowland, 1979 (S). P, pollution. Tributaries marked on left for display: most are on right

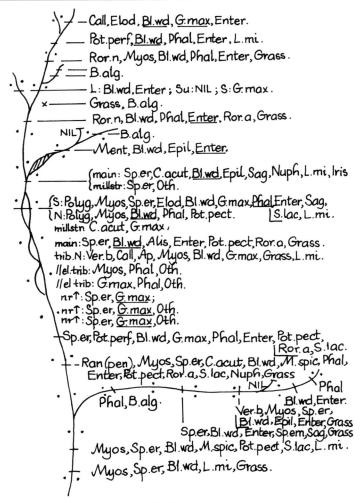

Figure 4.9 River map, Upper Welland, England. Pollution near source, slight recovery, then maintained by many small effluents. Clay (limestone), lowland, 1980 (S)

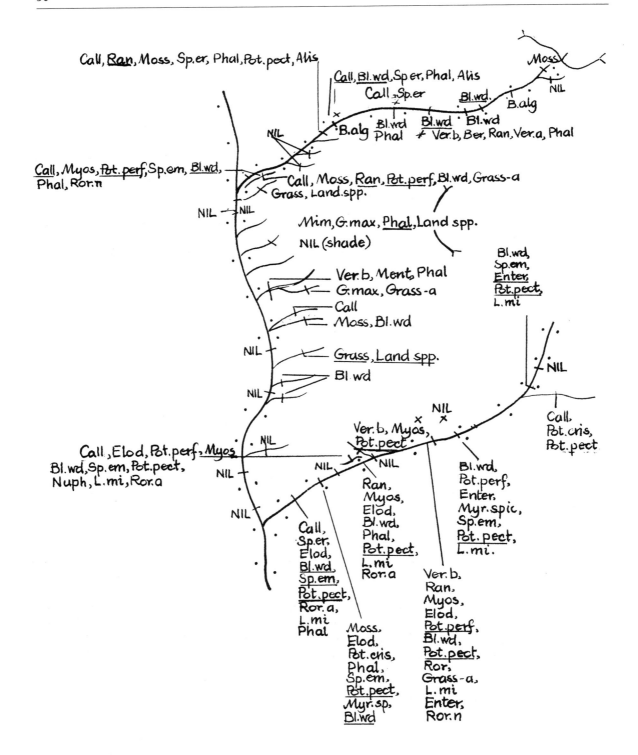

Figure 4.10 River maps. Emmer and Nethe, Germany (from Haslam, 1987a). Moderate pollution (Weser damaged by boats and erosion). Muschelkalk (mainly), lowland, upland, 1977 (S). Notes as for Figure 4.2

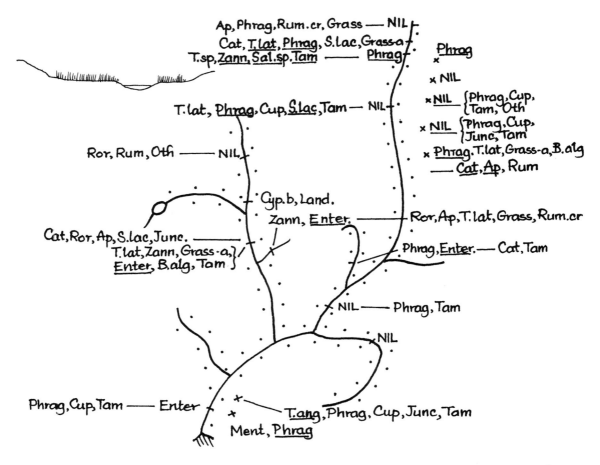

Figure 4.11 River map, Belise, Sicily (from Haslam 1987a). Polluted from near source. Where two species lists are given for the same site, the inner refers to the inner channel wet in early summer, the outer to the wide fringe, drying in early summer. Limestone, hilly, 1980 (M). Notes as for Figure 4.2

the other river maps here. The inner bed bears no, or few and tolerant macrophytes. (Note that *Phragmites communis* becomes more pollution-tolerant in the south.)

The last two in this series, the Dyle/Dijle and the Suir (Figures 4.12 and 4.13) illustrate well how man's opinion on river health depends on his perception of pollution, which again depends on what is considered normal. The Dyle is a typical Belgian lowland river. Apart from one clean tributary, its state is disastrous. Since most people there see nothing different, little concern is aroused (or was not, some years ago). Quite different to the Suir, renowned as shockingly polluted for Ireland, and with quite a literature (see, for example, Flanagan, 1974; Horkan, 1980;

1984). The contrast is great! Certainly the Suir is polluted; there is too much *Potamogeton pectinatus*; *Scirpus lacustris* and *Sparganium emersum* occur too far upstream for limestone; sewage fungus occurs by sewage works, and the angling may be poor below towns. The worst town already has a new sewage works, and more is in hand. But. . .! This shows the importance of EEC directives on river quality. Those who do not know the good have to be shown it, and the force of law is often needed for such improvements, often regarded as cranky, unnecessary and expensive, to be carried out.

Looking at ten European countries (Figure 4.14), pollution is one of the two most important factors reducing river vegetation. The other is the

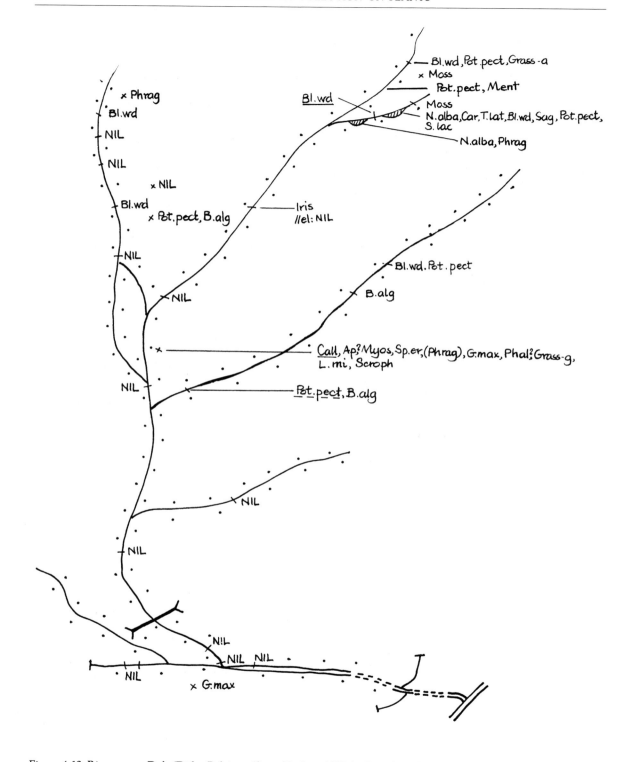

Figure 4.12 River map, Dyle/Dijle, Belgium (from Haslam 1987a). Grossly polluted. Mixed rock types, lowland, 1980 (L). Notes as for Figure 4.2

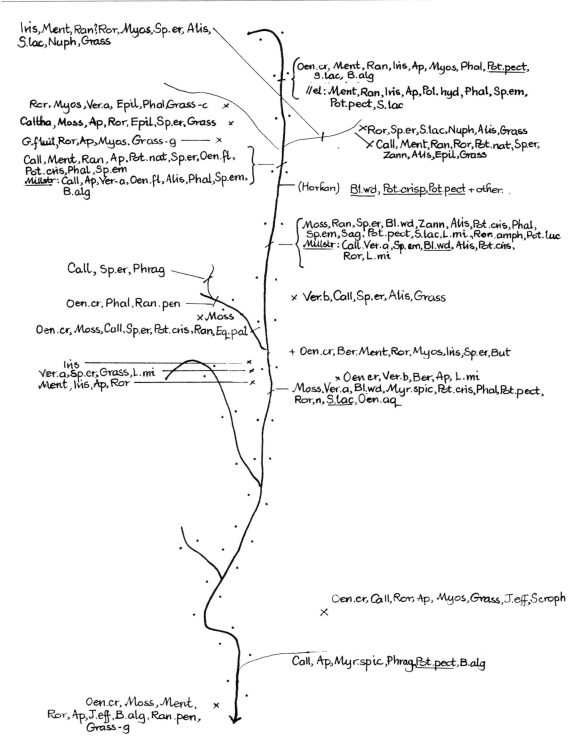

Figure 4.13 River map, Suir, Ireland (from Haslam 1987a). Some intermittent pollution. Limestone, upland, 1977 (L). Notes as for Figure 4.2

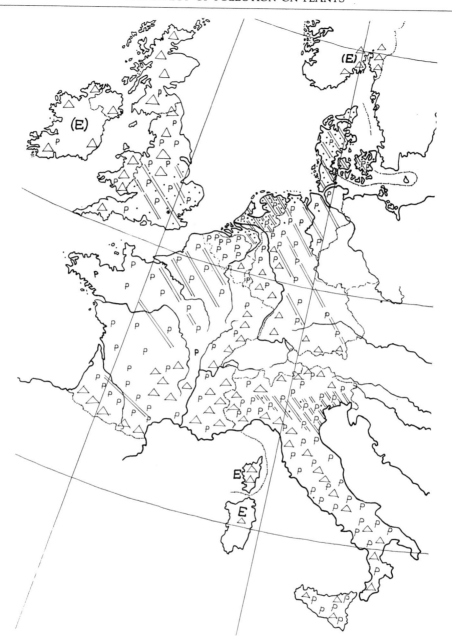

Figure 4.14 Distribution of the main habitat factors reducing vegetation in part of Europe (from Haslam 1987a)

As these factors vary so much in intensity, and in local distribution within regions, the areas marked are approximate. Triangles: main mountain regions, where vegetation is likely to be reduced by high water force, making much weed control unnecessary, and where straightening is usually both difficult and unnecessary. P, pollution. The density of the symbols indicates the relative density of pollution — dots; large areas with the most intensive weed control; lines; (large) areas in which the streams have been most straightened; E. early historical phase, parts with exceptionalllly low human interference; (E) less human interference than other comparable regions (Ireland with grassland, Norway still with much woodland).

Table 4.3 List of typical sewage fungus organisms

Bacteria
- [a] *Sphaerotilus natans*
- [a] *Zoogleal bacteria*
- [a] *Beggiatoa alba*
- [a] *Flavobacterium* sp.

Fungi
- [a] *Geotrichum candidum*
- [a] *Leptomitus lacteus*
- *Fusarium aquaeductum*
- *Penicillium fluitans*

Ciliates
- *Colpidium colpoda*
- *Colpidium campylum*
- *Chilodonella cucullulus*
- *Chilodonella uncinata*
- *Cinetochilum margaritaceum*
- *Trachelophyllum pusillum*
- *Paramecium trichium*
- *Paramecium caudatum*
- *Uronema nigricans*
- *Hemiophrys fusidens*
- *Glaucoma scintillans*
- [a] *Carchesium polypinum*
- *Carchesium lachmanni*
- *Carchasium spectabile*

Algae
- [a] *Stigeoclonium tenue*
- *Navicula* spp.
- *Flagilaria* spp.
- *Synedra* spp.

[a] Indicates a common 'sewage fungus' organism.

Source: adapted from Hellawell (1986)

high water force found in mountains. (Straightening comes next, reducing vegetation only where water is, or becomes, shallow or eroding. Intensive weed control is more local.) The cleanest parts have the lowest human interference (Ireland, Norway, Corsica, Sardinia). The dirtiest, in regional terms, are not the most industrial parts, but those with least control and treatment, as in lowland Belgium, South-West France.

How macrophyte habit is affected by pollution

When exposed to pollution, a plant may grow poorly — particularly when roots are affected. It

Table 4.4 Pollution tolerance of diatoms in Belgium 4, tolerant species; 1, sensitive

Species	Tolerance
1. *Amphipleura pellucida*	4
2. *Amphora ovalis*	4
3. *Caloneis alpestris*	4
4. *Caloneis silicula*	4
5. *Cymatopleura solea*	3.5
6. *Cymatopleura elliptica*	4
7. *Cymbella affinis*	4
8. *Cymbella cistula*	3
9. *Cymbella lanceolata*	4
10. *Cymbella prostrata*	4
11. *Cymbella tumida*	4
12. *Cymbella ventricosa*	4
13. *Diatoma vulgare*	3.5
14. *Diatoma vulgare* var. *productum*	2.5
15. *Diploneis elliptica*	4
16. *Eunotia pectinalis*	2
17. *Frustulia vulgaris*	3.5
18. *Frustulia rhomboides* var. *amphipleuroides*	4
19. *Gomphonema abbreviatum*	3
20. *Gomphonema augur* (et var. *gautieri*)	3.5
21. *Gomphonema constrictum*	3.5
22. *Gomphonema olivaceum*	4
23. *Gomphonema parvulum*	1
24. *Gomphonema tergestinum*	2
25. *Gyrosigma attenuatum*	3.5
26. *Gyrosigma spenceri* var. *nodiferum*	4
27. *Hantzschia amphioxys*	2.5
28. *Melosira varians*	3
29. *Navicula cryptocephala* (+ *N. rhynchocephala*)	2
30. *Navicula cryptocephala* var. *veneta*	2
31. *Navicula gracilis*	4
32. *Navicula menisculus*	4
33. *Navicula mutica*	1
34. *Navicula viridula*	2.5
35. *Nitzschia amphibia*	2
36. *Nitzschia commutata*	3
37. *Nitzschia dissipata*	4
38. *Nitzschia filiformis*	3
39. *Nitzschia linearis*	3
40. *Nitzschia palea*	1
41. *Nitzschia sigma*	3
42. *Nitzschia sigmoidea*	4
43. *Pinnularia bredissonii*	1
44. *Surirella biseriata*	4
45. *Surirella ovata*	3
46. *Synedra affinis*	3
47. *Synedra pulchella* ar. *lanceolata*	2
48. *Synedra ulna*	2
49. *Synedra vaucheriae*	2

Source: Descy (1976d)

Table 4.5 Effect of sewage works effluent on root development of *Berula erecta* and *Myosotis scorpioides* Plants were suspended on black polythene sheets above water containing nutrient solution and above water collected 100 m downstream of the entry of Cambridge (Britain) sewage works effluent.

Results obtained after 3 months, when the more polluted *Berula erecta* shoots were yellow and the *Myosotis scorpioides* upper roots were much less bushy (so would anchor less firmly) than those of the control.

The root length/shoot ratio compares the relative amounts of root growth.

	% pollutant (effluent)	Root length/ shoot length
Berula erecta	0	0.40
	50	0.28
	100	0.20
Myosotis scorpioides	0	0.84
	25	0.50
	50	0.50
	100	0.43

Source: Haslam (1987a)

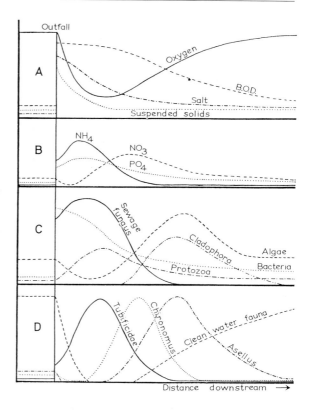

Figure 4.15 Downstream effects of organic pollution. A and B physical and chemical changes, C changes in micro-organisms, D changes in invertebrates.

Source: Hynes (1960)

Elodea canadensis is plastic, varying greatly with different chemicals (Figure 4.20).

is more easily washed out by storm flow, and so in streams pollution acts twice, affecting growth, and loss.

Decreased root growth is important. Species with very long roots or deep rhizomes are still well anchored even if the parts are considerably smaller. In short-rooted species, however, even a small decrease in the size of the roots will much decrease anchorage. Partly-rooted plants will then be washed out irrespective of whether the rest of the plant is in good condition. In fact if the shoots are normal and the roots small, the effect on the plant is greater, and erosion is easier, than if the shoots were small as well, and the plant is more likely to be washed out. Table 4.5 shows one example. *Ranunculus* spp. are more tolerant to pollution on consolidated gravel: where roots entangle and shorter roots have least effect. *Apium nodiflorum* is semi-tolerant to pollution only in brooks with low water force and solute-rich rock type. Polluted plants anchor less well. Some variation in shoots can be summarised as shown in Table 7.6.

Within one site yellowing will occur on the shoots on the siltiest substrate furthest into the water: i.e. receiving the most pollution.

Sewage fungus

This is quite a different kind of — partly plant — community, defined (as in Sanders, 1982) as excess and readily visible microbial growth attached to surfaces in watercourses which may receive polluting discharges. The type of micro-organism is determined by the effluent type. Its typical distribution is shown in Figure 4.15, some of its common constituents in Fig. 4.16, and a fuller list in Table 4.3. The Biological Oxygen Demand (BOD) is usually over 5 mg l^{-1} (but may be less than 1 mg l^{-1}); sewage fungus therefore normally indicates the presence of pollution. Sewage fungus may smother invertebrates — and indeed macrophytes — fish eggs, etc., and cause

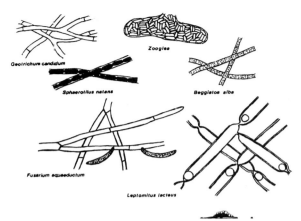

Figure 4.16 Common bacterial and fungal constituents of sewage fungus.

Source: Hellawell (1986)

deoxygenation. However, sewage fungus removes dissolved organic matter, particularly short-chain carbohydrates. Growth can be very quick.

Microphytes

River surface communities are all different: those in the water, on sand, on silt, on stones, on macrophytes. They are usually dominated by diatoms, though other algae are always present, too. Most work has been done on diatoms (Figure 4.18). These can live in the water, as phytoplankton, and on all river surfaces. Species may be numerous, even reaching several hundreds. As well as the algae, each of these microhabitats also bears bacteria, fungi and small animals.

On firm substrates, diatoms are likely to develop in layers, or 'forests'. The bottom layer is of flat forms, attached to the substrate. The second layer, growing above, is of upright forms, usually stalked and tree-like. These diatoms are non-motile. The top layer is a 'canopy' in mucilage, and the algae here are motile. In swift upper reaches, the bottom, attached layer is well developed, but the others are sparse or absent. In medium flow the middle layer develops, and in slow, silted places, the top one. There are usually only a few abundant species (see Chapter 7) (see Methods for the Examination of Waters and Associated Materials, 1989).

Diatom communities are influenced, like macrophyte ones, by their environment: by flow, rock type, grazing, pollution, etc. Those in the Belgian Ardennes vary with rock type and water chemistry. And, of course, they have the deformation of community imposed by pollution.

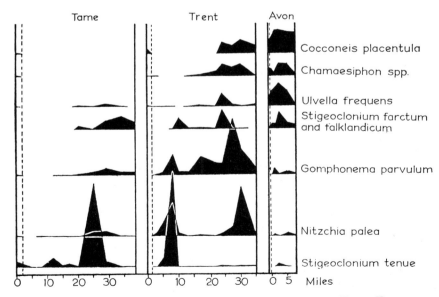

Figure 4.17 Changes in algae downstream of organic pollutions, Tame, Trent, Avon, England. Algae grown on grass slides in the river. Discharges shown by dashed lines. 5 miles = 8 km. *Source:* Hynes (1960)

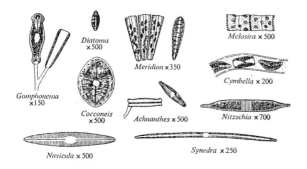

Figure 4.18 Diatoms which are common in the attached algal community of streams.

Source: Hynes (1960)

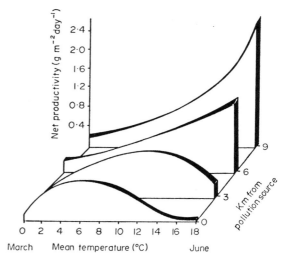

Figure 4.19 Periphyton productivity and stream temperature downstream of domestic pollution. Red Cedar River, USA (from Ball and Bahr, 1970). Data are for seasons of increasing photoperiods

Some species flourish at a particular — and mild — level of pollution, when more sensitive species disappear (Table 4.4). This means reduction of community occurs only with more severe pollution (Descy, 1976a, 1976c, 1976d, 1979).

Other patterns of algae with pollution are shown in Figures 4.18 and 4.19. Algal blooms of phytoplankton in the water developing because of pollution, may poison cattle, sheep, pigs, ducks, and other animals.

Pollution and the organism

The reaction to toxins is governed by the nature and concentration of the toxicant, the time of exposure, the other features of the habitat, the characters of the organism, and the other toxins present (Cairns *et al.*, 1972). Most toxins are considered to act by combining with enzymes, cell membranes or other functional components of cells (Warren, 1971).

In the field studies of Kohler and his associates (for example Kohler and Zeltner, 1981; Figure 4.19; Table 4.6), phosphate and ammonium ions are the two measured water constituents most closely associated with species distribution. Some species have wide ranges, some narrow. Since numerous chemicals are present, each in variable quantities, there are no precise ranges for any species. Detergents, for instance, increase ammonium uptake by roots and where ammonium is high death would occur with detergents in concentrations otherwise tolerable (Litav and Lehrer, 1978).

Transplants to test such findings are shown in Table 4.7. Transplanted species do develop best in — and so indicate — their own habitats. Sewage works effluent tests show remaining toxicity when diluted to below the level at which the individual analysed chemicals are toxic, and laboratory tests show that high levels of groups of, for example, nutrients are more toxic than are

Table 4.6 Macrophyte response to pollution ranked in order of damage (from most to least sensitivity)

a) Marlon A, 0.5 mg l^{-1} — % net loss of Photosynthesis

Potamogeton crispus	77 sensitive
P. coloratus	45
P. alpinus	39
P. lucens	39
Groenlandia densa	31
Elodea canadensis	27 tolerant

(Low to moderate damage with 0.5 mg l^{-1}. At 5 mg l^{-1}, damage was heavy).

b) Zinc, 5×10^{-5} mg l^{-1}

Groenlandia densa	sensitive
Elodea canadensis	
P. coloratus	
Ranunculus spp.	tolerant

c) Chloride — concentration for significant damage (mg l^{-1})

P. alpinus	50 sensitive
Elodea canadensis	250
Myriophyllum alterniflorum	1,000 tolerant

d) Boron — Concentration for significant damage (mg l^{-1})

M. alterniflorum	5 sensitive
P. alpinus	10
Ranunculus spp.	10
Elodea canadensis	10 (killed in 11 days with 100 mg l^{-1}) tolerant

e) Phosphate

P. alpinus	damage 0.01 mg l^{-1}, kill 5 mg l^{-1} sensitive
Elodea canadensis	damage 0.1 mg l^{-1}, kill 5 mg l^{-1} tolerant

f) Toluene

Juncus effusus	sensitive
J. bulbosus	
Comarum palustris	
Eleocharis palustris	
Carex riparia, Iris pseudacorus	
Menyanthes trifoliata, Scirpus lacustris	
Phragmites communis	
Typha latifolia	
Sparganium erectum	
Potamogeton natans	
Lemna minor	
Ranunculus aquatilis	
Sagittaria sagittifolia	
Ceratophyllum demersum	
Callitriche stagnalis	
Polygonum hydropiper	
Alisma plantago-aquatica	
Mentha aquatica	tolerant

Sources: Labus and Kohler (1981); Nobel (1980)

separate ones. (The same is found for fish, as in Alabaster et al., 1972.)

Many poisons accumulate in organisms, so their concentrations become greater than those in the surrounding water or soil. So, in rivers with concentrations which are tolerable for a short time, long exposure can build up much higher levels in the plants and animals. This bioaccumulation happens in two ways (Baudo, 1985). The first is bioconcentration or the uptake of the chemical from the environment. This may be active (metabolically controlled) or passive. Metals vary in the form (species) in which they occur, and this determines their effect, since only the form compatible with the uptake mechanism is absorbed. The second, which is for animals only, is biomagnification through the food chain. Cadmium in food leads to cadmium in the eater. The relative importance of the two pathways, for animals, depends on the relative supply coming from, on the one hand, water and soil, and, on the other, from food. The degree of bioaccumulation depends on the difference between the pollutant's ability to penetrate the organism and on the organism's ability to excrete it (Boudou et al., 1971).

The remainder of this section, up to and including the discussion of waste heat, except where separately referenced, is summarised from the review by Kohler and Labus, 1983.

Nutrients

Studies of nitrogen metabolism show complex interactions. Several species can use ammonium or nitrate, but prefer ammonium if both are present. (In one, the moss Fontinalis antipyretica, ammonium suppresses nitrate reductase activity, inhibiting nitrate uptake. Reactivation is light-dependent.) Some carbon sources are shown in Table 4.8. Increased organic loading to Potamogeton pectinatus reduces starch, hexoses, malic acid, and several amino acids, while (with increased phosphate and nitrogen) asparagine and sucrose contents rise.

Heavy metals

Some heavy metals are essential trace elements (Co, Cu, Fe, Mn, Mo, Zn), which can occur in

Table 4.7 Transplant tests between community zones, Germany

Species transplanted	Development of transplanted spp. in zone characterised by a specific sp.			
	Potamogeton coloratus	Berula erecta	Good Callitriche obtusangula	Poor Callitriche obtusangula
Potamogeton coloratus	Good	Good	Fair	Poor
Groenlandia densa	Fair	Good	Good	Fair
Ranunculus trichoides	Good	Good	Good	Poor
Hippuris vulgaris	Good	Fair–good	Good	Fair
Callitriche obtusangula	Fair–good	Fair	Good	
Ranunculus fluitans	Poor	Fair	Good	Good

Source: Kohler et al. (1974)

Table 4.8 Macrophytes using bicarbonate for photosynthesis

	Maximum pH	HCO$_3^-$ using		Maximum pH	HCO$_3^-$ using
Callitriche hamulata	9.5	↙	Juncus bulbosus	7.9	x
Elodea canadensis	9.8	↙	Myriophyllum alterniflorum	9.5	↙
Fontinalis antipyretica (hard water)	6.6	x	Potamogeton alpinus	9.6	↙
Fontinalis antipyretica (soft water)	7.6	x	P. coloratus	9.6	↙
Groenlandia densa	7.4	↙	P. crispus	9.5	↙
Isoetes echnospora	7.7	x	P. pectinatus	9.8	↙
I. lacustris	7.7	x			

Source: Kohler and Schoen (1984)

quantities much exceeding those needed for growth and reproduction. Others (such as Cd, Cr, Hg, Pb, V) have no known function in the plant. Algal growth (photosynthesis) may be inhibited at 1–2 parts per billion (ppb) of Cu^{2+}, though with variations: *Chlorella* is more sensitive than *Scenedesmus*. Sensitivity varies with metal, also, *Chlorella* being retarded more by silver than by cadmium, mercury or nickel, and cell division being reduced more by cadmium than by copper or mercury (at 0.32 parts per million (ppm)). *Lemanea* is especially resistant to zinc and lead Within-species variation occurs also. The Cyanophyceae (blue-green algae), Zygnemaceae and diatoms all include strains varying from sensitive to resistant, so variation probably occurs elsewhere also. At least some Bryophytes appear to be highly resistant.

Toxicity varies with speciation (for instance, chelators may decrease toxicity, methylated forms may increase it), organic matter (in which metals are absorbed), oxygen (reduced oxygen may increase toxicity), and temperature (which may increase accumulation and hardness). Joint toxicities may be additive or synergistic (enhancing) or even antagonistic (vanaduim and rubiduim). Tables 4.9–4.12 illustrate some concentrations, accumulations and tolerances.

Acute toxicity by metals in freshwater is less common, and therefore less important than chronic toxicity, that is, than long-continued sublethal levels. Concentration factors in organisms here vary from 100 to 110 000 between different species and metals. Accumulation tends to be more in mosses than in angiosperms, and so these are more likely to be useful for assessing river quantities (see Chapter 7). Within the plant, more tends to accumulate in lower than in upper parts. Species differ, and metals differ. Three algae, one liverwort and three mosses were compared for accumulation of zinc, cadmium and lead, and marked variations were found, e.g. for zinc (at 0.01 mg/e) the bryophytes were the highest, the two green algae showed the greatest change in plant concentration in response to water concentration, and the red algae had a change closer to the

Table 4.9 Maximum concentrations of heavy metals in aquatic bryophytes (ppm)

Species	Lead	Zinc
Fontinalis squamosa	–	5 430
Philonotis fontana	5 965	7 023
Rhynchostegium riparioides	–	6 705
Scapania undulata	14 825	1 950
Scapania undulata	8 902	3 558

Source: Hellawell (1986)

bryophytes but at a much lower concentration (Kelly and Whitton, 1989).

Impact is reduced in hard, well-buffered freshwater systems (see also the discussion of fragile waters in chapter 3); for example, zinc toxicity is less with high calcium for *Stigeoclonium* and *Hormidium*, less with high pH for *Hormidium* (pH 6.1–7.6 rather than 3.5–4); cadmium uptake is less in hard water in *Nitella* and *Elodea*. Acid-tolerant diatoms, similarly perhaps, are more tolerant to copper and zinc.

Table 4.10 Chemical composition of aquatic macrophytes from Lake Como (mean values, mg kg^{-1} dry weight, and percentage dry weight for Ca)

	No. of samples	Al	Ti	Fe	Mn	Ca	Mg	P	Zn
P. perfoliatus	43	1012	149	785	58.5	4.04	7218	3319	93.5
P. gramineus	16	1531	353	1283	118	6.06	3419	2820	101
P. lucens	12	643	144	439	51.7	3.78	3936	2890	75.9
P. crispus	10	1208	112	952	81.3	3.36	3124	3615	100
P. pusillus	1	6712	2336	6675	415	1.45	4855	6384	137
M. spicatum	35	2079	196	1410	124	2.87	4776	3372	125
V. spiralis	14	3404	362	2676	293	2.79	8227	3265	200
R. trichophyllus	6	2054	138	1135	205	3.11	4980	3376	121
R. circinatus	3	3718	439	2317	230	5.26	7028	4074	182
E. densa	6	2790	323	1656	417	2.95	3671	3663	276
E. canadensis	2	2404	171	1094	76.5	12.6	3080	1070	49.2
N. marina	2	616	45	1600	121	1.91	7048	2904	88.9
C. demersum	1	807	49.4	2395	1504	2.24	6330	7603	1394
	No. of samples	Ni	Cr	V	Cu	As	Hg	Cd	Pb
P. perfoliatus	43	31.7	15.70	1.43	11.9	4.68	0.11	0.84	5.70
P. gramineus	16	23.1	6.92	3.21	11.8	7.86	0.14	0.28	9.24
P. lucens	12	23.7	4.22	1.79	15.5	4.02	0.09	0.44	2.04
P. crispus	10	24.9	6.22	1.47	12.7	3.37	0.12	0.66	5.37
P. pusillus	1	21.4	18.10	15.20	21.8	5.80	0.30	0.27	27.70
M. spicatum	35	21.6	7.85	3.80	12.4	6.62	0.12	0.57	12.30
V. spiralis	14	34.5	9.34	7.72	25.4	5.49	0.15	0.56	20.20
R. trichophyllus	6	29.0	5.31	2.71	13.3	17.9	0.08	0.75	7.37
R. circinatus	3	32.2	10.40	8.26	15.0	7.93	0.09	0.86	22.90
E. densa	6	49.6	5.48	3.49	19.8	9.29	0.12	1.36	10.50
E. canadensis	2	18.1	5.00	2.70	8.0	6.75	0.05	0.22	8.56
N. marina	2	23.9	4.25	12.70	4.94	4.79	0.85	0.25	22.40
C. demersum	1	49.4	20.60	3.31	35.0	20.00	0.12	11.00	33.90

Source: Baudo, Canzian, Galanti, Guilizzoni and Rapetti (1985)

Table 4.11 Tolerance towards zinc of some of the more abundant species in the study area, as determined by field observations in old mine workings, N. England

Very resistant >10 mg l⁻¹ Zn	Resistant 2.0–10.0 mg l⁻¹ Zn	Moderately resistant 0.5–2.0 mg l⁻¹ Zn	Low resistance 0.1–0.5 mg l⁻¹ Zn	Very low or no resistance <0.1 mg l⁻¹ Zn
Euglena mutabilis	Chamaesiphon	Chrysonebula holmesii	Navicula radiosa	Diatoma vulgare
Caloneis bacillum	polymorphus	Ceratoneis arcus	Synedra ulna	Tabellaria flocculosa
C. lagerstedtii	Homoeothrix varians	Cymbella ventricosa	Drapamaldia	Microspora amoena
Eunotia exigua	Phormidium mucicola	Fragilaria intermedia	mutabilis	Ulothrix zonata
E. tenella	Plectonema	Gomphonema		
Neidium alpinum	gracillimum	parvulum		
Pinnularia borealis	Hydrurus foetidus	Meridion circulare		
P. subcapitata	Achnanthes	Staurastrum		
P. viridis var. sudetica	minutissima	punctulatum		
Cylindrocystis	Diatoma hiemale var.			
brebissonii	mesodon			
Mougeotia sp.	Microspora sp.			
(8–12 μm)	(8–12 μm)			
Hormidium rivulare	Stigeoclonium tenue			
Microthamnion				
strictissimum				
Ulothrix moniliformis				

Source: Say and Whitton (1981b)

Table 4.12 Accumulation of Zn and Pb by an alga, moss and a (semi-aquatic) angiosperm from a moderately polluted stream (Rookhope Burn, Co. Durham, England) and an unpolluted river (R. Tweed, Scotland). All sites have similar pH values (c.7.5–8.2); values for water chemistry (filtrable through No. 2 SINTA funnel) are means of various analyses. The level of the heavy metal in the organism is expressed as mg g⁻¹ dry weight, and accumulation is indicated by the ratio of this value to that in the water

Condition of river	Site	Water chemistry (mg l⁻¹) Ca	Zn	Pb	Lemanea fluviatilis Zn	Ratio	Pb	Ratio	Hygrohypnum ochraceum Zn	Ratio	Pb	Ratio	Mimulus guttatus Zn	Ratio	Pb	Ratio
Polluted	A	24.2	0.056	0.046									485	870	307	665
	B	27.0	0.18	0.072									170	940	103	1 430
	C	29.3	0.15	0.063	1 514	10 000	842	13 500	6 730	45 000	5 032	79 000				
	D	32.2	0.08	0.063	1 197	15 000	306	8 100	4 109	51 000	2 254	63 000				
Unpolluted		16.5	0.02	0.001	331	150 000	16	16 000	251	120 000	110	110 000	54.1	27 000	13.1	13 000

Source: Whitton (1970)

Primary production decreases, as Baudo *et al.* (1981) record. However, they also note that heavy metals seldom occur singly and alone, but with, for example, pollution from biocides, fertilisers, farming hormones, and the like, so care is needed when interpreting field results.

Field data are recorded by Say and Whitton (1981a). Algae showed different patterns with high and low levels of zinc (associated with high cadmium and other elements), and also with water source: whether stream, or mine adit. Bryophyte distributions were less marked (varying more with water source, and metal accumulation being influenced by water chemistry).

Biocides

These include many substances with different patterns of toxic action, decomposition pathways and environmental persistence. Among herbicides, phenoxycarboxylic acid derivatives (2,4-D, MCPA) are used in greater quantity than all the rest (triazines, carbonates, chlorinated carboxylic acids). Among insecticides, persistent organochlorine compounds (DDT, dieldrin, endrin, toxaphene) have largely been substituted by organophosphorus compounds which are more easily degraded. Polychlorinated biphenyls (PCBs) are similar to organochlorine insecticides, but are mainly of industrial origin. Phenols and their derivatives come from coke and coal conversion works, oil refineries and petrochemical works. Such poisons therefore come either from intentional use for farming, or as waste products from industry.

Biocides in the water may remain there, or be absorbed on to particles (suspended solids, sediments) in the stream, depending on the amounts of absorptive humic substances and clay, on the solubility of the components, and on water factors. Compounds with acidic functional groups are influenced by pH.

A few pesticides degrade better in anaerobic sediments (e.g. DDT, Lindane), but more degrade better in aerobic ones. Plants are of much service in degrading many poisons (see Chapter 3).

Aquatic herbicides are used on potentially or actually choked drainage and other channels. These herbicides may be short-lived or persistent. Sensitivity varies between plant groups, for example, 20 ppm of 2,4-D may kill angiosperms while not influencing Cyanophyceae. (Blue-green algae)

The most common toxic effect is to impair photosynthesis directly, but others may, for instance, uncouple oxidative phosphorylation, or inhibit nitrate reductase. Bioaccumulation can be great, up to 84 000 times being recorded. This is partly because of their low solubility in water.

Herbicides alter the vegetation directly. Insecticides may do so indirectly, for instance by allowing algal blooms to develop. Blooms cause turbidity, some cause toxicity, and, since zooplankton, etc., feed on algae, blooms may lead to alterations in predator populations.

Chlorine

Chlorinating drinking water is common and widely approved. Chlorination elsewhere, of cooling water (see Chapter 6) or sewage, affects the river directly and is less acceptable. The same applies to chlorinated hydrocarbons used in farming. Residual chlorine below sewage works decreases bacteria and phytoplankton and therefore decreases self-purification. Chlorine also makes highly toxic derivatives of organic materials.

Surfactants

Detergents are contact poisons coming largely from household waste. The biologically non-degradable alkyl benzene sulphonates are increasingly, in more enlightened parts, being replaced by the degradable linear-chained forms. At the time of writing (1989), Britain still uses the phosphate-rich detergents while countries such as Germany and Italy now use phosphate-free ones. (This is an immediately obvious way in which British pollution could be decreased. There are no signs of this happening, for the usual reasons. Comment is by the writer, not by Kohler and Labus.) Even quite low concentrations interfere with gas exchange. Algal growth is hindered, depending on the alga and the surfactant, at between 0.1 and 10 ppm of cation-active compounds, and at between less than 1 and around 10 000 ppm of non-ionic ones. They interact with cellular and sub-cellular membranes, and penetrate internally.

Macrophytes are the most sensitive indicators of damage by anionic surfactants (see Table 4.13).

Radioactive wastes

Since the nuclear test ban treaty of 1963, fall-out

Table 4.13 Macrophyte tolerance to turbid water

1. Most tolerant of turbid water

Ceratophyllum demersum	*Polygonum amphibium*
Lemna minor agg.	*Sagittaria sagittifolia*
Nuphar lutea	*Scirpus lacustris*

2. Intermediate

Callitriche spp.	*P. pectinatus*
Myriophyllum spicatum	*Sparganium emersum*
Potamogeton natans	*Sp. erectum*

3. Least tolerant of turbid water

Elodea canadensis	*Ranunculus* spp.
Potamogeton perfoliatus	Mosses

Source: Haslam (1978)

from weapons has been insignificant and the focus of concern has shifted to nuclear power plants. The average annual amounts reaching freshwaters are (barring accident) levels considered acceptable. Attention is moving to naturally radioactive streams, those with radon in the water or uranium, thorium, etc., in the sediment (Chapter 1), whose widespread distribution is only now becoming apparent.

The elements with the longer halflives (the time taken to degrade half the radioactivity) are considered the more toxic (^{90}Sr, ^{137}Cs, ^{144}Ce of the fission products, and ^{54}Mn, ^{60}Co, ^{60}Zn of the activated nucleides). They can remain in solution or be in suspension or sediment. Sediment levels can be high. Ordinary river reactions determine their movement, and so the decontamination of specific areas. As with ordinary forms of metals, plant uptake is greatest at low pH when most is in solution, and is in proportion to the total present. Concentration factors vary (from 100 to 500 000 for ^{90}Sr), algae accumulating more than angiosperms. Even so, algae (Table 5.22) are relatively tolerant compared with fish and mammals. Radioactive springs in Japan have been recorded with algae (e.g. bearing twelve diatoms, two other algae).

Salinity and chlorides

Salinity increases near estuaries, with rock type (salt deposits), the balance of evaporation over precipitation, and pollution. Chloride is always present in sewage. Clean waters usually have 2–20 ppm, while at the other extreme deicing salt can give temporary peaks of 3000 ppm. Coal mines can give 1700 ppm river chloride in Germany.

Macrophytes vary in sensitivity (Table 4.13). *Myriophyllum spicatum* is all right at 1.33 per cent salinity (NaCl) and occurs in estuaries (as does *Potamogeton pectinatus*). *Ceratophyllum*, on the other hand, has 50% inhibition of growth at 0.38 per cent (and *Potamogeton perfoliatus* at 3.5 per cent). Anion ratios and indeed other ratios, may affect tolerance. Seawater is tolerated better than freshwater with added sodium chloride.

Acidification

Water plants may obtain their carbon supplies from dissolved carbon dioxide, or from the larger molecules of biocarbonates etc., in the water. Below pH 4 macrophytes are limited and CO_2-source species are favoured, e.g. *Juncus bulbosus*, *Sparganium emersum*, *Glyceria fluitans*, *Eleocharis acicularis*, *Typha latifolia*, *Juncus effusus*, and spp. of *Polytrichum*, *Anisothecium*, *Fontinalis*, *Catharina* (see Table 4.8.). The next-less restricted group include *Isoetes lacustris* and *Lobelia dortmanna*. Some algae are more tolerant, *Chlamydomonas acidophila* going down to pH 1, and *Euglena* and others to 1.6. In acid mine drainage, bacterial and fungal counts decrease between pH 4 and pH 2.

Acidification (the causes of which were discussed in Chapter 1) lowers productivity and diversity, restricting distribution to tolerant species. Acid rain includes the effects not just of pH but of the sulphur and nitrogen compounds which cause the acidity, and all the other pollutants (hydrocarbons) which may be present.

Waste heat (thermal pollution)

An 8–10°C rise in temperature is not uncommon, though a limit of 3–5°C is prescribed by law in various countries. The rise depends on the amount of heat discharged, the mode of release, the properties and quantity of the receiving waters, climate and weather. The temperature of water influences its viscosity (lower in hot water) density (same), ionisation, solubility of compounds (more soluble in hot water) and gases (less soluble in hot water: oxygen deficit possible). Then the microorganism metabolism probably increases with heat, changing the BOD and therefore self-purification from organic waste.

Death usually occurs only a few degrees above

Table 4.14 Suspended solids content of industrial and other effluents and rivers into which they are discharged

Substance/effluent	Typical range or mean value (mg l^{-1})
China clay (rivers)	28–91 268
	500–100 000
Paper manufacture	200–3 000
	1 180
Leather (tannery)	3 000–8 000
	2 000–3 000
Petroleum production	441
Metal finishing and plating	200–1 000
Cotton textiles	30–300
Laundry	2 000–5 000
Meat packing	99–3 200
	1 400
	500–5 000
Cannery	200–2 500
	1 350
Fruit canning	100–750
Vegetable canning	30–2 220
Sugar-beet	800–4 300
Poultry	200–2 500
Primary sewage treatment	252
Sewer storm overflows	2 968

Source: Hellawell (1986)

maximum growth temperature — which, of course, varies with species. *Myriophyllum spicatum* growth increases in the range 15–25°C, and decreases at 25–33°C, in which range *Najas guadalupensis* is still increasing in growth. Among angiosperms *Potamogeton perfoliatus* grows well at 35°C. Cyanophyceae typically tolerate temperatures above this, green algae 30–35°C, and diatoms 20–30°C. Fluctuating temperatures hinder algal growth more than constant ones, even when averages are considered (see also Chapter 6).

Suspended solids

Suspended solids reduce light and so reduce low-growing plants. Apart from pollution, this is due both to silt from the land (varying with rock type, clay streams bearing the most), which increases in storms, and, in large rivers, to phytoplankton. The phytoplankton can also be increased by (suitable) pollution. Suspended

Table 4.15 Macrophyte rooting level in relation to sedimentation
Species vary considerably in their behaviour, and thus classification is on typical behaviour only. There are no species which never alter their rooting level.

1. Species whose rooting level commonly varies
   a) Shallow-rooted
   Apium nodiflorum   *Rorippa nasturtium-aquaticum* agg.
   Berula erecta   *Veronica anagallis-aquatica*
   (*Callitriche* spp.)   *V. beccabunga*
   Myosotis scorpioides   *Zannichellia palustris*

   b) Deep-rooted
   Nuphar lutea   *Sparganium erectum*
   Potamogeton pectinatus

2. Species whose rooting level commonly remains constant
   a) Shallow-rooted
   (*Callitriche* spp.)   *Ranunculus* spp.
   (*Potamogeton crispus*)   (*Elodea canadensis*)

   b) Deep-rooted
   Carex acutiformis agg?   *Scirpus lacustris*
   Phragmites communis   *Sparganium emersum*
   (*Potamogeton pectinatus*)

solids — particularly the larger particles — may abrade plants during storms. If toxic, their toxicity acts as described above. Suspended solids are also deposited as sediment (see below, and Table 4.15).

Sediments

Sediments are particles deposited from suspension. The discussion of farming in Chapter 2 describes soil deposition. Table 4.14 lists some depositing pollutants, and Table 3.1 gives a summary of effects (see also Chapter 11). Toxic responses are described above. Sediment with no nutrients is, of course, poor for plant development. To stay in place, flowering plants, angiosperms, must be well anchored. The species most harmed by some sediment deposition are shallow-rooted plants whose rooting level adjusts itself to remaining on top of the soil (Table 4.15). These become entirely rooted in new, unstable sediment and are easily washed out. Those most harmed by deep sediment deposition are species which maintain their rooting level at the old ground level, and/or are short. Both are likely to be smothered.

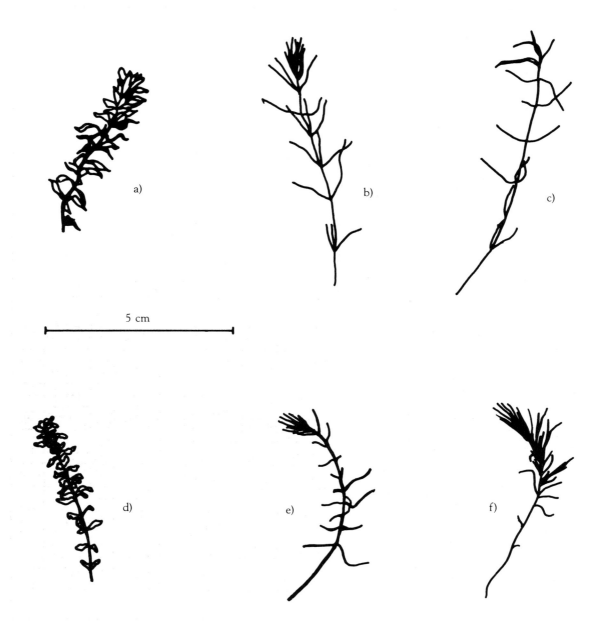

5 cm

Figure 4.20 Morphological changes in *Elodea canadensis* with different chemical treatments (from Haslam, 1987a).
(a) Control, (b) Tannic acid (1.25 mg l^{-1}), (c) Simazine (0.07 mg l^{-1}) and NPK, (d) Agrochemical factory effluent, (e) Sewage works effluent, (f) Ioxynil (K Salt, 0.006 mg l^{-1}) and NPK. (b) was, and remained for the six years tested, externally (including florally) similar to *E. nuttallii*, identified as such by the Institute of Terrestrial Ecology and the University Botanic Gardens, Cambridge. Soil as well as water was changed. (c) was similar but died after four months. (d) and (e) retained the form for the period (nearly two years) of the test. (d) had, in life, strongly curved leaves, (e) had flaccid leaves wider at the base than (b). (f) was unchanged for the first two months, then developed narrow, long, green and flaccid leaves. (The NPK solution was, in mg l^{-1}, NO_3-N, 6; PO_4-P, 4; K, 2; Mg, 0.66; Fe^{3+}, 0.01.)

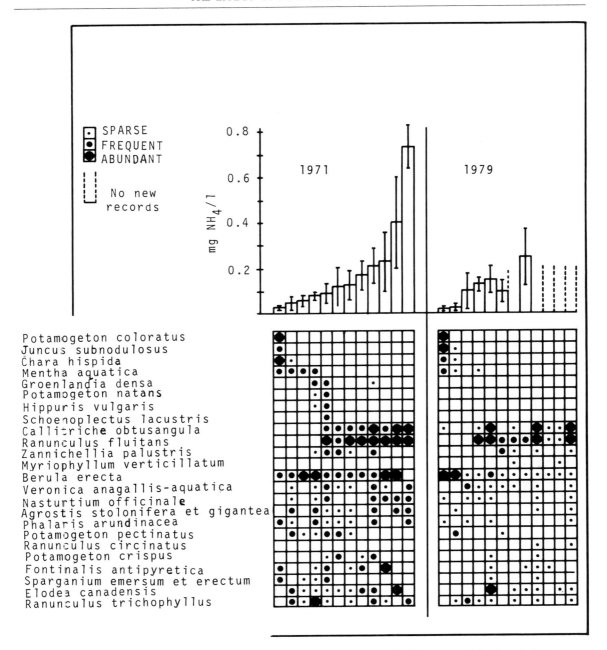

Figure 4.21 Distribution of macrophytes and of ammonium in the Moosach, Germany, in 1971 and 1979. Redrawn from Kohler and Zeltner (1981)

5
Effect of pollution on animals

Fish

Many people, who have no knowledge of biology, hydrology, or any other 'ology', enjoy looking at a river. Some of these, the writer hopes, will read this book and enjoy river-gazing more. Of those who look with knowledge, most know fish. And it is the death of fish that arouses the most public concern about pollution. 'Thousands of fish die' is a headline which clamours for remedy.

River fishing is, and was, a sport. It was, but in European waters, rarely is, a livelihood (Figure 5.1; landscape strip above). Many settlements on larger rivers, such as the Rhine, were fishing villages. In England, as described in *Domesday Book*, all settlements on rivers had at least one fishery. Down the ages much of the table fish — indeed table flesh — in inland areas came from the river.

River table fish at present are few, and are often expensive trout from a trout farm or expensive salmon from a salmon farm or non-industrial highlands. It was Modern Times and the Industrial Revolution which spoiled the river fish. Salmon used to be cheap and ordinary food e.g., from the Rhine (Friedrich and Muller, 1989). The change from the nineteenth to twentieth-century is great.

Now, though, there is another concern. Excessive abstraction is causing too low flows. Both would need to be remedied before table fish were once again plentiful — as also, in some parts, would other types of damage. (Table fish are not just fish but fish all are prepared to eat and that may be eaten with safety.)

In the Rhine, 1765 was the start of the decline. River engineering damaged the spawning grounds, steamships disturbed the fish and destroyed the fry — and serious pollution began. Following town and industrial development, 1915 saw a sudden decline of fish. In 1924 the rents of the lower German Rhein fisheries were cut to

Figure 5.1 Mill and fishing, sixteenth century, Germany (C. Pencz). Three methods of fishing, all downstream of the mill.

half or one-third, and with more heavy industry the river declined further. There was a short post-war improvement when it was industry that was in decline, but this did not last: 1969 saw the worst fish kill on record. Massive fish kills also occurred in the Dutch Rijn (van Urk, 1984): it is at least something that the fish could still be killed in large numbers! Some recovery occurred after 1979, and the lower Rijn in 1983 had a sparse population of *Barbus barku*, *Chandrostoma nosus*, *Leuciscus leuciscus* and *L. cephalus*, like other similar Dutch waterways. Bad, that is, but less atrocious than formerly. Coarse fish like these species are the most

tolerant to pollution.

Some say fish are the best indicators of river health, since, of those organisms living within the water, fish are at the top of the food chain (they eat other organisms; other organisms do not eat them). One snag, however, is that fish can swim away from some forms of localised pollutions (which they detect), and return, making interpretation of parts, rather than the total river, impossible. Another is that sampling is at best disruptive, at worst destructive. (Hynes, 1960, considers fish the least satisfactory group for indicating pollution for these reasons.) Few researchers study fish by watching them and few

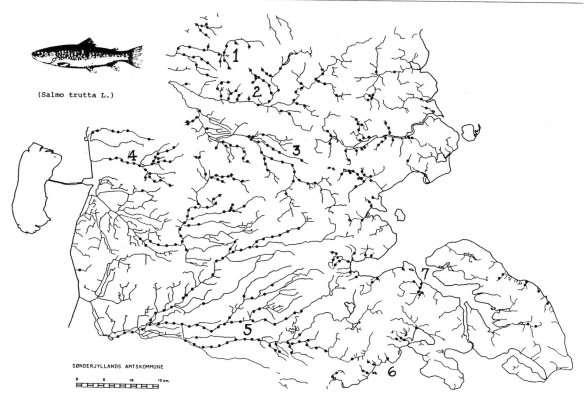

(Salmo trutta L.)

SØNDERJYLLANDS AMTSKOMMUNE

Figure 5.2 Distribution of trout in South Denmark (from Sønderjyllands Amstkommune, 1986). Trout occurrence is associated, in part, with saprobic invertebrate zones. I = oligosaprobic; II = beta-mesosaprobic; III = alpha-mesosaprobic; IV = polysaprobic (see Table 7.15). 1. Cleaner river, mostly zone II and better. 2. Cleaner downstream, worse upstream. 3. Trout in upper river, almost confined to zone II. 4. Trout in zone II, absent in II/III. 5. Upper reaches with sparser trout and lower grades. 6. Trout frequent in zone I/I–II, sparse in zone II. 7. Trout absent in zone III and worse.

anglers record their observations. Work has concentrated more on fish zones within individual rivers than on fish communities, many schemes, such as that of Huet (1954), being proposed:

- Trout zone, rapidly-flowing, well-oxygenated cool water, steep-sided valley.
- Grayling zone, less rapid, some slow reaches, well oxygenated, steep-sided valley with flat bottom.
- Barbel zone, moderate current, long slow reaches, oxygen levels may fall in warm weather, a variety of fish present. Valley wide and flat-bottomed.
- Bream zone, slow flow, canal-like, oxygen may fall to low levels, fish hardy, like bream and tench.

Each scheme works well in its own region, but does not transfer (for example, grayling are sparse in Britain), and good Europe-wide schemes are absent. It is difficult to assess ecological effects on communities when communities are not studied. Salmonids indicate clean and not solute-rich conditions. Where there are only a few, sparse species of coarse fish, such as bream and eel, pollution is probable.

Fish have home ranges, territories within which they move, and outside which they rarely go. However, they will go outside to get away from pollutants (like ammonia) that they can detect and perhaps be distressed by. Many pollutants, however, are not detected (Hellawell, 1986) and the fish suffer — as they do from a very rapid pollution (for instance, spills) or if all the river is

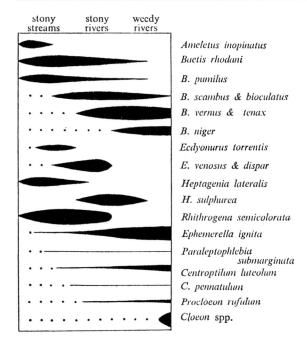

stony streams	stony rivers	weedy rivers	
			Ameletus inopinatus
			Baetis rhodani
			B. pumilus
			B. scambus & bioculatus
			B. vernus & tenax
			B. niger
			Ecdyonurus torrentis
			E. venosus & dispar
			Heptagenia lateralis
			H. sulphurea
			Rhithrogena semicolorata
			Ephemerella ignita
			Paraleptophlebia submarginata
			Centroptilum luteolum
			C. pennatulum
			Procloeon rufulum
			Cloeon spp.

Figure 5.3 Ephemeroptera in polluted water (from Hynes, 1960).

affected and there is no clean place to go to. (Hawkes (1975) notes that fish zones are more practically useful than those of invertebrates, and ascribes this to fish being at the top of the food chain.) Fish distributions are shown in Figure 5.2. If a fish population is present, conditions are tolerable for that population.

Fish requirements may greatly differ during their lifetime. Salmonids spawn on gravel, usually in upper reaches, and the sites may be damaged by only a little sewage fungus or suspended solids. (Trout eggs die under sewage fungus even if water and oxygen are suitable.) Fish lay many eggs, so pollution effects may be locally devastating but almost harmless to the population as a whole, or mild organic pollution may be a disaster. In addition, many fish hunt by sight. Organic pollution leads to burrowing rather than surface invertebrates, so may harm the fish through lack of food (Hynes, 1960).

Salmon find their home river, on their return from the Atlantic ocean, by smell. Pollutions affecting smell may thus be harmful.

Fish usually avoid severe organic pollution, except sometimes round sewage works outfalls, where they eat well, at least in cold weather. Fish

are repelled by low oxygen levels, particularly at high temperatures. Where carbon dioxide rises (with decaying organic matter), fish appear to need more oxygen — and are less sensitive to ammonium poisoning. Oxygen levels are also affected by weather (Hynes, 1960).

Macro-invertebrates

In Britain, Denmark, France and Germany in particular, routine pollution monitoring includes examining invertebrates. The appointments in Britain are termed 'biological' — the biologists of, for example, the water authorities — but almost all are just zoological. Which means a large body of invertebrate zoologists, all looking at pollution, all assessing rivers — and most, indeed most of those publishing on macro-invertebrates, are not ecologists. The significance will soon be seen.

Invertebrates, like the macrophytes described in Chapter 4, are, of course, affected by the habitat in which they grow, and so their communities can be classified by this habitat. These communities can be divided by landscape: alpine, mountain, highland (equivalent to the writer's 'upland' for macrophytes) and lowland; by the water chemistry derived from rock type (carbonate or silicate); and by drainage order (a variant on the writer's upstream–downstream position and size category). The lowland streams are further classified as shaded or open, and by land use. This allows equivalence with the writer's fourth factor for macrophytes: Man's activities (Otto, 1979; 1980; Kothé et al., 1980, U. Braukmann, pers. comm.). Unfortunately the scheme has not been extended Europe-wide. Detailed studies of communities in a limited range show these are related to water flow and source, nutrient states, water regime, organic content of the sediment, and size of the emergent zone (Castella, 1987), the emergent vegetation being of course determined by the habitat factors (Chapter 4).

Most workers have instead been interested in the single factor of pollution. Pollution imposes uniformity (Chapter 3). Absence of invertebrates occurs with excess of any type of pollution in any type of stream, and tubifex worms are only a little less limited. Figure 4.14 shows the characteristic pattern below sewage works in quite a range of stream types. When pollution

Table 5.1 Invertebrate distribution in different water qualities.
The percentage composition of the faunas of eroding substrata in various British rivers, arranged, from left to right, in order of decreasing water hardness. Some of the data are taken from published work as indicated. Figures to nearest whole number. * = less than 0.5 per cent

	Great Driffield, Yorks., gravel	R. Avon, Hants	R. Dee, Denbigh-shire	R. Ceiriog Denbigh-shire	Trout stream, Northum-berland	R. Towy, Cardigan-shire	Afon Hirnant, Merioneth
Season of investigation:	all	all	winter and spring	winter and spring	spring and and summer	summer	all
Flatworms							
Tricladida	1	—	1	*	2	—	*
Worms							
Oligochaeta	27	47	8	4	1	1	*
Leeches							
Hirudinea	1	*	2	*	—	—	—
Crustacea							
Shrimps (Amphipoda)	32	3	1	1	—	—	—
Asellus	*	*	—	—	—	—	—
Mayflies							
Baetis	3	1	24	15	22	26	5
Caenis	—	4	1	—	*	—	—
Ephemerella	*	—	22	6	2	19	4
Ephemera	—	*	—	—	—	—	—
Ecdyonuridae	—	—	6	17	15	1	19
Stoneflies							
Leuctridae and Nemouridae	—	—	4	14	23	9	47
Perlidae and Perlodidae	—	—	1	2	4	*	2
Chloroperla	—	—	*	2	1	—	2
Caddis-worms							
Caseless	*	1	3	6	3	*	9
with cases	30	2	*	*	*	5	*
Bugs							
Hemiptera	—	1	—	—	—	*	—
Beetles							
Helmidae (Elmidae)	1	2	3	1	6	8	1
Other beetles	1	—	*	—	*	2	1
Midges							
Chironomidae	2	22	7	33	16	5	4
Ceratopogonidae	*	—	1	—	1	*	—
Buffalo-gnats							
Simulium	*	—	—	—	1	3	6
Other flies							
Diptera	1	—	—	—	—	1	*
Water-mites							
Hydrocarina	*	—	1	—	*	23	—
Snails							
Gastropoda	*	16	1	—	1	—	—
Limpets							
Ancylastrum	*	1	14	*	2	*	*
Other animals							
(mostly insects)	1	—	—	—	—	—	*
Total no. in collections	3372	3258	4444	1620	13 745	2657	7788

Source: Hynes (1960)

Table 5.2 Invertebrate distribution in different altitudes, Afon Hirnant (percentage composition of the total fauna of each station)

	Altitude of station (feet)			
	700	800	950	1100
Baetis spp.	4.6	9.2	10.1	11.7
Ephemerella ignita	3.8	4.1	4.8	5.4
Rhithrogena semicolorata	17.6	11.7	6.9	2.7
Heptagenia lateralis	0.1	0.1	1.0	1.1
Amphinemura sulcicollis	24.1	23.1	25.8	7.6
Isoperla grammatica	1.0	3.3	5.3	11.6
Hydropsyche instabilis	7.3	1.8	0.2	<0.1
Stenophylax spp.	0.1	0.2	0.2	0.5
Helmis maugei (*Elmis aesisa*)	0.3	1.3	2.8	12.7
Orthocladiinae	2.5	3.5	3.6	7.4

Source: Hynes (1960)

becomes less overriding, however (when pollution is milder), all the other habitat factors come into play, influencing the nature of the community. And here is the difficulty of the sparsity of ecologists. Good schemes are drawn up to detect the amount of pollution, for the rivers in question — not necessarily for other rivers. Both heat and very fancy mathematics are generated. Observing each other's rivers might be more profitable. The Saprobic system, for instance (see Kolkwitz and Marsson 1908; 1909) was designed to show organic pollution below sewage outfalls. Extend it as a general monitoring scheme and, behold, effectively no lowland stream rates as clean. How odd.

Distributions of fauna are shown in Tables 5.1, for water hardness, 5.2, for altitude; 5.3 for season, and 5.4, for substrate. Season is

Table 5.3 Invertebrate distribution in different months. Afon Hirnant at four stations. Only those species are shown which made up at least 2 per cent of the fauna in one month. Figures are shown to the nearest whole number. * = less than 0.5 per cent

Date of sampling:	1 XI	29 XI	10 I	7 II	14 III	13 IV	17 V	21 VI	18 VII	29 VIII	28 IX	25 X	22 XI
Mayflies													
Beatis spp.	7	12	10	11	4	4	5	12	9	24	9	7	10
Ephemerella ignita	—	—	—	—	—	—	*	29	36	9	*	*	*
Ecdyonurus venosus	6	3	1	1	*	*	1	1	1	1	5	2	1
Rhithrogena semicolorata	16	14	16	12	12	10	13	2	*	—	4	10	8
Heptagenia lateralis	*	—	*	*	1	1	3	2	1	*	—	—	*
Stoneflies													
Brachyptera risi	—	*	4	7	3	2	—	—	—	—	—	—	*
Protonemura spp.	7	6	5	3	*	*	—	*	1	5	9	5	4
Amphinemura sulcicollis	24	27	27	26	22	11	—	1	*	—	20	39	32
Leuctra hippopus	2	1	*	*	*	*	—	—	—	*	2	2	1
Leuctra inermis	7	15	18	18	33	30	25	3	—	—	3	15	23
Leuctra fusca	—	—	—	—	—	—	—	3	13	11	1	*	—
Isoperla grammatica	3	8	7	8	7	8	8	3	1	*	1	1	3
Chloroperla spp.	1	1	1	1	2	3	5	3	*	1	2	2	3
Caddis-worms													
Rhyacophila spp.	1	1	*	1	1	1	1	2	2	5	3	1	1
Plectrocnemia conspersa	2	1	*	*	*	1	2	2	1	2	2	*	*
Hydropsyche instabilis	5	2	2	2	5	3	4	4	1	1	1	3	1
Beetles													
Helmis maugei	7	3	1	2	5	10	4	3	8	7	14	1	1
Midges													
Orthocladiinae	5	*	*	*	1	11	9	12	4	4	11	3	2
Tanypodinae	*	*	*	—	—	1	*	2	5	*	*	*	*
Buffalo-gnats													
Simulium spp.	*	2	5	5	1	1	5	15	13	23	5	2	8
Total no. in sample	2380	3543	3846	2885	1371	3438	1743	2218	1811	580	575	1671	4367

Source: Hynes (1960)

Table 5.4 Invertebrate distribution on different substrates. The percentage composition of adults of various groups of insects emerging into traps set over different types of substratum in streams in Algonquin Park, Ontario, and comparison of the total numbers emerging from the different types of substratum

	Rubble	Gravel	Sand	Muck
Ephemeroptera	35.5	4.6	9.3	20.3
Trichoptera	7.0	1.7	1.7	3.8
Plecoptera	4.1	2.1	0.7	0
Chironomidae	38.2	67.6	83.9	74.8
Simuliidae	10.8	21.4	0.9	0
Miscellaneous	4.4	2.5	3.5	1.0
Ratio in total numbers emerging to total numbers emerging from sand	4.6 in rapids 3.3 in pools	2.1	1.0	1.8

Source: Hynes (1960)

important. Many invertebrates are aquatic larvae in winter, but terrestrial or aerial adults in spring, and eggs in summer. Others are the reverse. More again never leave the river. So a river should not be classed as clean or dirty because some of the components happen to be eggs and are not sampled. The habitat variation within invertebrate groups is shown in Figure 5.3. These are just to demonstrate some aspects of non-pollution variation.

Variation with organic pollution, and common pollution-tolerant species, is shown in Table 5.5, and a characteristic pattern is shown in Figure 5.4. The polluted River Tame — by now an old friend — is shown in Figure 5.5 for its invertebrates, a selection of invertebrate pollution zones is shown in Figure 5.6 and a good example of recovery after stopping pollution in Figure 5.7. Recovery is, of course, quicker the less the pollution, and the fewer the pollution sources. Variation over time is well shown in the Lee. Below the sewage works the Blanket weed (plant!) zone was: 1950, 2–7 + miles; 1952, 0–12.5 miles, 1953 and 1954, 1–13.25 miles; the *Chironomous* zone in 1950, 1953 and 1954 was 1–2 miles, but in 1952 the river was cleaner and this zone was absent, though the next downstream zone, *Asellus*, was actually longer. Hynes (1960) records that these zones may be compressed, or be extended to, for

example, 400 km, as in the Illinois River downstream of Chicago.

Micro-invertebrates

The same general principles apply to the effects of pollution on micro-invertebrates. Being so small, they are difficult to sample, identify and analyse. Like the smaller algae, those in the water reflect the conditions in the upstream water: where they and the water have come down from. Where little time elapses between the water leaving the stream source and it joining the sea, little development can occur, swift-water species being few in kind and abundance, but large blooms of plankton can develop where the retention time is as long as say, twenty days. Most are Rotifers, with *Cladocera*, Protozoa, Copepods, etc., also present. Such may be an integral part of the river, or be transient, for instance washed in from lakes (Winner, 1975). With downstream recovery from (organic) pollution, those feeding on bacteria change to those with other habits, and this change is independent of where the water happens to be. The river-bed micro-invertebrates reflect present pollution more (Figure 5.8 and Table 5.5). The soil is full of them — and they stay put. So, water and soil pollution patterns are independent (Hynes, 1960).

Birds

Birds, of course, do not live permanently on and in water: they move elsewhere, and for varying times. Defining a river bird is thus less easy than defining a river fish (Table 5.7). They vary from resident waterfowl spending much time in the river, to migrants present seasonally, and birds living in bankside trees.

Birds may be polluted by being in the water, by poisons, or by surfactants like oil. They may, like swans, eat plants, invertebrates and gravel; the plants and animals may contain toxins, and the lethal lead weights dropped by anglers may be mistaken for gravel. They may feed on animals from invertebrates to fish, and the higher in the food chain they go, the more non-degradable poisons the birds eat — which are then further accumulated in the birds. Birds of prey, as already noted, who feed at the top of the

Table 5.5 Invertebrate distribution with organic pollution
Association of invertebrate taxa with different degrees of organic pollution in riffles.

Group	Clean		Dirty
Platyhelminthes	Planaridae		
	Polycelisnigra		
	Dendrocoelum lacteum		
Annelida			
Oligochaeta			*Limnodrilus hoffmeisteri*
		Lumbriculidae Enchytraeidas	
			Tubifex tubifex
Hiradinea	*Chaetogaster* —— *Piscicola geometrica* —— *Stylaria*		
	Glossiphonia complanata		
		Heleobdella stagnalis	
		Haemopsis	
		Eropobdella testacea	
		E. octoculata	
Arthropoda			
Crustacea	*Austropotamobius pallipes* —— *Gammarus pulex* ——		
	Cyclops		
	Asellus meridianus	*A. aquaticus*	
Insecta			
Plecoptea	Most		
	Isoperla grammatica		
	Leuctridae		
	Caprinidae		
	Nemouridae		
	Amphinemura sulciocollis		
Ephemeroptera	Heptageniidae		
	Leptophlebiidae		
	Caenidae		
	Ephemerella ignita		
	Baetis rhodani —————————— *Baetis rhodani*		
Megaloptera	—— *Sialis fuliginosa* —— *S. lutaria*		
Trichoptera	—— Hydroptila spp. ——		
	—— Agraylea spp. ——		
	Other cased caddis		
	—— *Rhyacophila dorsalis* ——		
	Other Rhyacophilidae		
	—— *Hydropsyche angustipennis* ——		
	—— Other cased non-caddis ——		
Coleoptera	Helminthidae		
Diptera	*Dicranota* spp. —— *Simulium ornatum* ——		
	Limnophora spp. —— *S. reptans*		
	Atherix spp. —— *Eusimulium auream* ——		
	Tipula spp.		
	Other Simulidae	*Chironomus riparius*	*Tubifera tenax*
	Other Chironomidae —————————		
Mollusca			
Gastropoda	*Lymnaea* spp. —— *Lymnaea* spp. ——————		
	Physa acuta ——————		
	Other gastropods ——————		
	Sphaerium spp. —— *Sphaerium* spp. ——		
	Pisidium spp.		

Source: adapted from Hawkes (1978b)

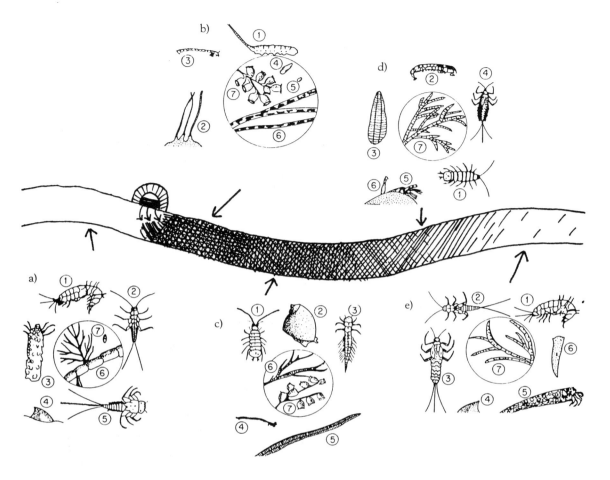

Figure 5.4 Succession of benthic riffle invertebrate communities associated with different stages of recovery from organic pollution (redrawn after Hawkes, 1978).

Key to organisms (micro-organisms shown within circles).

A. 1. *Gammarus pulex* (freshwaster shrimp), 2. *Nemoura* (stonefly nymph), 3. Limnophilid caddis, 4. *Ancylus*, 5. *Ecydonurus* (mayfly larva), 6. *Draparnaldia* (green alga), *Cocconeis* (diatom)

B. 1. *Eristalis tenax* (rat-tailed maggot), 2. *Tubifex* (sludge worm), 3. *Chironomus riparius* (blood worm), 4. *Paramecium caudatum*, 5. *Colpidium*, 6. *Sphaerotilus natans* (sewage fungus), 7. *Carchesium* (sewage fungus)

C. 1. *Aselllus aquaticus* (water hog-louse), 2. *Lymnaea pereger* (wandering snail), 3. *Sialis lutaris* (alder-fly larva), 4. *Chironomus riparius* (blood worm), 5. *Erpobdella* (leech), 6. *Stigeoclonium* (green alga), 7. *Carchesium* (sewage fungus).

D. 1. *Asellus aquaticus* (water hog-louse), 2. *Hydropsyche* (caseless caddis larva), 3. *Glossiphonia* (leech), 4. *Baetis rhodani* (may fly nymph), 5. and 6. *Simulium ornatum* (pupa and larva of buffalo gnat), 7. *Cladophora* (Blanket weed).

E. 1. *Gammarus pulex* (freshwater shrimp), 2. *Nemoura* (stonefly nymph), 3. *Ephemerella* (may fly nymph), 4. *Ancylus fluviatilis* (limpet), 5. *Stenophylax* (caddis larva), 6. *Dugesia* (flatworm), 7. *Cladophora* (Blanket weed).

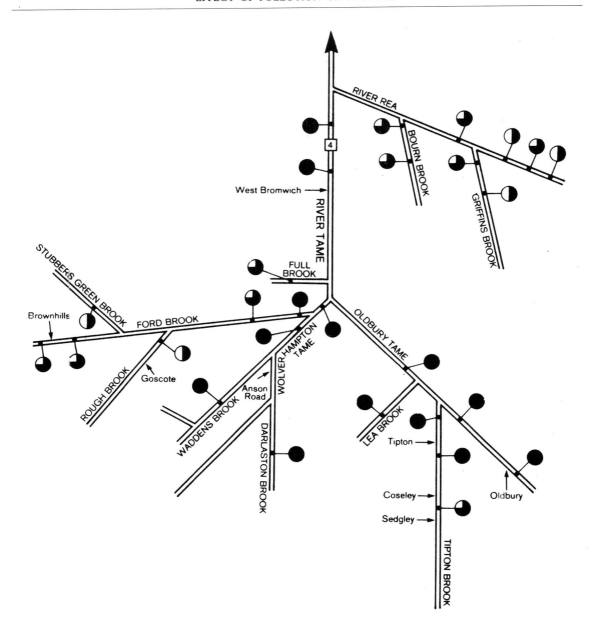

Figure 5.5 Invertebrate (NWC/BMWP) scores in the polluted upper Tame, England (simplified from Severn-Trent Water 1988).
Circles: black = bad, score 0–12; ¾ black = poor, score 13–35; ½ black = moderate, score 36–70 (Excellent = 150).

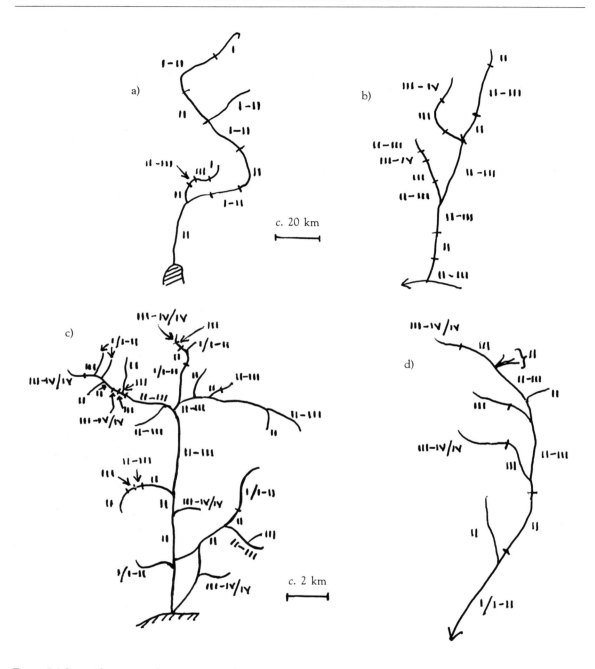

Figure 5.6 Invertebrate saprobic zones in polluted rivers, Germany and Denmark (after Staatsministerium des Innern, 1986; Sønderjyllands Amstkommune, 1986).
Zones as in caption Figure 5.2 (a) upper Ammer, (b) Kleine Laber, (c) Aller, (d) upper Sønder

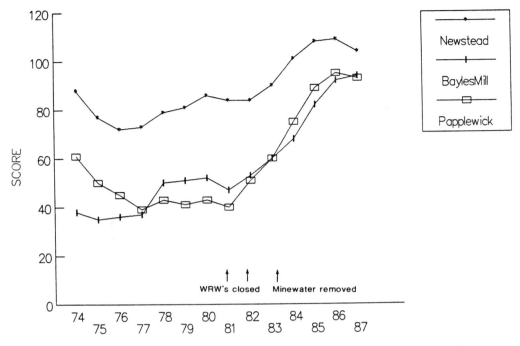

Figure 5.7 Invertebrate recovery after decreasing pollution, Leen, England (from Severn-Trent Water, 1988).
NWC/BMWP scores. Results of closing water reclamation works and removing mine pollution. The two downstream sites are affected by other pollution.

food chain, declined drastically as a result of ingesting organochlorine biocides, slowly returning when the use of these dropped. Fish are eaten by, for instance, cormorant; invertebrates by dippers; while moorhen and mallard are omnivores, and riverside birds in general often eat insects. Pollution eliminating the food removes the birds. Acidification decreases invertebrates. The dipper has declined in Scottish streams where pH has dropped. It can survive the pollution, but not the effects of pollution on its food. Likewise a loss of vegetation, through pollu-tion, may mean the loss of the invertebrates it supported — even if they could tolerate that particular pollutant — and so, again, mean the loss of the birds feeding on those.

Of the British populations of smew, gadwall, tufted duck and goosander, 5 per cent of the total occur within 30 km of St Paul's Cathedral, the heart of London. In the polluted waters of Birmingham, in our well-known River Tame, coots live on dark ponds with tyres and bicycles, where sparrows and pigeons drink (Lovegrove and Snow, 1984; and see landscape strip p. 1).

Table 5.6 Micro-organism distribution in relation to organic pollution

a) Zones of pollution
1. Severe organic pollution
   Bacteria, *Rodo* and a few other ciliates.
2. Where sewage fungus can grow
   Bacterial-eating Protozoa, e.g. *Colpidium colpoda*, *Glaucoma*, *Paramecium*.
3. Joined by *Carchesium*, *Vorticella*, algal-eating forms like *Chilodonella*, *Spirostoma*, *Stentor*.
4. Diversity increases much more, e.g. *Coleps*, *Didinium*, *Lionotus* and Rotifers.
5. Early recovery shown by addition of green flagellates, e.g. *Euglena*, *Procus*.

b) Rotifer distribution.
1. Moderate organic pollution, alpha-mesosaprobic.
2. Improved, alpha-beta-mesosaprobic
   Brachionus angularis, *B. calyciflorus*.
3. Beta-mesosaprobic
   Asplanchna brightwelli, *B. budapestinensis*, *B. falcata*, *B. quadridentatus*, *Hexarthra mira*, *Polyarthra vulgaris*, *Platyias patulus*, *Ploesoma hudsoni*, *Synchaeta oblonga*.
4. Near-clean, oligo-beta-mesosaprobic
   A. priodonta, *Filinia longiseta*, *Keratella cochlearis*, *K. crassa*, *K. earlinae*, *K. quadrata*, *K. vulga*, *S. pectinata*.

Sources: Hynes (1960); Sládaček (1973)

Table 5.7 River birds
The following groups in several orders of birds are fairly closely associated with rivers and streams

Gaviiformes	Divers.
Podicipitiformes	Grebes.
Pelecaniformes	Cormorants, shags, darters, and pelicans.
Ciconiiformes	Herons, egrets, bitterns, storks, and ibises.
Anseriformes	Ducks, geese, swans, and mergansers.
Falconiformes	Fish-eagles and ospreys.
Gruiformes	Rails, coots, and waterhens.
Charadriiformes	Waders and gulls.
Strigiformes	Fish-owls.
Piciformes	Kingfishers.
Passeriformes	Wagtails and dippers.

Source: Hynes (1970)

Where birds can feed on unpoisoned food, such as Man's waste food, water poisons are less harmful. Where, in Birmingham, the river-banks are lined, birds cannot nest, and populations will decline (Lovegrove and Snow, 1984). But that is another story (Chapter 8).

Mammals

Chanin (1985) has documented the distribution of otters. Britain has the best records of population, from the hunt records, but the mid-twentieth century decline came at much the same time across Europe. It came with the introduction of organochlorine insecticides, dieldrin, aldrin and the like, in 1955. Otters feed at the top of the food chain, so get the most of these unpleasant and undegrading compounds in their bodies. The otter numbers dropped between 1957 and 1967 in different parts. Breeding dropped in the late 1950s and early 1960s, but may have improved by 1969.

Otters were not analysed for dieldrin at the time, but the circumstantial evidence is very convincing. Dieldrin use decreased from the 1960s and fish levels were much lower by 1982 (Chanin, 1985). Not, be it noted, nil — they ought to be. But dieldrin is still allowed in Britain for specialist purposes, such as mothproofing woollens or growing daffodils, and Britain does not require those using these nasty chemicals to remove them from their waste water. (It also takes time for remaining land poisons to wash out). Affected fish are, of course, not knowingly eaten by Man, who has better means of knowing than does the otter. Unlike the birds of prey, otters have not recovered. Other pressures supervened (see Chapter 6).

The biocides are not the only poisons eaten by the otter. Fish may also contain dangerously high levels of heavy metals (Chanin, 1985) and no doubt other poisons in fish (see below) are also poisonous to otters. Otters eat mostly fish, but also crayfish, frogs, birds, worms, lampreys, etc. Plant-eating animals fare better!

In addition, otters hunt mainly by sight. So pollution making water turbid or coloured is, at best, discouraging (Chanin, 1985; see also Chapter 6).

Pollution in the organism

For the toleration and responses of animals to

Table 5.8 Water quality criteria for freshwater fish and aquatic life based on EPA (USA) and EIFAC (Europe) guidelines

Substance	Criteria for protection of fish		Criteria for protection of aquatic life
	EIFAC	EPA	EPA
Aldrin/dieldrin		0.003	0.003
Ammonia (un-ionised)	25	(20)	20
Arsenic			100
Cadmium hardwater	1.5[a]	1.2	12
softwater	0.9[a]	0.4	4
Chlordane		0.01	0.01
Chlorine	4	2	10
Chromium		(100)	100
Copper hardwater	112[a]	0.1 96 hr LC_{50} (c. 10)	0.1 96 hr LC_{50}
softwater	22[a]	0.1 96 hr LC_{50} (c. 2)	0.1 96 hr LC_{50}
Cyanide		5	5
DDT		0.001	0.001
Endosulphan		0.003	0.003
Endrin		0.004	0.004
Heptachlor		0.001	0.001
Lead hardwater		0.01 96 hr LC_{50} (c. 500)	0.01 96 hr LC_{50}
softwater		0.01 96 hr LC_{50} (c. 5)	0.01 96 hr LC_{50}
Lindane (HCH)		0.01	0.01
Malathion		0.1	0.1
Mercury		0.05	0.05
Methoxychlor		0.03	0.03
Nickel		0.01 96 hr LC_{50} (c. 100)	0.01 96 hr LC_{50}
Oil		0.01 96 hr LC_{50} (c. 100)	0.01 96 hr LC_{50}
Parathion		0.04	0.04
PCBs		0.001	0.001
Phenols	1 000	1[b]	
Phthalate esters		3	3
Selenium		0.01 96 hr LC_{50} (c. 20)	0.01 96 hr LC_{50}
Silver		0.01 96 hr LC_{50} (c. 0.1)	0.01 96 hr LC_{50}
Sulphide (undissociated H_2S)		2	2
Zinc hardwater	50[a]	0.01 96 hr LC_{50} (c. 25)	0.01 96 hr LC_{50}
softwater	20[a]	0.01 96 hr LC_{50} (c. 3)	0.01 96 hr LC_{50}

[a] 95 percentile values.
[b] Value selected to avoid tainting of flesh.
Values are in μg l^{-1}. Where application factors of the 96 hr LC_{50} are specified approximate values are given in brackets.
LC Median lethal concentration. Concentration at which half the population will die in a specified time. (LT, time taken for half of test organisms to die at a given concentration. ILL, Incipient lethal level. Concentration of toxic substance or other potentially lethal condition which can be tolerated indefinitely).

Source: Hellawell (1986)

Table 5.9 Variations in sensitivity to ammonia by different fish species, based on data in the literature (milligrammes of ammonia per litre)

Species	Acute toxicity	Chronic toxicity
Salmo salar, smolts	0.28 (24 hr LC_{50})	
Salmo trutta, fry	3.60 (10 hr LC_{50})	
Salmo gairdneri, fry	0.07 (24 hr LC_{50})	0.06–0.02
fingerlings	0.44 (96 hr LC_{50})	0.2
adults	0.1 (24 hr LC_{50})	0.49–1.8
	0.5	
Rutilus rutilus		0.42
Scardinius		
erythrophthalmus		0.24–0.44
Abramis brama		0.50
Perca fluviatilis		0.35–0.6
Cyprinus carpio	1.5–2.0	
Tinca tinca	2.0	

Source: Hellawell (1986)

Table 5.10 Comparison of main characteristics of synthetic insecticides

Characteristic	Organo-chlorine	Insecticide organo-phosphorus	Carbamate
1. Potential for entry into freshwater	strong	strong	moderate
2. Solubility in water	very low	low	low
3. Aquatic toxicity	high	moderate	moderate
4. Aquatic persistence	prolonged	short	short
5. Bioaccumulation potential	strong	weak	weak

Source: Hellawell (1986)

Table 5.11 Toxicity of organochlorine insecticides aldrin, dieldrin and endrin to fish

Species	Test	Value ($\mu g\ l^{-1}$)
Aldrin		
Rainbow trout	24 hr LC_{50}	50
(*Salmo gairdneri*)		
Bluegill	96 hr LC_{50}	13
(*Lepomis macrochirus*)		
Goldfish	24 hr LC_{50}	50
(*Carassius auratus*)	96 hr LC_{50}	28
Dieldrin		
Rainbow trout	24 hr LC_{50}	50
(*Salmo gairdneri*)		
Bluegill	96 hr LC_{50}	8
(*Lepomis macrochirus*)		
Goldfish	96 hr LC_{50}	37
(*Carassius auratus*)		
Endrin		
Rainbow trout	96 hr LC_{50}	0.41
(*Salmo gairdneri*)		
Coho salmon	96 hr LC_{50}	0.27
(*Oncorhyncus kisutch*)		
Coho salmon	96 hr LC_{50}	0.77
(*Oncorhyncus kisutch*)		
Bluntnose minnor	96 hr LC_{50}	0.27
(*Pimephales notatus*)		
Fathead minnor	96 hr LC_{50}	1.8
(*Pimephales promelas*)		
Goldfish	96 hr LC_{50}	1.96
(*Carassius auratus*)		
Carp	48 hr LC_{50}	140
(*Cyprinus carpio*)		
Bluegill	96 hr LC_{50}	0.6
(*Lepomis macrochirus*)		

Source: Hellawell (1986)

pollutants, Hellawell (1986) must be studied. Tables 5.8–5.18, 5.20, 5.21, 5.24 and 5.25 are but a tiny selection from those in this comprehensive work. It should also be consulted for details of chemical reactions and inhibitions. As explained in Chapter 3, there are reservations about removing and analysing animals both for their removal and for the interpretation of the analysis. At least, however, this can prove to all whether levels exist which are dangerous to human health. This is the most effective stimulus to action. Conservation of river life ranks a poor second, sometimes effective, other times not. Of course what is widely suspected, but as yet unproved either way, is whether there are any 'safe' levels of the accumulator poisons — whether, in fact, chronic, as yet unmeasureable, levels produce chronic ill health. Table 5.8 gives some current guidelines. It is noteworthy that fish are marginally the most worthy of protection: see the beginning of this chapter.

Standard toxicity tests have been invented because it is difficult to interpret laboratory results in terms of the river. A standard test will at least compare the relative toxicities of different compounds in the laboratory, and these relative data perhaps transfer to the river even if the actual toxic concentrations do not apply.

Table 5.12 Toxicity of several organophosphorus insecticides to fish

Insecticide	Species	Test	Value (μg l^{-1})
Guthion®	Fathead minnow	96 hr LC$_{50}$	100
	(*Pimephales promelas*)	96 hr LC$_{50}$	1 900
	Bluegill	96 hr LC$_{50}$	5.6
	(*Lepomis macrochirus*)		
	Goldfish	96 hr LC$_{50}$	2 400
	(*Carassius auratus*)		
Dursban®	Rainbow trout	96 hr LC$_{50}$	8.0
	(*Salmo gairdneri*)		
	Rainbow trout	96 hr LC$_{50}$	7.1
	(*Salmo gairdneri*)		
	Green sunfish	36 hr LC$_{50}$	22.0
	(*Lepomis cyanellus*)		
	Bluegill	96 hr LC$_{50}$	3.6
	(*Lepomis macrochirus*)		
	Golden shiner	36 hr LC$_{50}$	35
	(*Notemigonus chrysoleucas*)		
	Mosquito fish	36 hr LC$_{50}$	215
	(*Gambusia affinis*)		
	Fathead minnow	96 hr LC$_{50}$	203
	(*Pimephales promelas*)		
Malathion	Fathead minnow	96 hr LC$_{50}$	12 500
	(*Pimephales promelas*)		
	Bluegill	96 hr LC$_{50}$	20 000
	(*Lepomis macrochirus*)		
	Channel catfish	96 hr LC$_{50}$	760
	(*Ictalurus punctatus*)		
	Brook trout	96 hr LC$_{50}$	120–130
	(*Salvelinus fontinalis*)		
	Cut-throat trout	96 hr LC$_{50}$	150–201
	(*Salmo clarki*)		
	Rainbow trout	96 hr LC$_{50}$	122
	(*Salmo gairdneri*)		
	Coho salmon	96 hr LC$_{50}$	265
	(*Oncorhynchus kisutch*)		
Parathion	Fathead minnow	96 hr LC$_{50}$	1 600
	(*Pimephales promelas*)		
	Mosquito fish	48 hr LC$_{50}$	350
	(*Gambusia affinis*)		
Methylparathion	Fathead minnow	96 hr LC$_{50}$	7 500
	(*Pimephales promelas*)		
	Guppy	96 hr LC$_{50}$	819
	(*Lebistes reticulatus*)		
Dipterex®	Fathead minnow	96 hr LC$_{50}$	51 000–180 000
	(*Pimephales promelas*)		
Disulfotor.	Rainbow trout	96 hr LC$_{50}$	3 020
	(*Salmo gairdneri*)		
	Fathead minnow	96 hr LC$_{50}$	4 000
	(*Pimephales promelas*)		

Source: Hellawell (1986)

Table 5.13 Toxicity of aquatic herbicides to fish

Herbicide	Species	Test	Value (mg l^{-1})
Asulam	Harlequin (*Rasbora heteromorpha*)	24 hr LC_{50}	5 200
Chlorthiamid	Harlequin (*Rasbora heteromorpha*)	24 hr LC_{50}	41
Copper	Rainbow trout (*Salmo gairdneri*)	48 hr LC_{50}	0.14
Cyanatryn	Harlequin (*Rasbora heteromorpha*)	96 hr LC_{50}	7.5
Dalapon	Rainbow trout (*Salmo gairdneri*)	24 hr LC_{50}	350
Dichlobenil	Rainbow trout (*Salmo gairdneri*)	96 hr LC_{50}	6.4
	Rainbow trout (*Salmo gairdneri*)	48 hr LC_{50}	22
	Roach (*Rutilus rutilus*)	10 day LC_{50}	1.6
	Bluegill (*Lepomis macrochirus*)	24 hr LC_{50}	20
	Harlequin (*Rasbora heteromorpha*)	96 hr LC_{50}	4.2
	Harlequin (*Rasbora heteromorpha*)	96 hr LC_{50}	6.5
	Roach (*Rutilus rutilus*)	96 hr LC_{50}	8.0
	Bream (*Abramis brama*)	96 hr LC_{50}	7.5
Diquat	Rainbow trout (*Salmo gairdneri*)	24 hr LC_{50}	90
	Mosquito fish (*Gambusia affinis*)	24 hr LC_{50}	723
	Mosquito fish (*Gambusia affinis*)	96 hr LC_{50}	289
Diuron	Rainbow trout (*Salmo gairdneri*)	48 hr LC_{50}	4.03
	Harlequin (*Rasbora heteromorpha*)	48 hr LC_{50}	152
2,4-D	Carp (*Cyprinus carpio*)	96 hr LC_{50}	96.5
	Rainbow trout (*Salmo gairdneri*)	24 hr LC_{50}	250
	Fathead minnor (*Pimephales promelas*)	96 hr LC_{50}	8.23
Glyphosate	Harlequin (*Rasbora heteromorpha*)	96 hr LC_{50}	12
	Rainbow trout (*Salmo gairdneri*)	96 hr LC_{50}	54.8
Maleic hydrazide	Harlequin (*Rasbora heteromorpha*)	96 hr LC_{50}	25–30
Paraquat	Harlequin (*Rasbora heteromorpha*)	24 hr LC_{50}	840
Terbutryne	Rainbow trout (*Salmo gairdneri*)	96 hr LC_{50}	3.5
	Harlequin (*Rasbora heteromorpha*)	96 hr LC_{50}	1.8
	Harlequin (*Rasbora heteromorpha*)	96 hr LC_{50}	11
	Roach (*Rutilus rutilus*)	96 hr LC_{50}	5.5
	Rudd (*Scardinius erythrophthalmus*)	96 hr LC_{50}	6.7

Source: Hellawell (1986)

Table 5.14 Toxicity of several organophosphorus insecticides to macroinvertebrates

Insecticide	Species	Test	Value (μg l^{-1})
Fenthion	Chaoaborus sp.	48 hr LC_{50}	12
	Cloeon sp.	48 hr LC_{50}	12
	Gammarus pulex	48 hr LC_{50}	14
	Lymnaea stagnalis	48 hr LC_{50}	6 400
Guthion®	Pteronarcys californica	96 hr LC_{50}	22
	Pteronarcys californica	96 hr LC_{50}	1.5
	Gammarus lacustris	96 hr LC_{50}	0.13
	Gammarus lacustris	96 hr LC_{50}	0.15
	Gammarus fasciatus	96 hr LC_{50}	0.1–0.38
	Daphnia magna	26 hr LC_{50}	0.18
Dursban®	Pteronarcys californica	96 hr LC_{50}	10
	Gammarus lacustris	96 hr LC_{50}	0.11
	Gammarus fasciatus	96 hr LC_{50}	0.32
Malathion	Pteronarcys californica	96 hr LC_{50}	50
	Gammarus lacustris	96 hr LC_{50}	1.6
	Gammarus fasciatus	96 hr LC_{50}	0.76–0.9
	Daphnia magna	26 hr LC_{50}	0.9
Parathion	Peteronarcys californica	96 hr LC_{50}	32
	Pteronarcys californica	96 hr LC_{50}	5.4
	Gammarus lacustris	96 hr LC_{50}	12.8
	Gammarus lacustris	96 hr LC_{50}	3.5
	Gammarus fasciatus	96 hr LC_{50}	1.3–4.5
	Daphnia magna	26 hr LC_{50}	0.80
Methyl-parathion	Daphnia magna	26 hr LC_{50}	4.8
Trichlorophon (Dipterex®)	Pteronarcys californica	96 hr LC_{50}	16.5
	Pteronarcys californica	96 hr LC_{50}	35
	Gammarus lacustris	96 hr LC_{50}	50
	Gammarus lacusris	96 hr LC_{50}	40
	Daphnia magna	26 hr LC_{50}	0.12

Source: Hellawell (1986)

Organic pollution and nutrients

This has, in principle, already been discussed. Table 5.9 adds some detail on ammonia toxicity (ammonia is an important constituent of sewage). This is under 'standard' laboratory conditions: ammonia toxicity varies with temperature, dissolved oxygen and pH — and this is important in the rivers! Oxygen deficiency has many effects, ranging from decreasing the concentration of haemoglobin in body fluids (Wilhm, 1975) to lowering the hatching and early growth of fry (of coho salmon; see Warren, 1971). Its effects, as expected, vary with other river characters (such as water velocity, in the last named example).

Biocides

While blaming the organochlorine insecticides (dieldrin, etc.) for awful effects on the environment, it must be remembered that at the beginning — and when their use was limited and localised — they saved countless lives by killing malarial mosquito larvae. Hellawell (1986) points out that casualties in the First World War from insect-borne disease exceeded those in battle, and DDT stopped all that in the Second World War. A useful invention becoming dangerous on misuse is not all that rare.

The greater danger of the organochlorine compounds is seen in Table 5.10: they get into

Table 5.15 Toxicity of polychlorinated biphenyls (PCBs) to fish

PCB	Species	Test	Value (μg l^{-1})
Aroclor 1242	Pimephales promelas	96 hr LC_{50}	300
1242	Pimephales promelas (newly hatched)	96 hr LC_{50}	15.0
1254	Pimephales promelas (newly hatched)	96 hr LC_{50}	7.7
1248	Pimephales promelas (newly hatched)	30 day LC_{50}	4.7
1260	Pimephales promelas (newly hatched)	30 day LC_{50}	3.3
1221	Salmo clarki	96 hr LC_{50}	1 200
1232	Salmo clarki	96 hr LC_{50}	2 500
1242	Salmo clarki	96 hr LC_{50}	5 400
1248	Salmo clarki	96 hr LC_{50}	5 700
1254	Salmo clarki	96 hr LC_{50}	42 000
1260	Salmo clarki	96 hr LC_{50}	61 000
1242	Salmo gairdneri	10 day LC_{50}	48

Source: Hellawell (1986)

Table 5.16 Acute toxicity of mothproofing agents to fish

Substance	Species	Temperature (°C)	96 hr LC_{50} (mg l^{-1})
Eulan WA New (chlorophenylid)	Rainbow trout (Salmo gairdneri)	4	5.4
		14	1.5
		15	0.5–1.0
Mitin FF (sulcofenuron)	Brown trout (Salmo trutta)	13	3.9
	Zebra fish (Brachydanio rerio)	23	1.2
Mitin N (flucofenuron)	Rainbow trout (fry)	15	0.068
	Rainbow trout (fry)	15	0.8[a]
	Rainbow trout (yearlings)	15	0.13
	Rainbow trout (yearlings)	14	11.20[a]
	Brown trout	13	1.7–6.8
	Zebra fish	23	4.0

[a] Solvent not used to disperse formulation.

Source: Hellawell (1986)

Table 5.17 Comparison of Lindane (HCH) residues in tissues of dead roach (Rutilus rutilus), from an acute toxicity test and a river mortality

Origin of material	Number of individuals	Range of Lindane residues in tissue (mg kg^{-1})		
		Muscle	Liver	Brain
Acute toxicity test	30	1.6–4.7	2.9–9.2	2.8–7.5
River mortality	3	1.6–2.0	3.0–3.7	2.8–3.3

Source: Hellawell (1986)

Table 5.18 Tentative table of the approximate order of toxicity of metals based on published data

Highly toxic   Decreasing toxicity →

Hg
  Cu Cd Au? Ag? Pt?
      Zn
          Sn Al
              Ni Fe^{3+}
                  Fe^{2+}
                      Ba
                          Mn Li
                          Co K Ca Sr
                                  Mg Na

Source: Hellawell (1986)

water easily (wash-out, direct spray, and dropped containers — the latter applying equally to other sprays), are very toxic, persist, and bioaccumulate. When comparing individual biocides, though, another factor is important: the toxicity of breakdown products. Degradation means change to simpler compounds. It does not mean rendered harmless, and some degradation products are more toxic than the original. Tables 5.11 and 5.12 illustrate the greater danger of the organochlorines to the organophosphoruses — less organochlorine is needed to kill.

Since herbicides are used to kill plants, many people think they must be harmless to animals. But of course! How else? Table 5.13 shows how else, for some herbicides applied direct to water. The toxic levels can even approach those of the organophosphorus insecticides. (They, of course, are used to kill insects: how can they kill fish, and not even where they were applied??) Table 5.14 shows that the organophosphorus substances are more toxic to invertebrates than to fish. The

Table 5.19 Reported concentration factors (CF) of metals for various classes of freshwater organism

Element	Group	CF range	Mean CF
Cs	Plants	80–4000	907
	Fish	120–22000	3680
Sr	Plants	80–410	200
	Fish	0.85–90	14
Mn	Plants	1300–600000	150000
	Molluscs	1100–1600000	~ 300000
	Crustacea	1700–250000	125000
	Fish	0.1–400	81
Co	Plants	300–30000	6760
	Molluscs	300–85000	32408
	Crustacea	–	–
	Fish	60–3450	1615
Zn	Plants	140–15000	3155
	Molluscs	30–140000	33544
	Crustacea	300–4000	1800
	Fish	10–7600	1744
Fe	Plants	40–45000	6675
	Molluscs	20–80000	25170
	Crustacea	60–1800	930
	Fish	0.1–1255	191
Ce	Plants	200–35000	3180
	Molluscs	400–1500	1100
	Crustacea	300–1000	600
	Fish	2–160	81
K	Fish	340–18000	4400
Ca	Plants	64–720	350
	Fish	0.5–470	70

Source: Langford (1983)

same applies to the organochlorines. PCB toxicity to fish is shown in Table 5.15, and that of mothprocfing agents in Table 5.16. Lastly, Table 5.17 shows the relationship between death from a biocide in a river and in a laboratory. The river fish values (few in number) fall within the laboratory ones, showing that when there is a single known poison, laboratory tests can be satisfactory.

Biocides tend to accumulate in the lipids (fat).

Heavy metals

Table 5.18 usefully lists the relative order of toxicity of metals, heavy and otherwise, from mercury, the most toxic, to sodium, the least toxic. Concentration factors for various metals, and in various groups of animals and plants, are shown in Table 5.19. Toxicity of copper and lead to fish is given in Table 5.20 and 5.21. The variation in concentration factor is enormous. Molluscs reach some of the highest values, and manganese is concentrated to an exceptionally high degree. Variation, of course, also occurs with habitat. Rainbow trout become acclimatised to copper, at between 30% and 60% of the incipient lethal level, but those at 10% of that level lost 20 per cent of their tolerance. However, the acclimatisation mostly did not last after the fish were returned to clean water. Exposure to copper decreases tolerance to zinc (Hellawell, 1986). As already explained, heavy metals are more soluble, so more toxic, in acid waters (see, for example, Table 5.21). Lead poisoning can occur in acid waters, with a level harmless in alkaline ones. It may be that fish kills at low pH may be due to the natural aluminium changing from the insoluble hydroxide to the soluble aluminate (Hellawell, 1986).

In general, salmonids tend to be more susceptible than coarse fish, which is as expected from the solute contents of their characteristic river types. Fish are generally more tolerant than invertebrates. Among invertebrates, molluscs and malacostracan Crustacea seem to be the most susceptible, then Oligochaeta, with Diptera, Trichoptera and some other insects the most resistant. While it is not easy to find field examples where heavy metals are the only pollutants, one study of low continuous copper in the laboratory, and fluctuating copper, chromium and zinc in the field, did show similar effects on the invertebrates. Some insects, such as *Sialis* and naiad damselflies, were unaffected. Otherwise the pattern was like that with organic pollution, with tubificid worms and chironomids in the worst part, and Ephemeroptera (mayflies), where the river improved (Hellawell, 1986).

Ecotoxicology is the study of pollutant levels in organisms. Molluscs grow well in large, messed-about rivers with moderate pollution. Filter feeders receive suspended solids, so receive metals from these as well as from the water. *Dreissenia polymorpha* may be a particularly good indicator, other Lamellibranchs such as *Anodonta cygnea*, *Unio pictorum*, and Gastropods also being recommended. Metal industries are particularly suitable for this method of assessment (see, for example, Léglize and Crochard, 1987, U.E.R. d'Ecologie, 1979). Other organisms are, of course, also used.

Table 5.20 Acute toxicity of copper to fish: selected examples from the literature

Species	Hardness (mg l^{-1} as $CaCO_3$)	Test	Value (mg Cu per litre)
Brook trout (Salvelinus fontinalis)	45	96 hr LC_{50}	0.1
Atlantic salmon (Salmo salar) parr	10	96 hr LC_{50}	0.03
Rainbow trout (Salmo gairdneri)	250	96 hr LC_{50}	0.9
	42	96 hr LC_{50}	0.06
	250	8 day LC_{50}	0.5
	12	7 day LC_{50}	0.03
	42	7 day LC_{50}	0.08
Coho salmon (Oncorhynchus kisutch)	20	96 hr LC_{50}	0.046
Goldfish (Carassius auratus)	220	96 hr LC_{50}	0.46
Carp (Cyprinus carpio)	53	96 hr LC_{50}	6.4
	250	96 hr LC_{50}	0.6
Fathead minnow (Pimephales promelas)	31	96 hr LC_{50}	0.075
	360	96 hr LC_{50}	1.5
Rudd (Scardinius erythrophthalmus)	250	96 hr LC_{50}	0.6
Stone loach (Nemacheilus barbatulus)	250	96 hr LC_{50}	0.76
Pike (Esox lucius)	250	96 hr LC_{50}	3.0
Perch (Perca fluviatilis)	250	96 hr LC_{50}	0.3
Bluegill (Lepomis macrochirus)	360	96 hr LC_{50}	10.2
Eel (Anguilla anguilla)	53	96 hr LC_{50}	0.81
	260	96 hr LC_{50}	4.0

Source: Hellawell (1986)

Table 5.21 Toxicity of lead to fish

Species	Hardness (mg l^{-1} $CaCO_3$)	Test	Value (mg Pb per litre)
Fathead minnow (Pimephales promelas)	20	96 hr LC_{50}	6.5
	360	96 hr LC_{50}	482
Bluegill (Lepomis macrochirus)	20	96 hr LC_{50}	23.8
	360	96 hr LC_{50}	442
Goldfish (Carassius auratus)	20	96 hr LC_{50}	31.5
Guppy (Poecilia reticulatus)	20	96 hr LC_{50}	20.6
Rainbow trout (Salmo gairdneri)	28	96 hr LC_{50}	1.32
	50	96 hr LC_{50}	1
	353	96 hr LC_{50}	(1.38)

Values given are total lead but, where known, free lead is indicated in brackets. In soft water total and free lead may be considered to be the same.

Source: Hellawell (1986)

Table 5.22 Dose of X-rays or gamma rays required to kill 50 per cent of organisms (röntgen; 1 röntgen = 2.58×10^{-4} C kg^{-1})

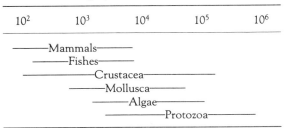

| 10^2 | 10^3 | 10^4 | 10^5 | 10^6 |

Source: Langford (1983)

Combined heavy metals (such as mercury, cadmium, chromium, nickel) may be more harmful to fish after a long exposure time than from additive toxicity for a shorter time. Sub-lethal effects, though, are less than additive, less than the added toxicity of each metal at that concentration (Hellawell, 1986). Metals are stored in different places in animals depending on the metal and on the species. Some species develop tolerance — good for the species concerned, but bad for their predators. Fish can regulate their content of some metals (such as copper, chromium, molybdenum and lead) over a fair range of concentrations (Bryan, 1976).

Radioactivity

The sensitivity of various groups is shown in Table 5.22. Higher organisms here are the more

Table 5.23 Summary of the effect of pH values on fish

Range	Effect
3.0–3.5	Unlikely that any fish can survive for more than a few hours in this range although some plants and invertebrates can be found at pH values lower than this.
3.5–4.0	This range is lethal to salmonids. There is evidence that roach, tench, perch, and pike can survive in this range, presumably after a period of acclimatisation to slightly higher, non-lethal levels, but the lower end of this range may still be lethal for roach.
4.0–4.5	Likely to be harmful to salmonids, tench, bream, roach, goldfish, and common carp which have not previously been acclimatised to low pH values, although the resistance to this pH range increases with the size and age of the fish. Fish can become acclimatised to these levels, but of perch, bream, roach, and pike, only the last named may be able to breed.
4.5–5.0	Likely to be harmful to the eggs and fry of salmonids and, in the long term, persistence of these values will be detrimental to such fisheries. Can be harmful to common carp.
5.0–6.0	Unlikely to be harmful to any species unless either the concentration of free carbon dioxide is greater than 20 ppm, or the water contains iron salts which are precipitated as ferric hydroxide, the precise toxicity of which is not known.
6.0–6.5	Unlikely to be harmful unless free carbon dioxide is present in excess of 100 ppm.
6.5–9.0	Harmless to fish, although the toxicity of other poisons may be affected by changes within this range.
9.0–9.5	Likely to be harmful to salmonids and perch if present for a considerable length of time.
9.5–10.0	Lethal to salmonids over a prolonged period of time, but can be withstood for short periods. May be harmful to developmental stages of some species.
10.0–10.5	Can be withstood by roach and salmonids for short periods but lethal over a prolonged period.
10.5–11.0	Rapidly lethal to salmonids. Prolonged exposure to the upper limit of the range is lethal to carp, tench, goldfish, and pike.
11.0–11.5	Rapidly lethal to all species of fish.

Source: Langford (1983)

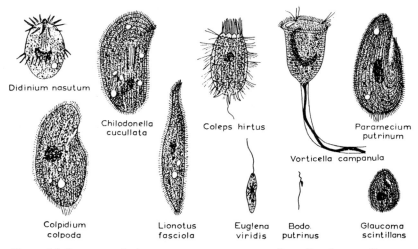

Figure 5.8 Protozoa which are important in organically polluted water (from Hynes, 1960).

Table 5.24 Final preferences of fish species in a temperature gradient

Species	Final preference (°C)
Roach, *Rutilus rutilus*	23
Carp, *Cyprinus carpio*	32
Tench, *Tinca tinca*	20.3
Salmon, *Salmo salar* (alevins)	14–15
Brown trout, *Salmo trutta*	12.4–17.6
Rainbow trout, *Salmo gairdneri*	13.6
Orfe, *Leuciscus idus*	17*
Bream, *Abramis brama*	19*
Crucian Carp, *Carassius carassius*	27*

* Estimated
Source: Hellawell (1986)

sensitive, unlike for heavy metals and biocides. Radio nucleides bioaccumulate also, so healthy invertebrates may be food for fish eaten unhealthily by Man.

pH

Effects can be seen in Table 5.23. In addition to the direct effect of pH, adding acid to waters may release sufficient carbon dioxide from the bicarbonate to kill the fish, etc. (This carbon dioxide effect may explain the recorded variations in lethal pH.) Values of pH from 5 to 9 are harmless to fish. Since toxicity of ammonia, cyanide and many others varies with pH, however, variations within the 5–9 range may be lethal by altering the toxicity of such compounds (Hellawell, 1986).

Waste heat

Tables 5.24 and 5.25 give temperatures preferred by, disturbing, and lethal to various species of fish. Fish are able to detect differences in temperature as small as and smaller than 0.05°C. They may be harmed by sudden and large temperature changes — both warmer, and colder. Hot water often stays at the surface of the river — if the river is large and slow enough — and then fish can stay unharmed in the lower water — where also are most macro-invertebrates. Short periods of warm water are tolerated (Hellawell, 1986).

Suspended solids

Turbidity affects animals both through its effect on plants (see Chapter 4) and directly. This may be by abrasion, clogging respiratory surfaces, interfering with feeding through collecting on feeding parts (for example, nets of filter-feeding organisms). Or their deposition may alter the

Table 5.25 Temperatures found to be disturbing and
lethal in European fishes

Species	Disturbing temp (°C)	Lethal temp (°C)
Rudd, Scardinius erythrophthalmus	30–33.6	38.2
Roach, Rutilus rutilus	29.2–33.0	36.4
Chub, Leuciscus cephalus	33.6–34.0	38.0
Ide, Leuciscus idus	–	37.9
Bleak, Alburnus alburnus	–	37.7
Gudgeon, Gobio gobio	30.8–30.9	36.7
Carp, Cyprinus carpio	32.3–32.5	40.6
Crucian Carp, Carassius carassius	34.6–36.0	38.5
Tench, Tinca tinca	32.1–33.7	39.3
Bitterling, Rhodeus amarus	30.5–31.8	36.5
Perch, Perca fluviatilis	30.5–32.0	?35.5
Pike-perch, Stizostedion lucioperca	33.0–33.3	37.0
Ruffe, Gymnocephalus cernua	29.4	34.5
Stickleback, Gasterosteus aculeatus	30.5–31.8	36.5

Source: Hellawell (1986)

habitat. Many invertebrates and some fish need a permeable substrate, and may suffer if gravel is coated with solids. Much deposition may prevent animals moving, and may eliminate their habitats and their food. Much deposition may also make the substrate unstable (Chapter 4).

If the particles are organic, only the most pollution-tolerant organisms can survive (Figure 4.14), and if they are toxic, the toxic effect follows likewise. The effect on fish of various solids is shown in Figure 5.9.

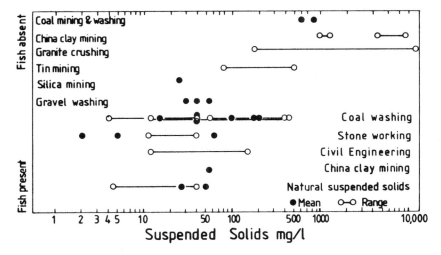

Figure 5.9 Distribution of normal and impaired fish population in relation to inert solids (from Hellawell, 1986).

6

When one thing is added to another

Until the last straw breaks the camel's back.

Macrophytes live in, and so react to, their total environment: the impact on their habitat, the total habitat, not just pollutants. Rock type, landscape, water force and size/downstream position are the basics, but others occur. Shade reduces vegetation. In earlier days, much of Europe was wooded. The natural forest has many glades, where streams have full vegetation, as now in North America. Planted woodland, hedges (mostly Britain) and tree-lines (mostly continental) more solidly prevent vegetation (Figure 6.1), but European rivers are now more open than shaded. Also, trampling is partly due to Man — domestic livestock — and partly not — vast herds, from elephant to deer, still roam wild in low-populated parts of, for example, Africa, need water, and cause much disturbance to river banks.

Understanding river ecology is like peeling an onion. The brown outer skin blocks all view of the onion within. This 'outside' may be the general aspect (is that rock important? or this bridge?) the mass of wriggly invertebrates in the net, or the absence of fish, etc. The writer came to rivers in middle life, and after two years was still saying 'I understand marshes, but I only see rivers'. With the literature now available it would no longer take two years — but neither can it be done in two hours. When one layer of the onion, say plant variation with rock type, is understood, this layer becomes transparent to the mind, leaving the next layer blocking the view, perhaps the effect of channel alteration.

This chapter tries to peel the onion, to show what are, and what are not, responses to pollution. All populations have checks and balances. Which are due to pollution?

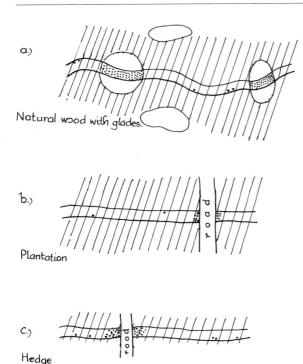

Figure 6.1 Shading trees and river vegetation. Crosshatching = shade, dots = vegetation. (a) Natural woodland, good vegetation in glades, (b) Plantation, vegetation very sparse, even by road, (c) Hedge, more vegetation by road, since hedge is short and narrow compared to (b).

How damage works

Actions lead to reactions, and the series of pictures in Figure 6.2 shows some reactions. Though diagrammatically presented, each occurs in the field. Similar amounts of damage have been posed for each. Pollution is but one of these damage factors, and the plants have a different pattern with each — and with the many more not here drawn.

Then the damage factors are added together. The effect may be simple addition. Or it may be less than that — if a shade-sensitive species has already been removed by pollution (being pollution-sensitive also), the addition of shade will have no further effect. Or more than that — shade weakens, so affected plants recover less well from repeated grazing.

Total damage, therefore, depends on:

- the damage factors present, and the intensity of each;
- the species present;
- the interactions between these.

The more one knows, the more one can interpret. The more one can interpret, the greater the fascination of the subject and the appreciation of river life.

In parts of France a distinctive feature is bed instability in lowland sandstone and Resistant rock streams. Instability leads to wash-out. So when water force slightly increases, less vegetation occurs than elsewhere (Table 6.1). Adding trampling compounds the damage: plants barely anchoring in an unstable bed will be lost when this anchoring is further disturbed. Another French example (Figure 6.3) has good vegetation where the banks are gently sloping and the water is clear (which here means clean). Vertical banks or pollution alone permit some vegetation; together they permit none. Bank shape in general determines how much, and in part what, grows on it (Figure 6.4). Fortunately, similar management produces similar banks, so within a regional stream type, bank and so bank vegetation, are typically similar. A typical effect of shallowing is shown in Figure 6.5. The species in and on the water (the 'water-supported' species) decline, and so total cover also declines. Edge emergents increase, in diversity as well as cover, so overall site diversity usually alters little.

Rivers under several influences

The German Tauber (Figure 6.6) is obviously badly polluted (Chapter 4). Why is there intermittent high cover and diversity — albeit only of pollution-tolerant species? Is it cleaner just there? No, old fords give shallow, gravelly and swifter water, where plants can grow and anchor much better than in deep, silted stretches, and where the turbid water is too deep to give adequate light to the bed. The ex-fords provide the same habitat as natural riffles, but are Man-made.

The upper Scottish Don (Figure 6.7) has various tributaries of similar landscape and size. One has a good water-supported vegetation: it alone is on sandstone. The rest are on Resistant rock, which is less fertile, and from which plants are more easily washed out.

a) undamaged

b) polluted

c) grazed

d) shallowed

e) shaded

f) disturbed

g) grazed and polluted

h) grazed, polluted
and shaded

i) disturbed, shallowed
and shaded

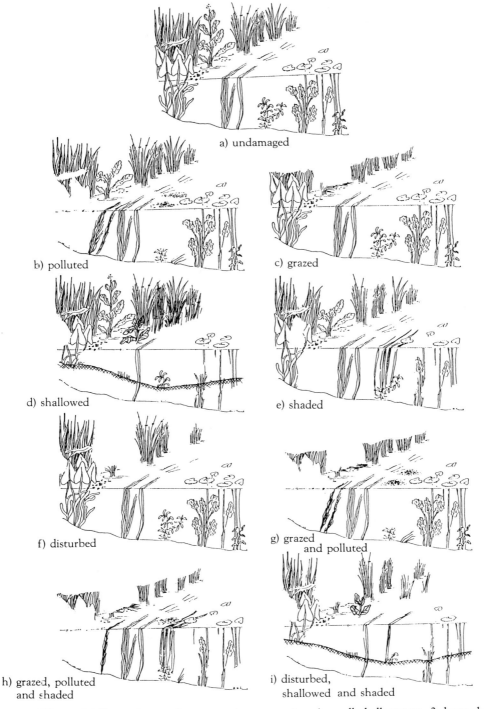

Figure 6.2 Effects of different damage factors on river vegetation. 1. small chalk stream, 2. large clay river. (a) Control: undamaged vegetation, (b) Polluted, (c) Grazed, (d) Shallowed, (e) Shaded, (f) Disturbed, (g) Grazed and polluted, (h) Grazed, polluted and shaded, (i) Disturbed, shallowed and shaded. Damage factor set to give equivalent change in vegetation. Diagrammatic but otherwise typical.

Table 6.1 Interaction between vegetation damage factors in small streams, France and Czechoslovakia. Selected from lowland areas of unsuitable substrate, where higher water force leads to wash-out (see also Figure 6.3)

a) Unstable sand, shade, cattle, France.

	Vegetation as % cover	
	Unstable sand	More stable sand
Open, no cattle	90	90
Moderate shade	50	70
Moderate cattle trampling	40	60
Shade and cattle	0	(30?)

Damage factors interact.
The same amount of damage removes more vegetation from less stable substrate.

b) Unstable sand, water force, France.

Water regime	Species
Mostly dry	Short *Glyceria* spp.
Summer-dry	Short *Iris pseudacorus, Juncus effusus*
Stable slow flow	" "
Moderate flow	Nil

(c) Resistant rock (which has easier wash-out), water force, France.

Topography	Species
Flattest, slow–moderate flow	*Glyceria fluitans, Iris pseudacorus, Juncus articulatus, Mentha aquatica*
Intermediate, moderate–fast flow	*Iris pseudacorus, Ranunculus* sp.
Lowland 'hill', fast flow but low water force	Moss.

This sequence is that expected between lowland and mountain.

(d) Small streams on Resistant rock, Czechoslovakia.

	Species
Polluted	*Phalaris arundinacea, Agrostis stolonifera*, Blanket weed
Shaded	*Veronica beccabunga, Rorippa nasturtium-aquaticum, Groenlandia densa, Agrostis stolonifera.*
Polluted and shaded	Nil.

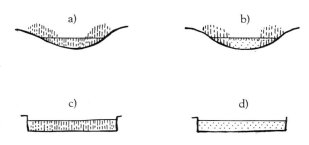

Figure 6.3 Interaction between vegetation damage factors in small streams, France (from Haslam, 1987a). Lowlands. (a) gently sloping banks and clear water: much vegetation, (b) gently sloping banks and turbid water: emergent and a little water-supported vegetation, (c) vertical banks and clear water: no emergent but much water-supported vegetation, (d) vertical banks and turbid water: no macrophytes. See also Table 6.1.

The English canal in Figure 6.9 has vegetation — except where boats queue and jostle at the locks.

The Corsican Golo (Figure 6.10) is alpine, with extreme water force reducing vegetation, which vegetation is most abundant in the lower-force tributaries, and least in the gorge.

In the German Lippe (Figure 6.11) the polluted river flows through an alluvial plain with chalk hills beyond. The river starts bad, and improves where diluted by a large clean tributary (note that where river water backs up into the tributary, diversity is cut from twelve to seven species. The river species are poor in quality (Table 4.2). In the little tributaries vegetation varies with rock type, whether the water is of calcareous or alluvial origin and, superimposed on this, with pollution. (*Berula erecta*, for instance, indicates lime; *Lemna minor* agg. and *Glyceria maxima*, alluvium; and low diversity and Blanket weed, pollution.)

The Luxembourg Attert (Figure 6.12) has an unstable bed, slumping banks, and pollution (villages, roads, and downstream factories). Cover is unduly low. Water-supported species are sparse upstream and absent downstream. Species quality is moderate (Table 4.2), so diversity and cover should be adequate: except that the unstable substrate forbids it.

The Welsh Rheidol and Ystwyth (Figure 6.13) were from ancient times polluted by lead mines

The Dutch canal in Figure 6.8 has considerable pollution and no vegetation — except where a fence excludes boats, and reduces their waves.

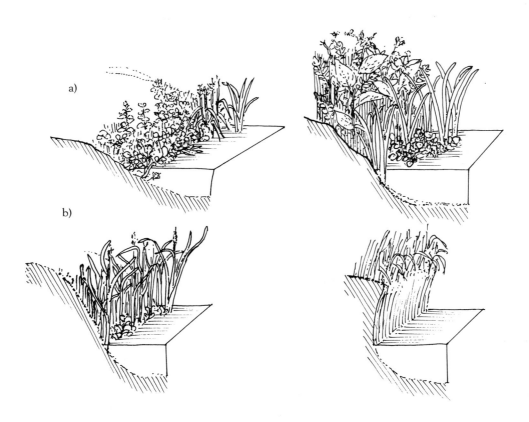

Figure 6.4 The effect of bank type on macrophytes (P.A. Wolseley from Haslam, 1987a).

(a) Sandstone stream. Left, gentle bank, lightly grazed. Right, steep bank, ungrazed. (b) Clay stream. Left steep bank, right, undercut bank, without emergents. In both streams, most of the nutrient-medium (mesotrophic, Blue band (see Tables 7.4 and 7.5)) species occur on the bank. Altering banks therefore alters site diversity and, perhaps, nutrient status band (see Haslam, 1987a, for interpretation).

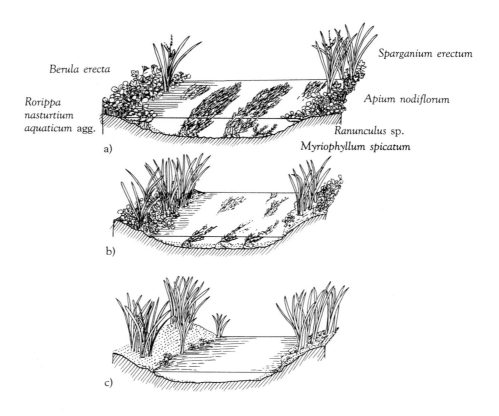

Figure 6.5 Changes in a chalk stream due to shallowing (P.A. Wolseley from Haslam, 1978). (a) Original state (with several species omitted), (b) First stage after the fall in water level, or stable stage after a lesser drop, (c) Second and stable stage after this fall in water level (with several species omitted).

— by lead, zinc and general mess. Use ceased during the twentieth century, and plants returned: Bryophytes and red algae with 400 parts per billion (ppb) of lead in the water, then as metal levels decreased, green algae, and two vascular plants (Carpenter, 1924; 1926; Jones, 1940; Newton, 1944). But the vegetation remained sparse — and still is sparse on the Ystwyth. On the Rheidol a reservoir was built *c.*1970, cutting the water force in the river below. Vegetation rapidly increased, and is typical of a good mountain stream: reasonable cover, and a *Ranunculus* zone below. Water force was the second reducing factor.

The English Pent (Figure 6.14) rises in nutrient-low chalk springs, with appropriate vegetation — even discoloured or distorted species. Then comes the motorway, whose minimum run-off (Chapter 3) is overpowering to this fragile water type, removing all characteristic species, cutting diversity, and introducing a nutrient-rich species. Downstream there are sparse nutrient-medium species, then an improvement, finally a reduction to benthic algae. What else happens, as well as motorway pollution? Public access allows disturbance upstream, then a town gives more disturbance, much road and roof run-off and walling (and intermittent culverting). No wonder such a little, fragile-water stream cannot retain its proper vegetation under so much stress.

The Czechoslovak Luznice (Figure 6.15) has interacting pollution, flow, unstable substrate and shade. Its sand is stable in low water force, unstable in high. Man controls flow, and pollutes. Turbidity (effluent and silt) reduces underwater growth, so enhancing the reducing

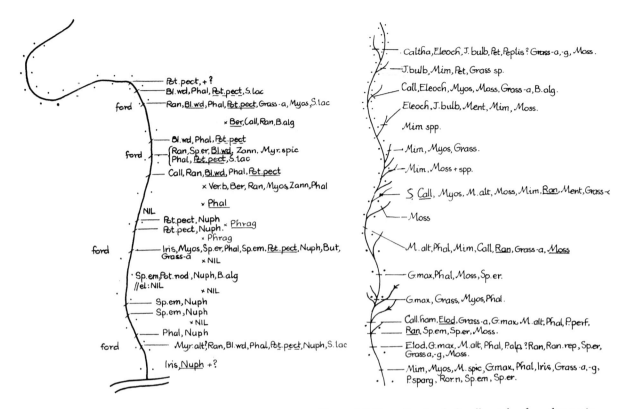

Figure 6.6 River map, Tauber, Germany (from Haslam, 1987a). Fords give wider, shallower, swifter water. Bridges now built at these points. Muschelkalk (sandstone below). Lowland, upland above, 1980. (M) Notes as for Figure 4.2.

Figure 6.7 Don, Scotland. Small patch of sandstone in upper river gives abundant water-supported vegetation in a stream otherwise similar to those nearby on Resistant rock, with little or no water-supported vegetation. Downstream eutrophication shows: nutrient-poor species (e.g. *Juncus bulbosus*) frequent upstream, nutrient-rich ones (e.g. *Glyceria maxima*) frequent downstream. Mostly Resistant rock. Mountain-rising with long lowland stretch downstream, 1986 (L) Notes as for Figure 4.2 (see also Figure 10.2).

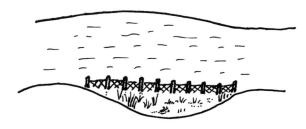

Figure 6.8 Fence protecting portion of Dutch canal from boats and allowing vegetation to develop (from Haslam, 1987a)

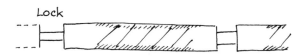

Figure 6.9 Canal with vegetation between locks, not where boats congregate at locks. Kennet and Avon Canal, England, 1980

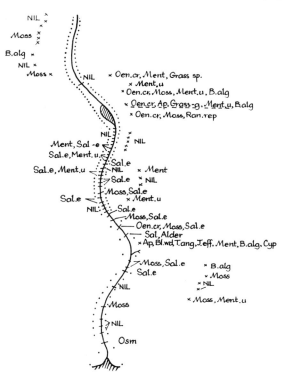

Figure 6.10 River map, Golo, Corsica (from Haslam, 1987a).
Extremely steep Alpine river, appreciable vegetation only in tributaries on flatter ground with lower water force. Even mosses in the main river are frequent only with both less water force and fairly deep water. Resistant rock, 1982 (M) Notes as for Figure 4.2.

effect of unstable sand. One site shows no vegetation in a swirling mill stream, five species in intermediate flow, and eight, with 30 per cent cover, in quiet water. In a diversion channel, polluted semi-eutrophic water is cleaned of turbidity and turned towards the nutrient-low habitat expected in its acid peat forest. Remarkable, since it is much easier to add than to remove pollution!

Small streams may have typical polluted vegetation, or typical shade vegetation — but when pollution and shade combine, vegetation fails.

A stream with a history: the upper Wylye
(Figure 6.16)

This English chalkstream should be bright and sparkling with the good (*Ranunculus*-based) vegetation of Figure 6.17. It is not. Why? What has been happening to the river?

1. Water taken away (abstraction, see Myatt, 1982; Stratton, 1982). The combination of depth and flow is often too little for large well-grown chalkstream plants, though this is partly due to:

2. Loss of flow controls and flushing. Many hatches (flow gates) allowed detailed control: heads of water for turning mill wheels, flooding water meadows, and flushing water down hatches in sequence to wash out deposited silt and mess. The mill in the centre used to pond water for sixteen hours, and grind for eight, so regularly flushing the now-bad reach below the village. Only one hatch remains, below this bad reach, and with a gauge, so not opening as it used to.

3. Loss of the cleaning action of the mill and water meadows. The mill ponded water. Most silt settled, and much organic cleaning would have occurred, as *Phragmites* grew here too (Chapter 3). The storm-carried silt which was used to fertilise the water meadows was,

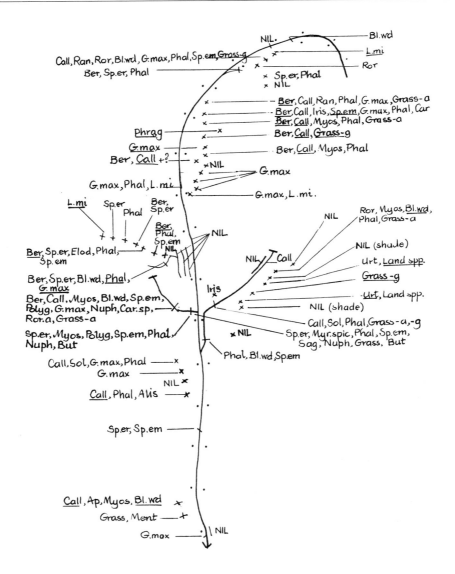

Figure 6.11 River map, (part), Lippe, Germany (from Haslam, 1987a).
Main river grossly polluted in middle of Map ('Nil'), improved by large clean
tributary. Alluvial valley with chalk outcrops above: *Glyceria maxima* and *Lemna
minor* agg. associated with alluvial dykes, *Berula erecta* and *Callitriche*, with chalk
brooks. Tributaries may also be dry (Land spp.), shaded (nil) or polluted
(Blanket weed). 1978 (L). Notes as for Figure 4.2.

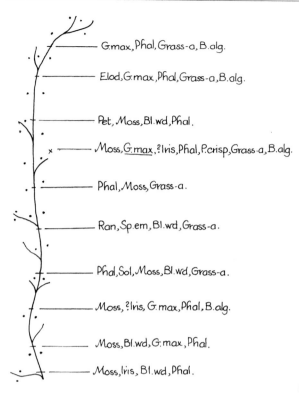

G.max, Phal, Grass-a, B.alg.

Elod, G.max, Phal, Grass-a, B.alg.

Pet, Moss, Bl.wd, Phal.

Moss, <u>G.max</u>, ?Iris, Phal, P.crisp, Grass-a, B.alg.

Phal, Moss, Grass-a.

Ran, Sp.em, Bl.wd, Grass-a.

Phal, Sol, Moss, Bl.wd, Grass-a.

Moss, ?Iris, G.max, Phal, B.alg.

Moss, Bl.wd, G.max, Phal.

Moss, Iris, Bl.wd, Phal.

Figure 6.12 River map, Attert, Luxembourg. Unstable bed, slumping banks and considerable pollution, 1985 (S). Notes as for Figure 4.2.

necessarily, removed from the river.

4. Loss of the working — or rather, the digging and planting — water bailiff. Even in the late 1940s surplus silt was dug out annually, preventing any build-up. And the plants wanted by anglers (chalkstream species like *Ranunculus*) were planted, keeping them present and healthy in doubtful (for example, ponded) habitats.

In recent decades, therefore, four factors aiding river health have been lost. What has been gained?

1. Increased silt. Much more silt is eroded from ploughed and bare hill slopes than from permanent grass. After 1939 the predominantly grass catchment (some of it arable since 3000 BC, see Hunt, 1982a, 1982b) has become predominantly cereal monoculture; not good for the soil. The checks which used to remove silt have gone, and this new silt just accumulates, most where water movement is

impeded, backing up from the hatch.

2. Fertilizer and biocide (see Chapters 2, 4, 11).
3. Livestock units and other organic effluent from farms.
4. Sewage from houses. Instead of (old) privies, houses near streams have their septic tanks overflowing to the streams. The blue colour of effluents means they can all be neatly located. Mains water, but not mains drains, came to the village in 1963. Piped water always leads to more use — easier baths, for instance. Again, septic tank effluent also contains the tank cleaner, detergent, bleach, soap, toothpaste, hair shampoo, fragrances and goodness knows what. (Let no one think medicines are harmless outside the body. The writer spilled some antibiotic into a sink: twenty years later the mark was still etched into the supposedly stainless steel.) The worst effect is where there are most houses: the village of Longbridge Deverill. The immediate influence here is obvious. The hamlet and farms next upstream are responsible for the pollution of the Wylye in the old mill area.
5. Cress bed (*Rorippa nasturtium-aquaticum*) and trout farm both on the old mill area. The former started in 1933, as the mill was declining, the latter in 1978. The cress farm therefore started while there was still control of water by the hatch, and digging-out of silt. The Wylye vegetation is unchanged by the entry of these effluents; the existing pollution overrides the incoming. In the separate exit channels, however, patterns vary. A clean cress exit channel has good chalk vegetation with *Ranunculus* dominant. Where it is cress plus six cottages and a farm with a cow unit, then the vegetation changes, and *Callitriche* is dominant. As it also is where cress and fish farm effluents join: good chalk water with mild organic pollution leads to vegetation more like sandstone. Fish farm effluent alone has a badly-polluted vegetation. Both farms add some silt to the river during cleaning: that from the cress farm clean and healthy, that from the fish farm containing more organic solids. Neither influences Wylye vegetation, or adds significantly to the silt from the hills. (Badly-managed fish farms can cause serious pollution, much more than the worst cress beds. see Chapter 9.)

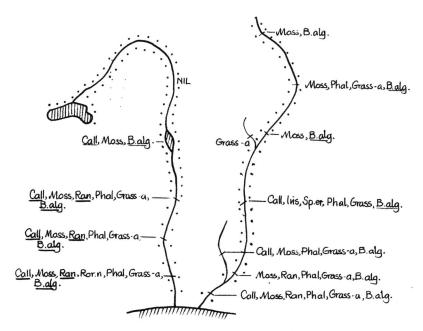

Figure 6.13 River maps, Rheidol and Ystwyth, Wales (from Haslam, 1987a). Former mines. Reservoir on Rheidol reduces water force and permits better vegetation. Resistant rock. 1980 (S). Notes as for Figure 4.2.

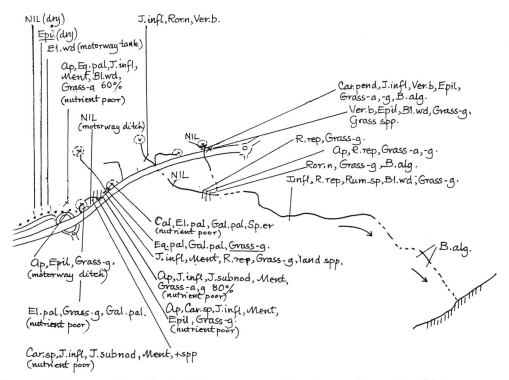

Figure 6.14 River map. Pent, England. Nutrient-poor and nutrient-medium springs. Motorway run-off, disturbance. Chalk 1985 (T). Notes as for Figure 4.2.

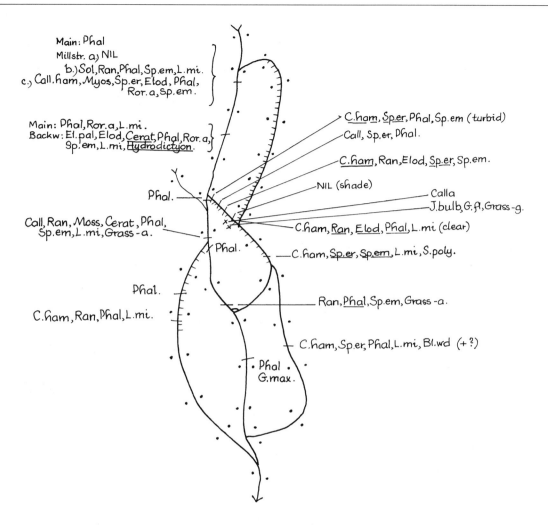

Figure 6.15 River map, part of Luznice, Czechoslovakia. Pollution, including turbidity, unstable sand, controls and variable water force, also acid forest peat, with oligotrophic tributaries and a cleaning action on the main river. Resistant rock. 1987 (M). Notes as for Figure 4.2.

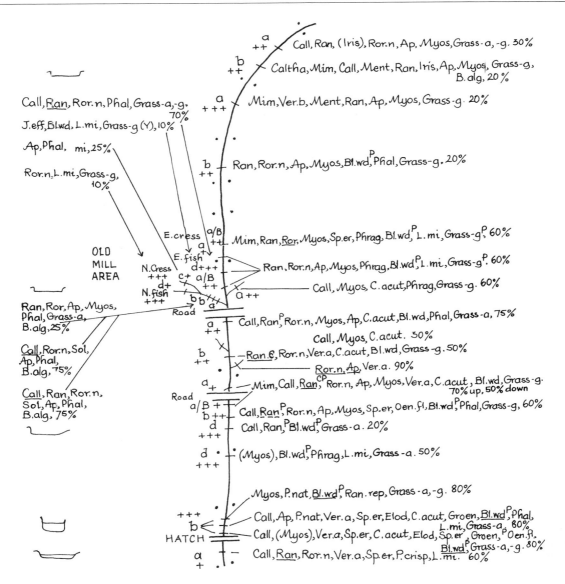

Figure 6.16 River map, Wylye, England. Pollution (village, farming, etc.), disturbance, lowered management, lowered water table. Chalk (a little sandstone downstream). 1984 (T). Notes as for Figure 4.2. Lower case letters = damage rating. *a/B* = rating of *a*, but a poor-quality indicator lowering Pollution Index to *B*. + = little silt, +++ = much silt.

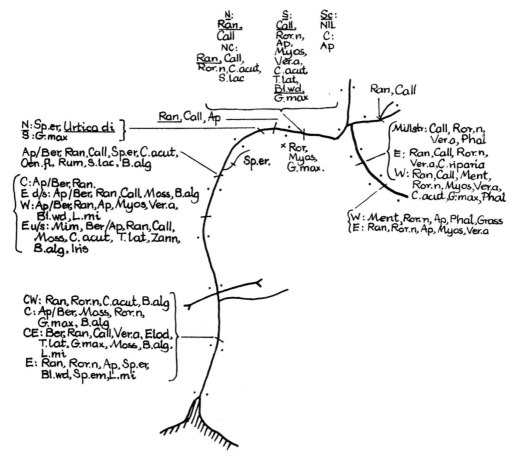

Figure 6.17 River map, Itchen, England (from Haslam, 1987a). To compare with Wylye, Figure 6.16. Good condition. Chalk (except below). 1977 (M). Notes as for Figure 4.2.

6. Disturbance. Past disturbance can be guessed at, but only the present is known (1984). It operated above and below the village. Above, a swan chewed and pulled the plants, lowering biomass, and appearance, and making the usual annual cut unnecessary. It hardly needs saying that the mess was attributed to pollution by some! (What else could make a river look like that?) Swans can be remarkably destructive, locally. A little pollution damage still continues. Below the village cattle drank, dogs swam, and children paddled. Less in all, but more damaging than the swan, since poor and polluted plants, in semi-consolidated silt, are easily uprooted with trampling.

Complaints abounded. The healthy, bright green *Ranunculus* community that the local people expect, and had been used to, is there no longer. Instead, nasty-looking ever-thickening silt, and nasty-looking plants, occur where people walk below the village. But complaint is one thing; identification and cure another.

Without knowing the history outlined above, what can be deduced solely from the vegetation and channel? Knowledge of the 'proper' vegetation for the channel is crucial. The *Callitriche*-based community here is quite a good one. In a chalk stream it means pollution. On sandstone or Resistant rock it does not. Excess silt throughout, particularly bad where water is ponded by the hatch, tells its own tale of input exceeding

Table 6.2 Distribution of macrophytes in dykes, Somerset Levels, England
Groups identified by TWINSPAN computer programme, species with a constancy of at least 40%

Species	1	2	3	4	5	6	7	8	9A	9B	10	11	12	13	14	15	16
Calystegia sepium																	1
Urtica dioica																	2
Solanum dulcamara														1			2
Agrostis stolonifera				1	1		1	1		1	1		2	1	2	3	1
Ranunculus repens																1	
Glyceria fluitans				3	1		1	1	1		2					3	
Glyceria maxima				2	1	2	1	2	2	3	2		2		4		
Phalaris arundinacea		2	2	1	1										1		
Phragmites australis														3			
Lemna minor		2	3	3	3	3	2	2	3	3	3	1	4	4			
Galium palustre										1	1	1	1				
Lemna gibba				3	3	3	2							3			
Sparganium erectum		1							2	1	2	3	2				
Mentha aquatica												1					
Oenanthe fistulosa												2					
Berula erecta											2	3					
Alisma plantago-aquatica								2	1		1	1					
Lemna trisulca								2	3	3	3	3					
Hydrocharis morsus-ranae					1	2	1		3	3	3	3					
Lemna polyrhiza					1	2	2		2	2							
Elodea canadensis							2										
Callitriche obtusangula				2	2			3									
Ceratophyllum demersum						3	2										
Potamogeton pectinatus							3										
Filamentous algae		2	3		3		3										
Elodea nuttallii					3												
Apium nodiflorum				2													
Sagittaria sagittifolia		3															
Nuphar lutea		3															
Enteromorpha	3																

Lemna minor (spanning subsets below): Fil. algae; Lemna gibba; Lemna trisulca / Hydrocharis; Agrostis

Species which characterise TWINSPAN subsets:

	Enteromorpha	Sagittaria/Nuphar	Glyceria fluitans	Elodea nuttallii/Algae	Ceratophyllum	P. pectinatus/Algae	Callitriche obtusangula	Elodea canadensis	Glyceria maxima	Berula	Berula/Sparganium	Sparganium/L. gibba	Phragmites	Glyceria maxima	Glyceria fluitans	Urtica/Solanum dulcamara
No of samples	1	11	19 20	21	46	10	20	51	78	76	31	39	19	32	16	22

Mean cover values: 1 = 0.1–2%, 2 = 3–9%, 3 = 10–29%, 4 = 30–69%
Descriptions of groups: 1. Pumped level ditches with variable water level and no aquatic macrophytes. 2. Wide straightened deep water channels in lowland region. Regularly cleansed and dredged. 3. Field channels on clays, unimproved pasture submergent cover up to 20%. 4. Field or throughflow drainage channels cleaned annually. Tree submergents absent. 5. Usually wide throughflow channels regularly cleaned with high submerged cover. 6. Mainly throughflow highly eutrophic drainage channels regularly cleaned – high cover of floating species low cover of submerged ones. 7. Coastal clay sites with high algal and submerged angiosperm cover. 8. Channels of catchment or improved pasture areas, regularly managed, nutrient-rich. 9A. Field and drainage broad channels regularly cleaned once a year, so with low cover of submergents. 9B. Field channels with variable cleaning but good submergent cover. Species-rich. 10. Field channels, cleaned at 2–5-year intervals, banks untrampled, so dominated by tall monocotyledons. 11. Field channels, cleaned at 2-year intervals, mainly on fen peat. Species-rich in all layers. 12. Field ditches infrequently cleaned. Dominated by tall emergents. 13. Infrequently managed channels on coastal clays, dominated by emergent monocotyledons. 14. Unmanaged improved pasture channels. Species-poor. Dominated by monocotyledons. 15. Unmanaged channels on unimproved pasture. Much grazing, and so with short emergents. 16. Dry, shaded channels.

Source: Wolseley (1986)

output. Little knowledge of our ancestors is needed to say that this is new: they understood, used and depended on, chalk streams, and would not have permitted this to happen. The depauperate community upstream is the classic response to shallowing (see Figure 6.5). In the old mill area the build-up of organic pollution is unmistakable (much Blanket weed, etc.; see Figure 6.16). The exit channels reflect their different origins (see above). (Recent dragging in the north channels upstream of their confluence means disturbance overrides all other influences.)

Severe pollution strikes in the village. Upstream of the village bridge, there are no houses and the vegetation has perhaps improved slightly, by self-purification, while flowing down from the old mill area. Just the other side of the village bridge there are houses, and vegetation drops — in cover, diversity and in quality (no chalk stream character and too many species from Table 4.2). The drop is rapid but not instantaneous. Since macrophytes are most harmed just by a discharge, where the pollution is strongest, this means effluents are added over a few tens of metres. (The pattern can be checked by inspecting the banks for septic tank effluents.) Vegetation slowly improves towards the hatch, but since the water is ponded, and silt is accumulated, no chalk community appears. At the hatch, *Groenlandia densa* and *Oenanthe fluviatilis* this far upstream indicate untreated sewage, *Lemna minor* agg. quiet water, *Potamogeton natans* plus *Elodea canadensis* quiet and silty, *Carex* plus *Sparganium erectum*, silty banks, and so on. The water here is fairly clean, and in the swift, gravelly water downstream of the hatch with a gravelly bottom the expected *Ranunculus*-based community appears. Here, for the first time, the water is sufficiently clean, deep and fast for healthy *Ranunculus*.

Management in the Somerset Levels

The Somerset Levels (Wolseley, 1986; Wolseley *et al.*, 1984) is an English alluvial plain with a dyke system. The original marsh has been drained over a period of centuries, and is now criss-crossed with dykes which lead to arterial drains, principally channels from the original rivers.

The results are summarised in Table 6.2 and Figure 6.18. Management recommendations are given in Appendix 2. The dykes, of course, form a continuum of species assemblages and behaviour. On this, a classification pattern can be imposed. Habitat factors are numerous. First, of course, comes the presence of water; and how much. Then vegetation is governed by subsoil, whether peat or clay, etc; by the land use — the most diverse (highest-quality) dykes were in permanent pasture with low grazing pressure; by the management (including size and shape) of the channels themselves, and by pollution — pollution from river water entering the dykes, from brackish water, or from other sources. **Satisfactory vegetation** here has many layers; high diversity; rare species present (rare either geographically or locally); and a long history of stable management. **Poor vegetation** has only one layer of vegetation (if any); low diversity; no rare species; and is found with changing chemical or physical conditions.

Drainage dykes, if let alone, will disappear — the water will flood and find new channels. Stable ecology therefore means stable management, not no management.

From the high-quality communities, placed in the centre of Figure 6.18, the changes in community radiate out with changes in channel size, water flow, brackish water, other pollution, bank grazing, fencing, too little management and too much management. River water is polluted, and bringing in more of it brings in *Elodea nuttalii* instead of *E. canadensis*. More direct sewage pollution leads to abundance of the tolerant *Apium nodiflorum* and *Glyceria maxima*. Herbicide use decreases diversity. More arablisation leads to increased *Lemna gibba*. Fen peat has more of, for example, *Scirpus maritimus* and *Sparganium erectum*, and coastal Wentloo clay has more *Carex riparia*, *Phragmites communis* and *Solanum dulcamara*.

Details of management are important. It is not what is done — whether or not vegetation is routinely removed — it is how it is done. Bucket-cleaning by an operator removing the vegetation without scraping the dyke bed gives high-quality vegetation structure and diversity. This is so even when the land is 'improved' for grassland. Taking this trouble slows invasive species, too. This is important. Explosions of one species or another are among the worst 'weed' problems: here we see that these troubles for Man may be created by Man. Toss Nature out — she will come back, in maybe a less wanted way. In another part of the

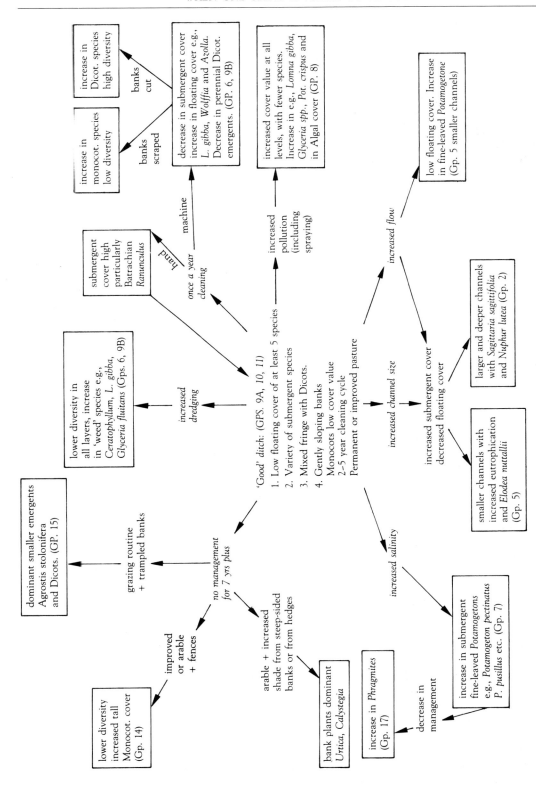

Figure 6.18 Habitat factors and vegetation in dykes, Somerset levels, England (after Wolseley 1986). Numbers refer to communities identified by TWINSPAN computer program (see Table 6.2).

Levels, careful cleaning by hand gives a flourishing water-supported vegetation. Cleaning by large machinery is generally much less satisfactory.

Additionally, when the bed is disturbed, chemicals are released into the habitat (for example, BOD up from 4.5 to 6.8, silicate from 4.46 to 5.30 ppm, ammonia from but a trace to 0.56 ppm, nitrate from 0.09 to 1.12 ppm). Lowering the water level in peat can give sulphurous by-products, which does not happen with clay. Management therefore affects dyke chemistry as well as dyke structure. That is, management can be polluting.

The Levels are mainly old-established grassland, improved grassland (with variations in both), and arable, with minor uses such as withy (*Salix*) beds and peat extraction, as well as some little-used marsh. Withy beds have the ecological advantage that their dykes must be hand-cleaned: and the disadvantage that they last only around twenty years. Arablisation includes changes in shape and depth of dykes (Figure 11.2) as well as in dyke chemistry, and much alteration of community results. Some marshes used to be acid, with typical bog flora. Acid conditions are now rare, after drainage and lowering of water level. Water passing through dykes is base-rich, and so alters and neutralises the bog conditions. Human interference, once more, has removed the most fragile original habitat. It has also introduced a more robust one: the large main drains made in the eighteenth century produced a habitat suitable for species like *Nuphar lutea* and *Sagittaria sagittifolia*. Man has created, over the centuries, a vast variety of aquatic habitats. They were created because of the variety of management. This variety is not found in either newly-drained marshes or in recently much-drained ones. The drainage has been done at different periods, in different ways, and is now maintained by different organisations and individuals, in different ways. Water may be penned, giving high summer and low winter levels (bad for water-supported species), or be related to normal water table, higher in winter. Channel size, shape and bank size and slope vary. Maintenance may be by hand, by dredger or intermediate. It varies in frequency, and intensity. Water source and so quality vary, land use around varies. And the vegetation varies with all these. How accurate are the macrophytes as indicators!

Truly, here, when one thing is added to another the mosaics and diversity of aquatic life are remarkable.

Otters (From Chanin, 1985)

Few animal studies are as complete as macrophyte ones, but the otters studies form a useful contrast, in that the main pollution came from the land, rather than as a discharge. Otters, the largest carnivorous European river mammals, have been killed for fur, for sport and as a pest for centuries. King John hunted otters in the thirteenth century. In the sixteenth century, they were officially designated pests, thought to be harmful to preserved grain, and no doubt also to fish ponds, eel traps, etc. Despite this, Man's impact was small until the nineteenth century, and not until the twentieth century were there 'game laws' to maintain the supply of furs. Later, some countries provided complete protection by law. In England, this came in 1978, though by 1975 few died in the hunt, and the law gave no benefit — except to the anti-bloodsports groups who wanted it. In the Netherlands, in contrast, the near-extinction in the 1940s was (partially and temporarily) reversed by a ban. In the Netherlands, though, the decline was attributable to hunting and trapping.

The British decline is shown in Table 6.3. Other countries, less documented, are similar. Organochlorine pesticides, spread on the land, reached fish and land animals, which were eaten by otters (Chapter 5). When the application of dieldrin and similar pesticides was reduced, otter numbers improved in Wales and South-West England, places badly affected by the 1950s decline, and with a low human population. In 1979–80, otters were, in general, most where people were fewest. By this time PCBs, heavy metals and the like had spread, and these, unlike the organochlorines, are associated with high population density.

Pollution, therefore, seems to be much more important than hunting in eliminating otters. But there is a third main reason for the otter's decline: loss of habitat. There is general disturbance by anglers and visitors, more specific disturbance and bank damage by boating, and the main culprit, the Protectors of the Waters, the legally appointed water authorities. These like

Table 6.3 Decline of otters, England and Wales

	Mean number of otters found per 100 days' hunting			
	100+	50–100	25–50	0–25
England				
Culmstock Otterhounds				
1900–50		+		
1950–57	+			
1957–65		+		
1965–71			+	
Dartmoor Otterhounds				
1900–50		+		
1950–57		+		
1957–62			+	
1962–71				+
Courtney Tracey Otterhounds				
1950–60		+		
1960–71			+	
Eastern Counties Otterhounds				
1950–60		+		
1960–71			+	
Wales				
Pembroke and Carmarthen Otterhounds				
1950–57		+		
1957–65			+	
1976–71				+

Source: Chanin (1985)

Table 6.4 Otters in Europe

1. Extinct or very sparse over much of the country: England, France, Italy, the Netherlands, Switzerland, West and East Germany, Austria.
2. Substantial decline, but still abundant in some areas, and present in most: Wales, Spain, Yugoslavia, Denmark, Sweden, Norway.
3. Decline in part of the country only: Scotland, Portugal, Greece.
4. Abundant throughout: Ireland.

Source: Chanin (1985)

nice neat channels, trapezoid-shaped, with no trees to fall on or obstruct machinery (Chapter 8, Fig. 12.2). Otters, however, need cover, places to rest and make their holts, nests and burrows. They want trees overhanging the water with good spreading tree roots (especially of ash, *Fraxinus excelsior*; and sycamore *Acer pseudoplatanus*). If the breeding sites are removed, the otters necessarily are lost, too.

On the continent, also, otters become sparse where people — and so channelisation — increase. The details vary from country to country (Table 6.4). Otters in Sweden suffer from heavy metals and fish nets; those in France and Spain from general pollution; the populations in Portugal and Greece are better, but with a potential threat from pesticides; south Italy is bad, with pollution, bank 'improvement', fishing, hunting and gravel extraction. The Dutch ban on killing led to a partial recovery, which is now in danger from pollution, land reclamation (habitat loss) and fisherman's nets. In West Germany, a contributory danger is the change in material of fish traps, which means otters cannot escape from them, and so drown.

To preserve the otter, the habitat must be preserved — or restored. Fortunately, the low-disturbance river with good waterside trees is beneficial to most other groups also! Otters move over ranges varying from c10 km to c15 km, the shorter ranges being for family groups, the longer for adult males. Nature Reserves seldom cover even half this length of river. It is therefore the water authorities rather than the conservation trusts in whose hands the otter's fate primarily lies. Not entirely, since the otter is not confined to the river and its banks, and the land around must be adequate, and the otter's food must be safe to eat. The Italian nature parks along and beside parts of rivers are the beginnings of conservation of the total river habitat.

Birds, for some reason, attract more human friends in high places who exert much pressure for conservation. Otters, regrettably, do so rather less. And all the more regrettably since good otter populations mean good populations of other groups, too, groups whose lack of friends in places of power is more understandable (such as water voles and diatoms).

Brown trout

The brown trout (*Salmo trutta* L.) is common in much of Europe, extending to the Near East. Outside, it has been successfully introduced into lands of the same temperature regime (growth

4–19.5°C, survival 0.2 to 2.5–30°C, egg development 0–15°C (Elliott, 1989a)). In Britain, fisheries are absent in bad pollution, negligible flow (dykes and drains) and the most solute-rich lowland rivers, such as those on clay. In addition to the native populations (which differ much in behaviour and genotype), stocking is often common. Brown trout are, of course, a major game (sporting) fish, much in demand for centuries, and with the vastly increased number of anglers in the twentieth century, native supplies ran short. Stocking became of major importance in Britain after c.1950, and may be done in an effort to increase the general trout population in the long term, or for catching during the next month or two ('put and take'). Even with this, there has now been widespread concern about trout decline since the mid-1970s, and while it is difficult to relate fish stocks to catch records by anglers, the sparse evidence supports this decline (Giles, 1989).

So, what has happened? Many factors seem to have contributed, but they can be summed up as *too much human interference*. Some rivers are overfished: direct human interference with the population. If this were all, the remedy, however objectionable, would be both obvious and possible. But it is not all. Trout avoid badly-polluted rivers, but can recolonise and are now returning to the rivers of south Wales from which industry is retreating (see the section on the Ebbw in Chapter 10). Mild domestic and industrial pollution is spreading over Europe, and must be causing chronic sub-lethal effects. Low oxygen is particularly harmful to young and adult trout. When under stress — as from pollution — the cortisol levels in the blood go up, and when up for weeks, disease (such as *Saprolegnia* infection and bacterial fin rot) spreads, killing some fish. Reproduction may also be suppressed (Pickering, 1989). Trout spawn in shallow, gravelly water, typically upstream and in small tributaries. These little streams are often not monitored for water quality, or if monitored, are considered of little account, and are left dirty. The loss of one spawning ground out of fifty is not important. With increasing interference, though, a couple may be lost to pollution, a couple more to channelling, and so on, until most are lost, and most of the population is lost also.

In small streams channelling may destroy spawning grounds, and in any it may lead to decreased vegetation. Less vegetation decreases cover for both trout and invertebrates, and may alter algae, on which many invertebrates feed. Channelling, other drainage processes, abstraction, regulation of flow and flow controls in the river may all alter discharge (flow and water depth). These alter habitat, and so fish — the change in the fish depending on the kind of alteration. Loss of streams and general lowering of water table lessen the available habitat (see Chapter 8).

Trout spawning success depends on dissolved oxygen, pH, depth, flow, clean gravel and water chemistry (Crisp, 1989). Activities increasing land erosion and river sedimentation damage spawning grounds, and may cause other habitat deterioration. Land drainage, afforestation and deforestation are all important here. If dams create stratified lake upstream, release of water from the 'wrong' level may give oxygen, temperature, etc., regimes less than optimal. Floods after drought, on peat, particularly in areas affected by acid rain, may kill trout from a combination of low pH, low calcium and high aluminium. Heavy rain may also wash in toxic heavy metals from old mine workings, and oxygen-deficient waters from reservoirs may have toxic levels of iron and manganese (affecting eggs, young and invertebrates). Biocide pollution is dangerous (see Chapter 5). Mechanical disturbance may also be dangerous, disrupting spawning grounds, and killing eggs either by washing them out, or by mechanical shock (Crisp, 1989). The cumulative effect of all of these, and more, appears to be — and is only too likely to be — causing the decline (not yet loss) of British trout. Trout, unlike otters, do not lack interest — but those wishing to conserve are also those wishing to catch. Those wishing to catch who have no interest in conservation are, at present, much less responsible for the decline than those who pollute, channel, and disturb.

Electricity generation

The examples above have illustrated how many factors affect one community or species. This section (based mainly on Langford, 1983) examines how one process, electricity generation, influences habitats, and so communities, in many ways.

Table 6.5 Environmental implication of reservoirs, heated water and power generation

Factor	Principal environmental, physical or chemical effect	Potential ecological consequences	Severity	Remedial or ameliorative action	Comments
A. Construction of reservoir	Barrier to normal river flow. Replace river by lake habitat	Prevention of upstream migration of fish and certain invertebrates. Effective barrier to upstream reproductive flights of some insects and downstream drift of many inverts: may interfere with normal downstream movement of fish.	Very severe; Moderate	Provision of effective fish passes, ladders or lifts. Non-flying migrating invertebrates may need help	Downstream effects less severe where other tributaries enter river short distance below dam, but headwaters above dam may suffer badly. Filter feeders may be more abundant below dam.
	Physical changes in impounded river (i) depth increase	Drown spawning grounds. Loss of normal riverine community, survival and sometimes explosive increase of a few species. If stratified, downstream O_2, temperature, etc., vary with zone released.	Severe	None	
	(ii) flow changes (a) upstream	Reduction in flow changes habitat from lotic to lentic	Severe	None	
	(b) downstream	Modified flow pattern ('compensation flows'); habitat modification	Moderate	Provide adequate variable flows of surface 'compensation' water	
B. Operation of power station 1. Abstraction of cooling water	Local increased water velocity	Attraction of and damage to fish species (and invertebrates)	Variable, usually not serious	Good design of screens ensure as low velocity as possible	
2. Anti-fouling protection of pipes, etc.	Possibility of escape and loss of biocides, especially chlorine	Toxic effects on normal river biota; death of organisms passing through system	Low	Careful design and operation of system	
3. General wash	Flue gas (S and N oxides), boiler water, drainage and spills (including minor radioactive) ash, sulphuric acid, salt, heavy metal, suspended solids, fuel store drainage, coal wash, oil store, sewage	Increased toxicity of many poisons. Attraction of mobile species (esp. fish) to thermal plume, repulsion of others	Variable; Variable		
4. Return of heated water	1. Elevation of water temperatures	Heat stress or death of sensitive species; acceleration of growth, shifts in timing of life cycles, increased rates of feeding of fish, invertebrates, etc. Enhanced micro-organism respiration, especially in organically enriched waters, reducing dissolved oxygen levels	Low; Moderate; Moderate	Minimise temperature differences	Heat stress may render some species more prone to disease. Cooling towers often provide net benefit through aeration of organically enriched water
		Survival of exotic species (especially or deliberately introduced, to compete with indigenous species)	Usually not serious	Complete removal rarely possible	Critical temperature for repulsion of many temperature species is 30°C. Populations may adapt to lower temperatures by selection, and spread further
	2. Lowering of dissolved gas solubility (i) Oxygen	Effects on sensitive biota (especially important when biochemical oxygen demand is elevated). Alter micro-organisms	Moderate	Minimise temperature differences; provide aeration or mixing facilities	
	(ii) Nitrogen	'Gas bubble disease' in fish and invertebrates	Low		

Effects on organisms may be direct, or via effects on prey or predator (e.g. waterfowl diversity may decrease through lack of food)

Source: Adapted from Hellawell (1986)

Electricity generation is the greatest single user of water in all developed countries. In England, half the water supplied (three quarters of industrial water) is for electricity, and in the United States it is 33 per cent, being 20 per cent of the total freshwater run-off of the country. The effects are summarised in Table 6.5. Water is abstracted for cooling and washing, rivers are dammed to make lakes, and electricity is generated by steam from coal, oil, gas or nuclear energy, or by a head of water.

At present, 60 per cent of heat generated is wasted, though with better use of existing technology this could be 40 per cent. Heated water returns to the river, where it may cause death, deformities, alteration to growth or breeding, or lesser distresses, directly, or by altering dissolved gases, etc. If a population of a predator species is diminished, its food species may increase, and *vice versa*. Higher temperatures may skew the existing population, or even permit the appearance of exotics.

Fish kills from heat are rare (less than 0.1 per cent of all reported fish kills in a US survey). This may be partly because fish can move from or under hot water. In suitably heated water, fish may actually move to hot currents — leading to overcrowding which, in hot water, is a recipe for parasites, diseases, and attracting predators. Heat can alter eating habits. Roach, for instance, ate *Dreissenia polymorpha* and algae instead of insects, though rudd, in the same observation, showed no change in food.

Waterfowl diversity decreases with heat, because food supply decreases (duck may, however, congregate). Phytoplankton develop best at around 25°C, decreasing if temperatures of 30°C or over are often reached. Zooplankton are less tolerant. Different groups, therefore, and different plants and animals within these groups, have different ranges of lethal, disturbing and preferred temperatures (Chapters 4, 5).

The hot water can be made useful where it is clean enough to be used for growing fish or plants. To use resources efficiently means not wasting heat in one place, while using it unnecessarily in another. Such should happen with all pollution. Use may be reduced by employing chlorine and other biocides to keep cooling towers free of micro- and, indeed, macro-organisms, or by, for example, copper corroded from the towers. In such circumstances the river is, of course, also poisoned, though sub-lethal effects are more likely than lethal ones.

Chemical wastes are varied: flue gas washings, boilers, sewage, etc. (Table 6.5). Many are highly toxic. They usually occur, however, in small quantities, and as power stations are, necessarily, sited on large rivers, dilution is great, and little harm results. Near the discharge there may be a kill with ash effluent, followed by a restricted fauna of filter-feeders: restricted, as typical in pollution. If temperature is raised, too, this is a second harmful factor in itself, and does further harm indirectly, as heavy metals, etc., dissolve more in hotter water, increasing toxicity.

When air pollution from the works leads to a low water pH, mayflies, molluscs and *Gammarus* decrease, and trout grow more slowly (Chapters 4, 5).

Normal radioactive discharges are nowhere toxic enough to kill aquatic organisms (see Table 5.22) though the effects of chronic, continuous low doses are unknown.

When a dam is put on a river, to create a head for hydroelectricity or cooling water — or for other purposes, for that matter! — a pond is inserted into that river, a pond with a wall at one end. Table 6.5 summarises the effects. The river habitat, flora and fauna are replaced by those of a lake. If in the wrong place, the lake may drown spawning reaches, and so eliminate the affected fish species. The dam wall stops movement along the river, which is particularly important for migratory fish, and migratory non-flying insects. Salmon are usually provided with special ladders or channels, but these may be useless for invertebrates. In the new lakes, fluctuating water levels may make habitats inhospitable or impossible for macrophytes and bottom-living invertebrates. Changing conditions and stratification may alter the plankton population, and therefore the intake of plankton to the river downstream, and, perhaps, the water type. If plankton becomes more plentiful, filter feeders may accumulate by the dam to feed on them. Also downstream, water is stabilised, with less drought and spate flows. (This can have the unfortunate by-product of allowing channelisation and removal of natural near-stream habitat, once high storm flows no longer occur.)

From this one process, electricity generation, therefore, the effects on the river include:

1. Heat: death, distortions and distresses of various kinds, affecting both the present organisms and their posterity, and the food supply (for animals) and predators.
2. Effluents: added toxins (chlorine, etc.), and waste products from all parts of the works. Effects are necessarily varied.
3. Dams, and therefore lakes: drastic alterations in discharge, all physical and some chemical characters. These affect river life to varying degrees, and interact with the first two factors. (A plant growing at its optimum temperature will be more likely to withstand mild pollution, or inadequate flow, for instance.)

7
Diagnostic methods

The plant is always right (A.S. Watt), and the plant is not an index.

Invertebrates live in rivers for their own purposes, not to enable man to assess pollution.

River chemistry can be monitored chemically or biologically. Chemical analyses are, and in the foreseeable future will be, needed. When aluminium was dumped in Cornish water supplies in 1987, the main detection was biological; the human population developed symptoms ranging from skin rashes to paralysis. In the absence of good data on the combination of symptoms characteristic of toxic aluminium levels in drinking water, chemical analysis was necessary. This was so because the biological understanding was imperfect. Tomatoes, watered for years from the Essex Stour, suddenly died. What had changed? Upstream of Cambridge a factory was discharging amounts of herbicide too minute to be detected until new equipment was devised (Williams *et al.*, 1977). This was diluted in, and passed down, the Cam, along a new water transfer channel and down along the Stour. After tens of kilometres and much more dilution, it killed the tomatoes. Again detection was biological but identification of the pollutant was chemical. Some 20 km downstream of the factory the writer was monitoring, and attributing the sparse aquatic vegetation to the numerous pleasure-boats — until diversity increased after the clean-up!

This pollution was also detected by macrophytes near the discharge, while the invertebrates there were near normal. It was, therefore, also a good example of the value of using several different methods of assessment, (Hawkes, 1978b).

Why is pollution being assessed? This must be known before a method is chosen. Is it for vegetation conservation, or to see whether man can safely eat the fish? The next questions are as important (in practice, though less so in theory). Who is going to do it, and with what funds? No one person can be expert in all groups from viruses, through actinomycetes, phytoplankton and frogs, to macrophytes and mammals. Mr X next door cannot do biocide analyses in the river he worries about: the test requires very expensive equipment, and either the training to use it, or the money to pay someone else. Biological monitoring is often quicker and cheaper than chemical — especially so for macrophytes; it integrates the effects of discharges varying over the day or year; it is a continuous — and free — monitor; and it shows river health (which, after a clean-up, may lag behind water chemistry).

Whim is important in river biology. Whim is not necessarily bad, as it may be the subconscious expression of what can, or should, be done. It acts as a) the whim of the researcher (more zoologists than botanists emerge from university, so there are more workers on river animals); b) the whim of those allocating funds; c) public whim — the media concern of the moment; d) the whim of those siting research stations, possibly in the distant past. This determines the habitats studied intensively. (Few researchers study each others' rivers and many disputes are due to arguing from different premises.)

Different biological methods, and their respective advantages, are shown in Table 7.1. Which method should be chosen when funds are short? The writer, a botanist, so biased towards plants, recommends vegetation surveys for routine monitoring, full-year invertebrate checks every few years, as well as the full battery of tests on algae, fish, etc., when difficult trouble spots are identified. Macrophyte monitoring is much the quickest, so the cheapest; it requires the least training, and has the additional advantage that good vegetation permits the good development of other groups, other factors being equal. However, they may not be equal. Disturbance outside the river reduces larger animals, and chemical sensitivity varies also — heavy metal pollution, for instance, may leave vegetation satisfactory while killing fish (Table 7.2).

Looking at a full list of species, their abundance, and peculiarities takes a little time, and needs some expertise to interpret. Hence the development of indices, a single letter or number to describe a community. Their purpose is to change when the environment changes. When wanted to assess pollution, it should change with pollution. Much information is lost but clarity is gained. A naive idea, regrettably still around, is that a species indicates one factor at one level — such and such pollution, and nothing but pollution! Perhaps even, such and such BOD, and nothing but BOD! Any reader will realise that is absurd. If a community or species is present, a — discoverable — set of conditions must be met. If it is absent when these conditions apparently are met, then there is another necessary condition which must be found (a new biocide more dangerous than expected? bank recently piled? Mr Y who kills otters in the dead of night?). Community studies are more reliable than those of individual species, unless the species are at the top of the food chain or have some relevant peculiarity. The plants and animals are real. The indices are not, they are just Man imposing an arbitrary order and simplification. Most indices are good for the area and habitat for which they were devised. They may not travel well (for instance, if devised on animals of soft water, they may need correcting before being applied to rivers rich in calcium and solutes).

Assessment is wanted to discover 1) what is in the river; 2) discover what is there in different conditions (including minimum influence of Man); 3) discover what changes happen with alterations, whether or not made by Man; 4) determine what standards are needed to keep various types of river health; 5) check these standards are kept, and 6) check the standards providing the protection designed for the uses they were established to maintain (Warren, 1971).

Macrophyte monitoring

Figure 7.1 shows where macrophyte indices can be used, though parts of improbable areas may, in fact, be indexable. Since this is the easiest method, easily used by anyone who has read this far, it is given in the most detail (for other methods, and for the full description of this one, see the appropriate references). The instructions on how to do it are given in Table 7.3. The commonest species of rivers can be identified with the key in Appendix 7.

Apart from Man's activities, vegetation is, of course, governed by rock type, water force, and

Table 7.1 Comparison of different types of assessment

Method	Sewage etc.[a]	Herbicides	Pesticides	Other Micro-organics	Oxygen status	Heavy metals	Sediments (surface deposits)
1. Macrophyte survey	+ + + +	+ + + + (aquatic) + + +? (land)	+ +?	+ + + +	+	+	+ +
2. Macrophyte chemical analysis	+	(P+ + + +)	(P+ + + +)	(P+ + +)	–	+ + + +	–
3. Diatom survey	+ + + (P+ + + +)	?	?	?	+ +	+ + +	+ +
4. Invertebrate survey	+ + +	?	+ + + +	+ +	+ + + +	+ + +	+ +
5. Fish surveys	+ +	?	?	?	+ + + +	+ +	+
6. Animal analyses	+	+ + + +	+ + + +	+ + +	+	+ + + +	–

These are possible guidelnies only, as present data are inadequate and not beyond dispute.

Species response differs between members of each group, and pollutant effect differs between chemicals of each group, so the common response marked here is not always correct. Also, response may vary with exposure; macrophyte, for instance, may be poor at detecting – by survey – low or intermittent heavy metal pollution, but good at reflecting sustained or high levels, so the symbol + for overall response signifies a range from nil to + + +. Plants integrate the inanimate chemical environment, animals also their food organisms. Animals at the top of the food chain integrate the whole of this. These are the most relevant to human bioaccumulation.

upstream–downstream position. Within one region these latter can be substituted by land-scape and stream size. The amount of damage by Man is therefore measured by the deviation from the expected plant community — or, rather, from the community found in the better available streams; those which are clean enough to be commonly termed clean, and under traditional or other reasonably good management. The best available vegetation is taken as the reference or standard community. Damage is the reduction from this, by whatever means, pollution or boats or shade or channelisation, etc. This standard vegetation is summarised in Table 7.4 The nutrient status bands used here are listed in Table 7.5. (If the available vegetation does not reach these levels, then the nearest of these levels, in habitat and region, should be used.) The final necessary list, that of pollution-tolerant species, is Table 4.2. The reference data, on diversity, cover and nutrient status band, for the community in question can be obtained from these. (Better, however, are up-to-date and locally-correct data obtained by the researcher, Next best are the actual species lists in the references. Table 7.5 is only an over-general summary.)

The Damage Rating Table (7.6) is then used to calculate the total damage to the community, and, where appropriate, the pollution index. This is really quite easy to do after the first try! Cover is used instead of biomass, as easier to measure and less sensitive to short-term changes (with, for example, storms). The first three criteria assess plant quantity, the second three, quality: change in nutrient status and the proportion of pollution-tolerant species present, and the amount of pollution-favoured species present. (In the damage rating, which assesses vegetation reduction from all causes, *a* is good (*a* for 'all right'), *h* bad (*h* for 'horrible'). When converted to assess just pollution — where this is possible — capital letters are used (*A–H*).) The pollution assessed is that from sewage or similar organic effluent, perhaps plus mixed industry, but the rating can be converted to any specific pollutant

Table 7.1 contd.

Habitat (see text)	Ease of identification	Season	Locate source pollution	Non-specialist use?	Speed	Expense	Destructive?	Available classification
not very rapid	+ + + +	summer	+ + + +	+ + +	+ + + +	+ + + +	+	+ + +
with mosses best, i.e. swifter	-	any	+ + + + .		+	+	+ +	+ +
stones	+	any	+ + +		+ + +	+ +	+ +[d]	+ + (P + + + +)
Riffle[c]	+ +	Need each	+ +	+ +	+ + +	+ +	+ +[d]	+ + + +
Any	+ + +	Any; or migrant season	+	+ +[e]	Variable	Variable	Variable	+ + +
Any	-		+ / + + + +[f]		+	+	+ + + +	+ +

Symbols: Increasing number of crosses = increasing favourability of method. Dash = not applicable. P = potential, probably; not yet available.
[a] Bacterial assessment, with *E. coli*, perhaps the most sensitive
[b] Vast range of compounds and, consequently, of response.
[c] Other habitats with colonisation sampler
[d] Non-destructive with colonisation sampler
[e] Low accuracy
[f] Depends on mobility of animals

by using the species sensitive and tolerant to that pollution (and if needed, altering the points categories in criteria 5 and 6). Any pollutant affecting cover or diversity will alter criteria 2 and 3, of course. The parts of the damage rating altered by different habitat factors are shown in Table 7.7.

Minor pollution and other damage can occur within the best, *a*, grade rating: *a* is satisfactory, not excellent or good. High-quality sites may well have eighteen species present. The *a* rating may be given on perhaps seven, the remainder being lost by a too-uniform channel, mild pollution and grazing, perhaps. If, though, the *a* rating is accompanied by yellowing or really flaccid habits (Table 7.6), then the pollution index is lowered to B. If the nutrient status of a river is rather finely poised (say moderate/moderate-rich band), and has slow silty and fast gravelly reaches, then the siltier ones, being more nutrient-rich, may be classed as moderate-rich, and the gravelly as moderate. Silt similarly contains more pollutants, and in such cases has the lower diversity (more toxic), while gravel has the lower cover (more wash-out).

The more the sites, the greater the accuracy. Only an expert can safely say whether interpretation from a single site is possible. The Severn Stour (Figure 7.2) illustrates the importance of time when habitat changes are small, and particularly when species are few. Some clean-up done in 1973 gave a doubtful improvement by 1977. A further increase in diversity in 1979 showed that there was a genuine long-term change, though one too slight to be proved in short-term surveys.

To those who find this method too fearsome, a rough guide can be given. Be able to identify *Potamogeton pectinatus* and Blanket weed (Figure 4.1). Note the commonest site diversity of streams of size ii is 6; of size iii, 7; and of size iv, 8. Count the species present. If there are at least this number, cover in water up to 1 m deep is at least 60 per cent and neither *P. pectinatus* nor Blanket weed are prominent, there is not much wrong (probably *a* or *b*, perhaps *c*, rarely *d*). The

Table 7.2 Advantages and disadvantages of macrophyte monitoring

Advantages

1. Best method for assessing habitat for macrophytes, for general conservation (if only one method used: as macrophytes create habitat for other groups), and for the safety of irrigation water (though pilot tests should still be run on the relevant crops). Best (with diatoms) for the total impact at a site, as the plants are stationary and do not eat organisms which may have come from other habitats.
2. Quick and therefore cheap. After half a day's practise, 35–60 sites recordable per day (c.160–320 km, traffic not heavy). Usually c.5 minutes per site: the time taken for a zoologist to change into boots and set up equipment at waterside. This means macrophytes are uniquely cost-effective. (More time hardly increases accuracy.) This speed assumes a) survey is of representative sites, so including some quick to record as macrophytes are sparse or absent due to the site being dry, heavily shaded, grossly polluted, etc; b) eyesight able to distinguish and identify species seen from, for example, bridges.
3. Simple. Macrophytes are large, stationary and few in taxa. Recognition of 50 taxa per country is sufficient, and approximate results are obtainable by identifying only c.15 species. Recording can be done from points of legal access (such as bridges), making permissions necessary only in special circumstances.
4. Non-destructive. Once common species are known (collected where relevant, by wading, or by grapnel or weighted hook on rope from bridge), sampling is negligible.
5. Assesses substrate as well as water quality: indicating the overall stage of the river (important, for example, during recovery, when water, and so invertebrates, return sooner than substrate to clean conditions).
6. Possible in waters too nutrient-poor for invertebrate communities?
7. Summer sample alone is satisfactory (unlike invertebrates).
8. Locates site of discharge accurately. Macrophytes are most affected by toxicity, which means most actively by the effluent discharge. (Invertebrates, in non-gross pollutions, are most affected by oxygen deficiency some way downstream of the discharge.)
9. Using macrophyte as well as, for example, invertebrate indices usually provides more information than either separately.
10. Very sensitive, especially to minor organic pollutions.
11. Macrophytes are often important to fish and so anglers. Regular monitoring can detect change when remedial action is still possible.
12. The larger and more abundant macrophytes add greatly to the aesthetic and amenity value, in their presence and their diversity of habit and tone (colour). Aesthetic value may also affect commercial value. Again, regular monitoring can detect change while remedial action is possible.
13. In some circumstances macrophytes have detrimental effects (flood hazard, interference with fishing and navigation) and regular monitoring can, again, give early warning of developing problems.

Disadvantages

1. Possible only where vegetation is plentiful and predictable (see Figure 7.1. Not where vegetation is almost prevented, as in the unstable, high-force Appenine (rivers). When moss indices are more developed, these will extend into (stable) mountain rivers.
2. Restricted to summer.
3. Unsatisfactory for detecting low oxygen concentrations, especially where these are brief but enough to threaten fish.
4. Less sensitive than invertebrates to certain types of pollution in lowland rivers, such as heavy metals. (Though chemical analyses, unlike surveys, will detect low levels of heavy metals.)
5. Few people have the interest to learn!

Table 7.3 Assesment by macrophytes
(For method in full, see Haslam and Wolseley, 1981; Haslam *et al.*, 1987.)

1. Learn identification at nearby streams, using any book with good pictures of vegetative parts, or Haslam *et al.* (1982). (Wade and use grapnel or weighted hook on rope.)
2. Look up the reference undamaged vegetation for the site in Tables 7.4 and 7.5. For accuracy, refer to Haslam and Wolseley (1981) (Britain); Haslam (1987a) (West Europe) or, better, survey to obtain the current reference vegetation for the locality. (If books are out of print, the writer can probably send photocopies on request.)
3. Survey when the vegetation is up (central latitudes, mid-June to mid-September), and without storm flow or, for deep water, other causes of poor visibility.
4. Select sites where (mostly) the whole cross-section is visible (bridges, vantage points, both banks).
5. Record stream width, depth, clarity (approx.), flow (slow, moderate, fast, rapid), substrate (which prominent of: mud plus silt, sand, gravel, stone, rock).
6. For the stream bank, record height above water level, slope (approx.), habit and type of vegetation (grazed grass, shading trees, etc.).
7. Record (obvious) management and damage factors.
8. Record cover to the nearest 10% in water up to 1 m deep (or, if none, that in bands at sides).
9. Record species partly or wholly in water. (In turbid water, check with, for example, grapnel until experience shows which habitats tend to have non-visible spp. Enter stream to check on small and doubtful species until experience shows where this is necessary.) Record frequency. Convert to two-point scale (much, little) for analyses. N.B. Safety advice omitted: see Haslam *et al.* (1987).
10. Determine rock type either by observation or, for Britain, using the 1:250 000 Geological Survey or Haslam and Wolseley (1981); for other countries, use relevant maps (Haslam, 1987a, gives small-scale maps for Western Europe).
11. Using data from 2, 9, 10, determine the damage rating from Table 7.6. Using data from 6, 7, and any other available, convert to Pollution Index.

N.B. Reference vegetation fluctuates with weather and other medium-type habitat variables. It should be re-established by surveys every few years. Reference vegetation is the best available under the most nearly ancient management (subject, of course, to this being up to the standards of Table 7.4).

Width and depth are usually correlated for any one stream type, and width, being less variable, is preferable for general use. Vegetation is governed more by depth, and when depth is abnormal for width, the site should be classed as smaller or larger (as relevant) than that allocated by width.

For general assessment of a river on one rock type and on typical topography, beginners should record for a river x kilometres long, ⅔x sites spread over main river and tributaries. With experience, fewer are needed to classify stream type. More are always needed in special circumstances, such as recovery from pollution, study change in rock type, much physical damage. Beginners should *never* interpret from a single site.

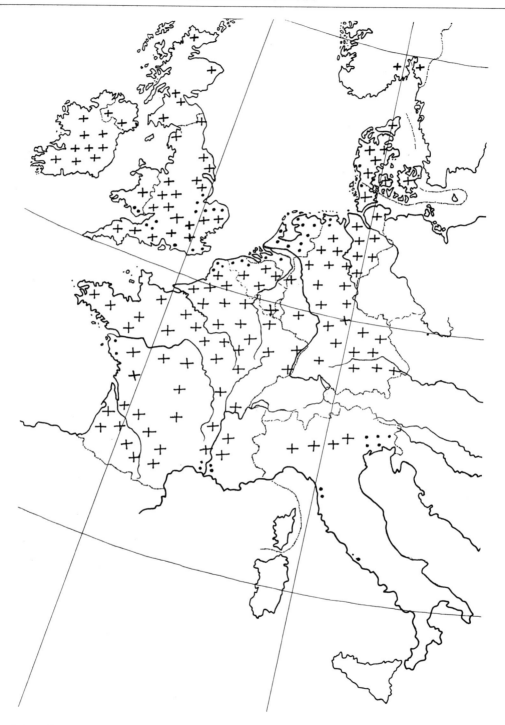

Figure 7.1 Map of EEC (excluding Greece, Iberia) showing areas where damage ratings can be used (from Haslam, 1987a). Dense crosses, lowlands which should be indexable. Sparse crosses, hills where a rating can be used locally. Dots, main areas where dyke ratings can be used.

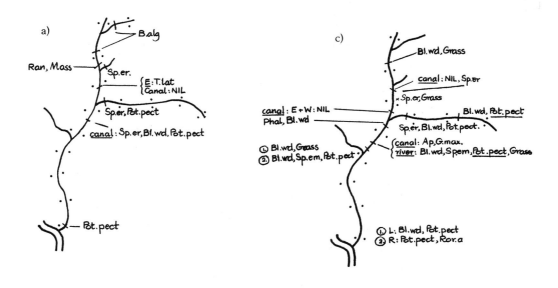

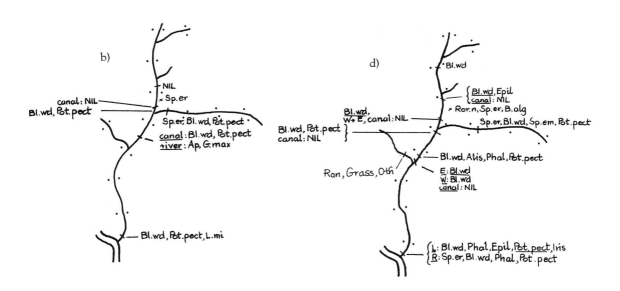

Figure 7.2 Changes in vegetation in a severely polluted river, the Stour, England (from Haslam, 1987a). (a) 1973, (b) 1975, (c) 1977, (d) 1979. Slight improvement in pollution status. E. east, W. west. Canal often parallel to river. Lay-out and species names as in Figure 4.2.

Table 7.4 Reference undamaged vegetation: minimum diversities covers and nutrient status bands of communities in different countries

When two figures are given, these apply to two variants (for example, variable hill steepness; with or without monodominant monocotyledons). Queries indicate communities with few records of undamaged communities. C, percentage cover; D, diversity. NSB, nutrient status band. Streamsizes ii, iii and iv, as in Figure 1.3. Nutrient status bands from low to high: Br, Brown; Or, Orange; Y, Yellow; G, Green; B, Blue (= T, Turquoise = P, Pink, see Table 7.5); M, Mauve; P, Purple; R, Red. Banding not applicable in the south, or in Dutch channels, see Tables 7.5 and 4.1

| | | Stream Sizes | | | | | | | | |
| | | ii | | | iii | | | iv | | |
Rock type	Landscape	D	C	NSB	D	C	NSB	D	C	NSB
Belgium										
Sands, N.E.	Lowland	6	80	(YT)T	8	80	T(TM)	9?	60?	M,MP
Sandstone, S.E.	Lowland	7	80	T	8	80	T,TM	8?	60?	T,TM
Clays and fertile sands	Lowland	6	70	TM,M	8	70	(TM)M	8?	70?	MP
Miscellaneous	Lowland	6	80	TM	7	80	(TM)M	8	80	MP
Resistant	Hill	4	60	T	7,3	80,50	T etc	8,3	60,50	T(TM)
Alluvium (dykes)	Plain	7	100	TM	9	100	TP	9?	100?	TP?
Britain										
Limestone (chalk)	Lowland	4	60	B	6	80	BP	9	80	BP,P
Limestone (oolite)	Lowland	4	50	B	5	50				
Clay	Lowland	2,6	20,50	BP	7	60	P,P(R)	8	60	PR,R
Clay	Upland	4	20	B,BP	6	20	B,BP			
Clay	Hill	0			0					
Sandstone	Lowland	3	30	B(BP)	6	60	BP			
Sandstone (New Forest)	Lowland	1,3	10,50	Or,B	5,7	60	B,P			
Resistant	Lowland bog	4	40	Br(Br,Or)	4	5	Br,BrOr	5	1	Or
Resistant	Lowland moor	5	40	Or,Y	6	40	Y(Or,Br)	3	5	OrY
Resistant	Lowland farmland	2,4	5,30	Y,B	4,6	10,20	B,BP	4,6	5,30	BP
Resistant	Upland	3	20	B(Y)	2	10	B etc	1	5	B etc
Resistant	Mountain	1	1	YB	1,3	1,25	Y,B(BP)	0,4	0,25	YBP
Resistant	Alpine	0		Y	0		OrY	0		Y
Sandstone	Upland	4(3)	40(20)	B	4	40(20)	B,BP	5,(7)	25(20)	BP,P
Sandstone	Mountain	0		YB	0,2		YB	0,5	0,20	B,BP
Sandstone	Alpine	0		Y,B	0		Y,B			
Limestone	Upland	3,5	40,80	Or(B)	4	50(20)	B(P)	5	40	B,BP
Limestone	Mountain and Alpine	0		YB	0		YBP	0,1		Y(BP)
Coal Measures	Low	2,4	20,40	B,P	3,8	10,25	Y,BP	3,8	10,25	
Coal Measures	Hilly	0		Y	3,8			3,8		
Alluvium (streams)	Plain	6	80	BP	6	80	P,PR			
Alluvium (dykes)	Plain	6	100	BP	8,9	100	P,PR			
Corsica										
Sand from Resistant	Lowland	6	80	–	7	80	–	8	25	–
Resistant	Mountain	6	80	–	4	5	–	2?	5?	–
Resistant	Alpine	6	20	–	3	5	–	2	5	–

Table 7.4 contd.

		Stream Sizes								
		ii			iii			iv		
Rock type	Landscape	D	C	NSB	D	C	NSB	D	C	NSB
Denmark										
Clay	Lowland	7	80	BM	8	80	M	8	80	M,MP
Moraine sand	Lowland	6	80	B	8	80	BM,M	8	80	M
Fluvial sand	Lowland	6	80	B	8	80	BM,M	?		
Alluvium (streams and dykes)	Plain	6	80	M	8	80	MP	9	80	MP
France										
Limestone (chalk)	Lowland	7	80	T	8	60	T	9	80	T,TP
Limestone (Jurassic)	Lowland	7	80	T	9	80	T,T(P)	10	80	T,TP
Limestone	Upland	7	80	(GT)T	8	80	(G)T,(TP)	9?	80?	T,TP
Limestone	Mountain	5	25	G,T	4	50	G,T	3	25	(GT)T(TP)
Limestone	Alpine	0,2		G	0,2	20	GT	0,3		GT
Muschelkalk	Lowland	6	60	T	7	60	T,TP			
Sandstone	Lowland	6	80	T	8	80	T,TP	9	50	TP
Sandstone (Jurassic)	Lowland	7	80	T	8	80	TP	9	80	P
Sandstone (Acid)	Lowland	6	80	Y,GT	7	80	(T)TP	8	60	P
Sandstone (Girondes)	Lowland	6	80	Y,YG	7	80	T,TP	8	80	T,TP
Sandstone S.	Mountain	0,5		GT	0,5		T?	0,4		P?
Sandstone	Upland	5	40	T	6	40	GT	7	30	T(TP)
Clay	Lowland	7	80	T,TP	8	80	(T)TP	9	80	P,PR
Clay	Hilly	2,4?		GT?	7?		T?	7?		TP
Alluvium streams (NC)	Plain	7	80	T,TP	8	80	TP			
Alluvium streams (EC)	Plain	6	80	T,TP	7	80	T,TP	8	80	T,TP
Alluvium dykes (Rhone)	Plain	6?	80	T,TP	6?	80	T,TP			
Resistant (Brittany)	Lowland-upland	7,6	60,30	GT	8,6	60,30	T	9,7	60,40	G(TP)P
Resistant (EC)	Lowland-upland	5,4	60,20	G,T	6,4	60,20	G,T	7,5	60,20	G,T,(TP)
Resistant (Massif)	Upland	3	10	T	4	10	T	5	20	T,TP
Resistant (Massif)	Mountain	5	10,50	GT,T	5	10,30?	T,TP	4	10,30	T(TP)
Resistant (Alp)	Alpine	0,1		G	0,1		G	0,1		G
Resistant (Alsace)	Alpine	0,2?	10?	YB	0,3	10?	G(GT)			
Resistant (Pyrenees)	Alpine	0,2		G	0,2		G			G
Germany										
Sands (acid)	Lowland	6	60	YT	7	60	TM	8	60	M
Sandstone	Lowland	5	80	T(TM)	6	75	TM	7	60	M
Sands/clays	Lowland	7	60	T	7	60	TM	8	60	M
Sandstone	Upland + etc	4,5	20,50	Or,T	5,6	20,50	T,TM	7?	60?	TM,M
Sandstone	Mountain	0,4	20	YT,T	0,5	20	YT,T	1,6	20	T,TM
Resistant	Alpine	0,5	5	YT	0,3	5	T	0,5	5	T,M
Resistant	Mountain	0,4	80	YT,T	0,5	20	YT(T)	0,6	10	YT,T
Resistant	Upland	0,4	75,20	(Or)Y,T	0,5	15	T,TM	0,6	50	T,TM

Table 7.4 contd.

Rock type	Landscape	Stream Sizes								
		ii			iii			iv		
		D	C	NSB	D	C	NSB	D	C	NSB
Germany contd.										
Resistant	Upland/ lowland	6	60,25	(Or)YT						
Resistant	Lowland	5	50	T	6	50	TM	7	50	M
Limestone	Upland	6	80	T	7	80	TM	8	80	TM(M)
Lime-moraine	Lowland	5	80	T	6	80	T,TM	7	60	TM,M
Muschelkalk	Mountain	0,4	50	YT,T	0,5	50	T,TM	0,6	60	M
Muschelkalk	Lowland	5	50	T	7	60	(TM)M	8	60	MP
Moraine (eutrophic)	Lowland	5	60	TM	6	60	M	7	60	MP?
Loess, etc.	Lowland	6	60	(T)TM	7	80	TM(M)	7	60	M?
Alluvium +chalk	Lowland	6	70	T(TM)	7	60	TM(M)	8?	60?	M?
Alluvium (dykes)	Plain	6	80	T(TM)	7	80	T,TM	7	80	T,TM
Ireland										
Limestone	Lowland	7	80	B	9	60	B	10	60	BP,P
Limestone	Hilly	0,3	0,30	B	0,4	0,10	B	0,5	0,10	B,BP
Resistant	Lowland	7	60	(YB)B	7	60	B	8	60	B
Resistant	Lowland boggy	6	50	OrY,YB	4	40	Y,YB			
Resistant	Hilly	0,3	0,10	Y(YB)	0,3	0,10	Y,B	0,2	0,10	YB?
Sandstone	Lowland	7	80	B	7	80	B	8	60	B,BP
Sandstone	Hilly	0,3	0,20	YB	0,3	0,20	Y,B	0,3	0,20	
Coal measures	Upland	5	20	Y,B	6	20	YB,BP	7	20	YB,BP
Italy										
Lime-alluvium	Lowland	7	80	B	7	80	B,BM	7	80	B,BM
Limestone	Hilly	10,4	0,80	0,3?						
Limestone (Carrara)	Hilly	1,7	100		1,8	80		0,7	100	
Limestone S	Hilly	0,10	50		0,10	50		0,10	50	
Alluvium (dykes)	Plain	6	100		8	100		9	100	
Alluvium (dykes)	Plain	1,7	60		1,13	60		1,6	30	
Clay-alluvium	Lowland	3			3					
Hard clay, N	Upland	0,3	80		0,4	20		0,5	80	
Hard clay, N	Mountain	0,1			0,4	0,25		0,6	0,50	
Hard clay, W	Coastal	0,7	100		0,9	60		0,7		
Hard clay, E	Coastal	0,7	100		0,8	100		0,5		
Hard clay, S	Hilly	0,7			0,2,12			0,1		
Resistant	Alpine	0			0			0		
Resistant	Hilly	0,10	90		0,10	90		0,10	90	
Sicily										
Resistant and Limestone	Hilly	0,8	0,100		0,8	0,100		0,8	0,100	
Sardinia										
Resistant	Hilly	1,12	10,90		1,12	10,90		1,12	10,90	
Limestone	Hilly	1,9	20,90		1,9	20,90		1,9	20,90	

Table 7.4 contd.

| Rock type | Landscape | Stream Sizes | | | | | | | | |
| | | ii | | | iii | | | iv | | |
		D	C	NSB	D	C	NSB	D	C	NSB
Luxembourg										
Sandstone	Lowland	5	60	T,TM	7	70	M	8	80	MP
Resistant	Hilly	5,2	60,5	(YT)T	7,1	50,5	T	7,1	40,1	T(TM)
Netherlands										
Sands	Plain	2,4	100		6	100			100	
Sands	Plain	8	100		8	100				
Clay	Plain	4	100		8	100		8	100	
Clay	Plain	7	100		8	100		8	100	
Clay	Plain	7	100		7	100		7	100	
Clay	Plain	4	100		5	100		6	100	
S. Norway										
Resistant, W	Hilly-low	6(0,12)	50	Br,M	7(0,11)	60	(Y)P(M)	7(0,8)	10	YPi,Pi
Resistant, E	Hilly-low	6(0,12)	70	Br,M	7(0,12)	50	YPiM	7(0,10)?	50?	YM

Source: Haslam (1987a)

Table 7.5 Nutrient status bands, macrophytes
In each country, species can be arranged in order according to their associated species, from nutrient-deficient to nutrient-rich habitats. The order can then be split, giving bands of about 5–20 species, named by a colour. The habitat conditions signified by Brown, Yellow, Blue, etc., and Red are the same in each. Intermediate bands are inserted when there are many species in that position. Bands are given for Belgium, Britain, Denmark, France, Germany, Ireland, N. Italy, and S. Norway (further south into Italy this system does not apply). Britain demonstrates the whole range of habitats. S. Norway is without the two most nutrient-rich ones. N. Italy has only medium to semi-rich habitats. See Table 4.1 for the approximate position of species.
    Nutrient analyses support the arrangement of habitats. (Surveys did not cover all rivers of all countries, and habitats may be have been missed.)

Trophic status	Colour band	Occurring in:	Example of habitat
Dystrophic, deficient	Brown	Britain, S. Norway	Blanket bog, i and ii
	Orange	Britain, France, Germany, Ireland	Moor streams, ii
Oligotrophic, poor	Yellow	All except N. Italy	Acid sands, ii
	Green	France	Semi-infertile sands, ii
Mesotrophic, medium	Blue[a] etc	All	Limestone, i, ii, iii
	Mauve	Denmark, Germany, N. Italy, S. Norway	Sandstone, iii
	Purple	All except S. Norway	Clay, iii
Eutrophic, rich	Red	Britain, France, Ireland	Nutrient-enriched clay, iv

[a] Different habitats are represented less often. Instead of calling all Blue, they may be:
1) *Blue.* Members of one species assemblage, often lowland limestone, without other chemical influences (Britain, Denmark, Ireland, N. Italy).
2) *Turquoise.* Species less likely to occur together (Belgium, France, Germany).
3) *Pink.* Species rank between nutrient-poor and nutrient-rich, but different to the above (S. Norway)

Table 7.6 Calculation of stream damage rating for British streams

1. Species diversity allowance*
   Number of species present**   0   1–2   3–4   5–6   7–8   9+
   Assign figure of: 5   4   3   2   1   0

2. Decrease in diversity: difference between expected and actual number of species[†]

3. Percentage decrease in percentage vegetation cover
   % loss in cover in water up to 1 m deep[‡] 100   80–95   60–75   40–55   20–35   0–15
   Assign figure of: 5   4   3   2   1   0

4. Change in Colour band

		One, or change to uncertain			
Change	Over one band	or nil	Half	Dubious	No change
Assign figure of:	4	3	2	1	0

5. Percentage of pollution-tolerant species. Add score of 1 for each tolerant species (and 1 if one or more land species are rooted in streams of sizes iii and iv), a half for each semi-tolerant species
   % tolerant spp.   Nil spp.   100   75–95   50–70   30–45   15–25   0–10
   Assign figure of: 5   5   4   3   2   1   0
   Assign 4 if only sensitive spp. are present but the number present is not over one-sixth of those expected, and 3 if one-quarter of those expected are present.

6. Weighting for special species

   | | Much *Potamogeton pectinatus* | Sparse *P. pectinatus* the only sp. | Much Blanket weed |
   | Assign figure of: | 4 (2 if intermediate) | 1 | 2 |

7. Britain only. Weighting for clay etc. in lower reaches and lowlands
   Clay, size iv ⎫   Clay, size iii, slower   ⎫
   Clay (mix), size iv ⎬ Subtract 2   Flatter sandstone, size iii–iv   ⎬ Subtract 1
                      ⎭   Clay-mix, size iv, and size iii if flat   ⎭

   Add the numbers for criteria 1–7. The damage rating is then:

Total	Damage rating
0–4	*a*
5–7	*b*
8–10	*c*
11–13	*d*
14–16	*e*
17–18	*f*
19–21	*g*
22+	*h* (*g* if channel has over 15% cover)

   * Use aggregates for *Ranunculus* spp, *Callitriche platycarpa/obtusangula/platycarpa* and mosses.
   ** Mosses restricted to man-made structures (bridge piers, concrete slopes, etc.) should be disregarded.
   [†] A negative number is recorded as 0.
   [‡] Or in bands at side if water all deep or turbid.
   Convert to Pollution Index of A–H if no physical damage present. If physical damage eliminates vegatation, irrespective of pollution, site is unclassable, *U*. If physical damage mild, a rating of *d* will have an index of C, B, or A, i.e. C+. Sites with *a* rating but yellow or flaccid species (below) are B.
   Yellow leaves: *Agrostis stolonifera, Apium nodiflorum, Callitriche* spp., *Catabrosa aquatica, Glyceria maxima, G.* spp., short-leaved, *Myosotis scorpioides, Phalaris arundinacea.*
   Flaccid or over-lush species: *Agrostis stolonifera, Apium nodiflorum, Glyceria maxima, G-* spp., short-leaved, *Phalaris arundinacea, Potamogeton crispus, Sparganium erectum.*

Source: Haslam (1982a)

Table 7.7 The effects of different types of damage on the components of the stream Damage Rating. The typical, not the sole type of response is shown. Symbols: + +, strong effect; +, mild effect; ?, doubtful effect; −, negligible effect.

Type of damage	Diversity	Cover	Colour band	% tolerant[a] spp.	Blanket weed	*Potamogeton pectinatus*	Notes
Dredging	+ +	+ +	?	?	+ +	−	
Cutting	+	+ +	?	?	+	−	
Shade	+	+ +·	−	−	−	−	
Herbicides, on emergents	−	?	?	−	?	−	
Herbicides, on channel	+ +	+ +	+ +	+ +	+ +	−	
Boats	+	+ +	?	+ +	−	+	Delicate spp. lost first
Trampling, etc.	+	+ +	?	−	?	−	
Unstable bed	+	+ +	−	+	?	−	Channel spp. lost first
Drought	+ +	+	+	+	+ +	−	Submergents lost first. Land spp. scored in % tolerant
Storm flow, etc.	+	+ +	?	−	?	−	
Eutrophication	−	−	+ +	?	?	−	
Turbidity	+	+	−	+	−	−	Submergents lost first
Salt	+ +	+	−	−	?	+ +	
Town effluent	+ +	+ +	?	+ +	+ +	+ +	Blanket weed and *P. pectinatus* irregularly affected

[a] These criteria have provisions for scoring even if all vegetation is lost. Such scores are not included in this table.

Source: Haslam and Wolseley (1981); Haslam (1982a)

Table 7.8 Calculation of the Structural Evaluation Number (SEN)

% cover of higher plants	Structural index			Diversity index		Index of filamentous algae	
	Scores						
	Emergent	Floating	Submergent	No. of species	Score	% cover	Score
76–100	1	1	2	<4	1	76–100	−4
[a]46(51)–75	2	3	4	4–5	2	51–75	−3
25–[a]45(50)	3	5	6	6–7	3	26–50	−2
[a]5(10)–25	4	6	5	8–9	4	5–25	−1
1–4[a](9)	5	5	4	10–11	5	<5	0
<1	3	3	3	12–15	6		
0	1	1	1	16–20	7		
	↓	↓	↓	>20	8		

● + ● + ●

(●   +   ●   +   ●)÷2.5

= SEN (rounded off to 1 decimal)[b]

[a] Depending whether cover measured to nearest 5% or 10%.
[b] SEN calculated as indicated by dots (which are allocated grades), the scores being added and divided as shown.

Sources: de Lange and van Zon (1977; 1983)

lower the diversity and cover, excluding *P. pectinatus* and Blanket weed, and the more the pollution-favoured species, the worse the rating. Except, of course, that no vegetation is the worst of all. (Remember that water force reduces vegetation in the hills!) It is important to remember that damage does not necessarily mean pollution. The sewage works must not be blamed for vegetation harmed by boats!

Channels with negligible flow have two indices available. The Structural Evaluation Number (SEN, Table 7.8) assesses the structure, the balance and total of the submerged, floating and emerged components, against an 'ideal' system. Its companion is the Geographic Rarity Number (species rarity listed for the well-mapped Netherlands, estimated elsewhere). Habitat affects the two differently. Structure is likely to be lost with grazing, rarity with mild pollution. The Damage Rating (Table 7.9) is better correlated to the SEN, except that polluted dykes with few species have a lower SEN (conservation value being negligible) than damage rating (some vegetation being present), and recently-cut ones will have a higher SEN (structure being present in miniature) than damage rating (most vegetation being gone). The SEN is consistently high in nature reserves, emphasising its use for conservation.

A promising new macrophyte method is being developed with computer definition of polluted communities (G. Bornette, pers. comm., method of Balocca-Costella, 1988). (See also Table 7.10 for a simpler method for Ireland.) The use of Bryophytes is being examined in Yugoslavia (see Vrhovšek et al., 1985), with some success.

Diatom indices

Diatom communities, like macrophyte ones, have been used for pollution assessment, and in much the same way, the more detailed studies taking into account the other habitat factors affecting them, such as rock type, nutrient status, grazing (Chapter 4). Diatoms are affected by the river water (and moving soil), and so provide a contrast to the rooted macrophytes which are also affected by the stable substrate. They are present all year in all rivers, and are the dominants on stones.

The number of relevant species is greater than that of macrophytes, but well below that of invertebrates. The community is dominated by but a few species, and these, plus some characteristic but sparser species, are all that need to be identified exactly. Of course the more the data, the fuller the interpretation — it is just that assessments can be made using only a small number of named species. Most papers report between one and five dominant species. In one instance forty-three species made up 85 per cent of the community.

Monitoring methods have concentrated on diatoms on stones (epilithic diatoms) rather than those on silt, macrophytes etc.; the communities varying with substrate. Collecting mixed samples from different microhabitats, without realising this, gives wrong results.

Sampling (collecting stones) is reasonably simple (but must be correctly done). Identification and counting are quick, given that microscope work is needed. There are several indices in use. Tables 4.4 and 7.13 show the tolerant species and water quality zones in Belgium, and Table 7.14 shows a generalised classification for Britain and elsewhere.

There is more literature on algae than on macrophytes; the invertebrate literature is more again. (Taken partly from Methods for the analysis of waters and associated materials 1987, which see for literature review, sampling methods, etc.)

Ecotoxicology, plants

Plants bioaccumulate heavy metals and other toxins, and therefore can be analysed chemically to assess their presence in the river (see, for example, Empain, 1976a; 1976b; Empain et al., 1980: Whitton and Say, 1975; Say and Whitton, 1981b). Bryophytes are used most. They are large enough, easily collectable, and non-rooted (so influenced by water chemistry alone). They have been used mostly for heavy metals (Tables 7.11, 7.12), but radionucleides, some biocides, etc., could also be monitored. The recommended species (Table 7.12, Britain) include a few algae and a flowering plant as well. This gives a range of plants, growing in a variety of habitats and seasons.

It is better to pick mosses than to kill fish, other factors being equal. Very satisfactory

Table 7.9 Dyke damage rating
a) Damage rating for British dykes and drains

1. Score for species present:
    1 for each tolerant sp.
    2 for each semi-tolerant sp.
    3 for each other sp.

2. Score for species diversity allowance:
    0 for 0–3 spp. present
    1 for 4–6 spp. present
    3 for 7–9 spp. present
    5 for 10+ spp. present

3. Score for cover allowance (not including tall edge plants):
    0 for up to 50% cover
    1 for 50–70% cover
    2 for 75–100% cover

Add the numbers for criteria 1–3. The Damage Rating is then:

Total	Damage Rating
27+	a_1
20–26	a_2
11–19	b
6–10	d
2–5	f
0–1	h

N.B. If the water is shallow or the dyke is covered with tall emergents, the rating cannot be used.

b) Damage- (including pollution-) tolerant species in dykes, different countries
4, tolerant species; 2, semi-tolerant species; 3 and 1 as appropriate, +, species occurring frequently but not tolerant to sewage and town effluent.

	Belgium	Britain	France	Germany	Italy	Netherlands
Agrostis stolonifera	4	4	2	4	2	4
Lemna minor agg.	2	2	4	4	2	4
Phragmites communis	4	2	2	4	4	4
Blanket weed	4	4	4	4	4	4
Glyceria spp., short leaves	2	4	+	2	2	4
Phalaris arundinacea	+	2	+	2	2	2
Potamogeton pectinatus	+	2	4	+	4	2
Enteromorpha sp.		4	2		4	4
Glyceria maxima	+	2		2		
Potamogeton crispus	2	+	2	+	2	
Sparganium erectum	+	2	+	+	2	4
Alisma plantago-aquatica	2				2	+
Callitriche spp.	+	4	+	+	+	2
Carex acutiformis agg.		1	4		+	
Elodea canadensis		2	+	+	+	2
Iris pseudacorus	2		+		2	+
Nuphar lutea		2	+	+	+	2
Scirpus maritimus		4			4	4
Apium nodiflorum		2	+			
Ceratophyllum demersum		1				+
Lemna polyrhiza						2
Potamogeton natans		2				+
Potamogeton perfoliatus		1				
Ranunculus scleratus	2					
Rorippa nasturtium-aquaticum agg.		2	+	+		
Sagittaria sagittifolia		1				+
Scirpus spp.						+
Typha latifolia		+			+	
Small land grasses					2	

Sources: Haslam and Wolseley (1981); Haslam (1982a; 1987a)

Table 7.10 Caffrey macrophyte index for Ireland
a) Indicator species

Sensitivity groupings	Macrophytes
GROUP A Sensitive forms	*Ranunculus* (not as below) *Callitriche hamalata*
GROUP B Less sensitive forms	*Ranunculus aquatilis* *Ranunculus peltatus* *Callitriche stagnalis* *Callitriche obtusangula* *Callitriche platycarpa* *Chara* spp. *Fontinalis antipyretica* *Potamogeton lucens* *Potamogeton obtusifolius* *Elodea canadensis* *Hippuris vulgaris* *Apium nodiflorum* *Rorippa nasturtium–aquaticum*
GROUP C Tolerant forms	*Zannichellia palustris* *Sparganium* spp. *Callitriche hermaphroditica* *Potamogeton crispus* *Potamogeton natans* *Potamogeton perfoliatus* *Nuphar lutea* *Lemna minor* *Lemna trisulca* *Enteromorpha* sp. *Scirpus lacustris* *Myriophyllum spicatum*
GROUP D Most tolerant forms (Pollution-favoured species)	*Potamogeton pectinatus* *Cladophora glomerata*

b) Index for water quality

Water Quality Class	Sensitivity grouping	Relative abundance
Q1 Bad Quality	Group A Group B Group C Group D	Absent Absent Emergents sparse Dominant
Q2 Poor Quality	Group A Group B Group C Group D	Absent Absent or sparse Abundant Dominant
Q3 Doubtful Quality	Group A Group B Group C Group D	Absent Common Dominant Abundant
Q4 Fair Quality	Group A Group B Group C Group D	Common Common or abundant Common Some algae
Q5 Good Quality	Group A Group B Group C Group D	Dominant Abundant Sparse Absent

Source: Caffrey (1986b)

Table 7.11 Advantages of using macrophytes to assess heavy metals in rivers

1. High bioaccumulation levels increase sensitivity of detection.
2. Moss samples can be collected by anyone suspecting contamination, and be sent for specialist analysis later.
3. Macrophytes reflect the total pollution impact, including intermittent pollutions.
4. Macrophytes can be harvested after a pulse of contaminated water has passed, so pinpointing pollution after it has occurred.
5. Metals in the macrophytes presumably reflect the metal fraction likely to affect the aquatic ecosystem better than most water or sediment analyses.
6. Dried plants are the easiest river samples to store and post.
7. Metal accumulation data may be helpful in designing systems for removal of heavy metals by macrophytes from industrial effluent.
8. Compared to sediments, macrophytes remain in place, so analysis reflects the history of just that place.
9. Compared to animals, macrophytes both remain in place, and reflect only the incident water (and sediment), not also that of the animals or plants used for their food.

Source: Whitton *et al.* (1981)

Table 7.12 Macrophytes recommended for routine use for heavy metal monitoring

Cladophora glomerata	(alga)
Enteromorpha	(alga)
Lemanaea fluviatilis	(alga)
Nitella flexilis	(alga)
Amblystegium riparium	(moss)
Fontinalis antipyretica	(moss)
F. squamosa	(moss)
Rhynchostegium riparioides	(moss)
Scapania undulata	(liverwort)
Potamogeton pectinatus	(angiosperm)

These are widespread in Europe, easy to recognise, and either easy to clean or usually have young material free of both dense epiphytes and a crust of manganese and iron oxides. They differ in their physical and chemical requirements and seasonality. While all will not all be present at any one site, a few should occur at most sites.

Source: Whitton *et al.* (1981)

Table 7.13 Diatom index grades for Belgium
The index is calculated from the sensitivity of species, their reliability (indicator) value, their relative abundance, and the number of species counted. A five-point scale results. (Calculated index values in brackets.)

1. Best biological quality, no pollution (over 4.5).
2. Almost normal quality (slight community change, slight pollution) (4.0–4.5).
3. Greater changes, decrease of sensitive species, moderate pollution or significant eutrophication (3.0–4.0).
4. Species resistant to pollution are dominant, decrease or loss of sensitive species, lowered diversity, severe pollution (2.0–3.0).
5. A few species resistant to pollution are dominant, many species are lost, very severe pollution (1.0–2.0).

Source: Descy (1979)

patterns of heavy metal distribution result, as in the River Team (see Chapter 10).

Fish monitoring

Fish communities are less useful for monitoring than macrophyte ones, as the fish can move. If fish are present conditions are favourable at that time: they may have been lethal a few months earlier. Fish do have home ranges, but can move from these. The degree of displacement has in fact been used to measure the intensity of sublethal effects — but only on pollutants sensed by the fish. Fish can, for instance, be distressed by toxic ammonia without responding to this by moving away (Hellawell, 1986). Greater disadvantages are that the number of species in any one place is usually small, and surveys of those fish that cannot be easily seen mean disturbance or destruction.

Invertebrate monitoring

Monitoring by benthic macro-invertebrates is the main biological method, used routinely in, for example Britain, Denmark and Germany. This has come about partly because there are more zoologists (see above), and partly because once one enterprising zoologist has, by much effort, got a method established, tradition follows.

Many methods have been devised, some of the best-known being listed in Table 7.18. A method may have been developed for particular rivers or for particular purposes, or may be wide-ranging. Comparisons can therefore be considered in a way not yet possible for macrophyte methods, which are still few. When Kolkwitz and Marsson (1908; 1909) first described pollution patterns they did so equally for plants and animals. This is known as the Saprobic system (Table 7.15). They were also describing a common pattern for organic pollution, in days when there were few household chemicals and much untreated sewage. When this is extended as a grading scheme for a country's rivers, anomalies can result, since invertebrates, like macrophytes, are not governed solely by river cleanliness (Chapter 5). This applies even though there have been later modifications (see, for example, Pantle and Buck, 1955). As well as pollution; rock type, water regime, etc., are important, in determining invertebrate distribution, and any lack of ecological understanding leads to much confusion. A method can be called the most satisfactory, even in the 1980s, if the grades are closely associated with only one pollutant, say, BOD. What about heavy metals, chloride and pesticides — let alone trampling or flow changes? There tends to be an assumption that the invertebrate data are 'right' only if correlated with simple, analysed, chemicals. The right assumption is that the invertebrates reflect the total environmental impact, and that the fauna of clean, healthy rivers is the more valid criterion. After the Saprobic system, in the number of countries and decades of use, comes the Trent Biotic Index (Table 7.16). A different British method, that most used in 1989, is shown in Table 7.17, and Table 7.18 sums up the essential features of a variety of methods. In this table 'basic data' speaks for itself — and the four symbol rating of 'ease of calculation' deserves notice. Saprobic and pollution indices give weight to whether species occur with or without pollution. This (as with macrophytes) may be flawed. Diversity indices are concerned solely with the number of taxa present, with the overt assumption that it must be satisfactory as it is objective, not subjective, and the covert one that many means clean. This may be even more flawed, since invertebrates decrease in purer (solute-deficient) waters. The diversity indices listed are to do with populations

Table 7.14 Diatom zones for Britain

Zone	Water	Dominants	Comment
1.	Clean. Uppermost reaches. Low pH and alkalinity	*Eunotia exiqua* *Achnanthes microcephala*	Small celled, firmly attached to stones. Little mucilage.
2.	Nutrient-richer and rather higher pH	*Hannaea arcus* *Fragilaria capucina* (var. *lanceolata*), *Achnanthes minutissima*	Common, *Tabellaria flocculosa*, *Peronia fibula*. Occurs along much of good-quality rivers.
3.	Nutrient-rich, pH 6.5–7.3, alkalinity 5.0–23.3	a) *Achnanthes minutissima* (upper); b) *Cymbella minuta* (mid); c) *Cocconeis placentula*, *Reimeria sinaata* or *Amphora pediculus* (lower)	Eutrophic rivers with mild pollution have these subzones over most of length. Varied by intermittent pollution. Silt/ mucilage component should be removed.
4.	Eutrophic, with moderate pollution	*Gomphonema parvulum*	Zone 3c species relatively absent. Low nitrogen; *Navicula mutica* common, higher: *N. seminulum*, *G. parvulum*, *N. accomoda*, *Nitzschia palea*. Sites often canalised, difficult to assess.
5.	Highly polluted	Small *Navicula* spp. (e.g. *N. atomus*, *N. pelliculosa*), small *Nitzschia* spp. (e.g. *N. palea*)	Exact identification unnecessary. Associated species: *N. goeppertiana*, *G. augur*, *Amphora veneta*, *N. accomoda*. *G. parvulum*. Characterisation depends on occurrence but rarely dominance of the first three of these.
6.	Saline	*Synedra pulchella* *S. tabulata* *Achnanthes brevipes* *Amphora coffeaeformis*	Salt mining/manufacturing. Not yet studied in Britain.

Source: Methods for the examination of waters and associated materials (1989)

and communities (see the relevant references for detail).

It is possible to devise very simple schemes, like that of WATCH (the junior section of the British Naturalists Trusts), but it is still necessary to have boots and collecting equipment, have legal access into the stream, have a stream suitable for wading, and disturb and remove the fauna. This may be more fun for children than macrophytes, but it is slower and less easily done.

All methods are satisfactory within their limits if they change when the environment changes.

It must be remembered that study of good field data is much to be preferred to any index. Indices are good to bring out a few salient points and to present complex data to busy decision-makers. After calculation, attention should revert to the field data. Equally important, no method is better than its field data. Invertebrates are sampled (removed) from only a tiny area, and if the sample is not representative of the community, the data are invalid. Invertebrate populations differ on different substrates (for example, gravel, silt, vegetation, see Chapter 5) with the type of sampling (taking a core, kicking, and for how long the substrate is kicked, month of sampling, etc.) and other factors (Chapter 5). Like must be compared with like if the pollution of two sites is to be compared. Sophisticated computer programs can look good, and be unchallengeable, but results are false if the field data are inadequate. How can one tell if they are

Table 7.15 Saprobic system (invertebrates)

Summary of classes

I   Oligosaprobic, clean and oligotrophic. Clean-water organisms (see Table 5.5), e.g. some caddis-fly larvae, stonefly nymphs.

II Beta-mesosaprobic. Dominance of *Gammarus pulex*, *Baetis rhodani* and various Caddis flies, not *Chironomus* or *Asellus*.

III Alpha-mesoaprobic. *Chironomus* and *Tubifex*, also with *Asellus*, *Herpobdella* and other pollution-tolerants.

IV Polysaprobic, severe organic pollution. *Tubifex* and *Chironomus* dominate in the worst pollution, *Eristalis* etc. may be present.

inadequate? This is a difficult question, but two guidelines can be given. Firstly analyse the data, then (and only then!) collect a great deal more under similar conditions. If the conclusions are the same, they are likely to be correct, within the environmental conditions of the survey (for rivers of the given rock type, flow, latitude, and at the given time of year). Secondly, investigate every discrepancy. If all can be explained in a way that adds to knowledge, the data are likely to be correct (discrepancy due to different season? unsuspected pollution? dredging? excess Blanket weed?). These guidelines do not guarantee adequate field data, but will help towards it. One water authority decided that its public image would be improved if the town river were cleaner. The river was disturbed, and short of

Table 7.16 Trent Biotic Index
(a) Classification of biological samples

Key indicator groups	Diversity of fauna	Total number of groups (see (b)) present				
		0–1	2–5	6–10	11–15	16+
				Biotic index		
Plecoptera nymphs present	More than one species	–	VII	VIII	IX	X
	One species only	–	VI	VII	VIII	IX
Ephemeroptera nymphs present	More than one species[a]	–	VI	VII	VIII	IX
	One species only[a]	–	V	VI	VII	VIII
Trichoptera larvae present	More than one species[a]	–	V	VI	VII	VIII
	One species only[a]	IV	IV	V	VI	VII
Gammarus present	All above species absent	III	IV	V	VI	VII
Asellus present	All above species absent	II	III	IV	V	VI
Tubificid worms and/or Red Chironomid larvae present	All above species absent	I	II	III	IV	–
All above types absent	Some organisms such as *Eristalis tenax* not requiring dissolved oxygen may be present	0	I	II	–	–

[a] *Baetis rhodani* excluded. [b] *Baetis rhodani* (Ephem.) is counted in this section for the purpose of classification.

(b) Part 2 Groups
The term 'group' here denotes the limit of identification which can be reached without resorting to lengthy techniques. Groups are as follows:

1 Each species of Platyhelminthes (flatworms)
2 Annelida (worms) excluding *Nais*
3 *Nais* (worms)
4 Each species of Hirudinea (leeches)
5 Each species of Mollusca (snails)
6 Each species of Crustacea (hog-louse, shrimps)
7 Each species of Plecoptera (stone-fly)
8 Each genus of Ephemeroptera (mayfly) excluding *Baetis rhodani*

9 *Baetis rhodani* (mayfly)
10 Each family of Trichoptera (caddis-fly)
11 Each species of Neuroptera larvae (alder-fly)
12 Family Chironomidae (midge larvae) except *Chironomus thummi* (=*riparius*)
13 *Chironomus thummi* (blood-worms)
14 Family Simuliidae (black-fly larvae)
15 Each species of other fly larvae
16 Each species of Coleoptera (beetles and beetle larvae)
17 Each species of Hydracarina (water mites)

Source: Woodiwiss (1964)

Table 7.17 The National Water Council (NWC) or Biological Monitoring Working Party (BMWP) Score

Families	Score
Siphlonuridae, Heptageniidae, Leptophlebiidae, Ephemerellidae, Potamanthidae, Ephemeridae Taeniopterygidae, Leuctridae, Capniidae, Perlodidae, Perlidae, Chloroperlidae Aphelocheiridae Phryganeidae, Molannidae, Beraeidae, Odontoceridae, Leptoceridae, Goeridae, Lepidostomatidae, Brachycentridae, Sericostomatidae	10
Astacidae Lestidae, Agriidae, Gomphidae, Cordulegasteridae, Aeshnidae, Corduliidae, Libellulidae Psychomyiidae, Philopotamidae	8
Caenidae Nemouridae Rhyacophilidae, Polycentropodidae, Limnephilidae	7
Neritidae, Viviparidae, Ancylidae Hydroptilidae Unionidae Corophiidae, Gammaridae Platycnemididae, Coenagriidae	6
Mesovelidae, Hydrometridae, Gerridae, Nepidae, Naucoridae, Notonectidae, Pleidae, Corixidae Haliplidae, Hygrobiidae, Dytiscidae, Gyrinidae, Hydrophilidae, Clambidae, Helodidae, Dryopidae, Eliminthidae, Chrysomelidae, Curculionidae Hydropsychidae Tipulidae, Simuliidae Planariidae, Dendrocoelidae	5
Baetidae Sialidae Piscicolidae	4
Valvatidae, Hydrobiidae, Lymnaeidae, Physidae, Planorbidae, Sphaeriidae Glossiphoniidae, Hirudidae, Erpobdellidae Asellidae	3
Chironomidae	2
Oligochaeta (whole class)	1

Source: Hellawell (1986)

macrophytes and because of both these, short of invertebrates. This meant a low index. So they implanted artificial substrates. Behold! the invertebrates grew! So the index improved; the river was "cleaner". How nice! Improved ecological interpretation at the outset would, of course, have made the exercise unnecessary.

Comparisons of German Saprobic Indices with macrophyte ratings showed an interesting pattern, in that the plant index never showed better conditions than the invertebrate Saprobic Index, suggesting a similar upper chemical control in both, but the macrophyte rating can also be much worse: affected by other factors and toxicities. Severn-Trent Water (1988) compared chemical and invertebrate assessments. Small streams in urban areas may have lower invertebrate indices because of intermittent storm run-off which influences invertebrates, but, being short-term, rarely happens to be sampled by chemists. Other factors depressing invertebrates more than would correspond to measured chemicals include lead mine pollution, bed instability, biocide pollution, diatom blooms and eutrophication. Various sites are likewise better in their invertebrates than in their chemical classification, but these are not accounted for.

Table 7.18 Summary of results of comparisons of effectiveness, taxonomic demands, and ease of calculation of several invertebrate indices and methods of data treatment based on experience with data derived from a spatial survey of the polluted River Cynon and a temporal survey of the unpolluted River Derwent. Number of symbols indicates rating. Open symbol indicates very low rating

Index or method	Expected effectiveness on theoretical grounds	Usefulness for water management purposes	Taxonomic demand	Ease of calculation	Subjective assessment of actual performance with test data	
					R. Cynon	R. Derwent
Basic data						
Total number of individuals	●	●	○	●●●●	●●●	●●●
Numbers of individuals in given taxa	●●●	●●	●●	●●●●	●●●●	●●●●
Numbers of taxa						
(i) Species	●●	●●	●●●	●●●●	●●	●●
(ii) Higher taxa	●	●	●●	●●●●	●	-
'Species deficit' (Kothé, 1962)	●●	●●	●●●	●●●	●●●	-
Saprobien system						
	●●	●●	●●●	-	-	-
Relative purity (Knöpp, 1954)	●●●	●●●	●●●	●●	-	-
Saprobic index (Pantle and Buck, 1955)	●●●	●●●	●●●	●●	-	○
Saprobic index (Zelinka and Marvan, 1966)	●●●●	●●●	●●●	●●	-	○
Pollution indices						
Trent biotic index (Woodiwiss, 1964)	●●	●●●●	●●	●●●●	●●	●
Biotic score (Chandler, 1970)	●●●●	●●●●	●●●	●●	●●●●	●●●●
Diversity indices						
Lognormal distribution (Preston, 1948)	●●●	●	●●●	○	○	●
Williams α index (Fisher et al., 1943)	●●●	●●	●●●	○	●●	●●
Margalef (1951)	●●	●●	●●●	●●	●●	●●
Menhinick (1951)	●●	●●	●●●	●●●	●●●	●●
Simpson (1949)	●●●●	●●●	●●●	●●	●●●	●●●●
Information theory index (Shannon, 1948)	●●●●	●●●	●●●	●	●●	●●●
McIntosh (1967)	●●●●	●●●	●●●	●●	●●●	●●●●
Sequential comparison index (Cairns et al., 1968)	●●	●●●	●	●●●●	-	-

Source: abridged from Hellawell (1986)

Other animal survey methods

Various methods using other, or single, groups have been devised, for example a Chironomid Index (Wilson, 1980). Something is known about pollution-tolerant species in rotifers, protozoa and other taxa, but the greater difficulty in sampling and analysis has meant less investigation.

Ecotoxicology, animals

This is a far more advanced study in animals than in plants. Results of analyses are given in Chapter 5, and, in quantity, in Hellawell (1986). Animal ecotoxicology is, unfortunately, destructive, but sometimes the data do help to preserve other populations. Rivers can be classed by the pollutants occurring in their animals, unsatisfactory effluents can be identified, and pollution changes monitored.

A diagnosis is, by itself, of little use. What does it mean? How accurate is it? Is the water still fit for its present use? Should any action be taken?

8
Structural damage: physical damage not due to pollution

Think a little, while ye hear
 Of the banks
Where the willows and the deer
Crowd in intermingled ranks
As if all would drink at once
Where the living water runs! —
 Of the fishes' golden edges
 Flashing in and out the sedges
Of the swans on silver thrones
 Floating down the winding streams
. . .And the lotos leaning forward
To help them into dreams.

Elizabeth Barrett Browning

Never in his life had [the Mole] seen a river
before — this sleek, sinuous, full-bodied animal,
chasing and chuckling, gripping things with a
gurgle and leaving them with a laugh, to fling
itself on fresh playmates that shook themselves
free, and were caught and held again. All was a-
shake and a-shiver — glints and gleams and
sparkles, rustle and swirl, chatter and bubble. . .

As he sat on the grass and looked across the
river, a dark hole in the bank opposite, just
above the water's edge, caught his eye, and he
fell to considering what a nice snug dwelling-
place it would make for an animal. . .
It was the Water Rat!. . .

'And you really live by the river? What a jolly life!'
[said the Mole]
'By it and with it and on it and in it,' said the Rat.
'. . .It's my world, and I don't want any other. What
it hasn't got is not worth having, and what it
doesn't know is not worth knowing. . .Whether in
winter or summer, spring or autumn, it's always got
its fun and its excitements. When the floods are on
in February, and my cellars and basement are brim-
ming with drink that's no good to me, and the
brown water runs by my best bedroom window; or
again when it all drops away and shows patches of
mud that smells like plum-cake, and the rushes and

weed clog the channels, and I can potter about dry-shod over most of the bed of it and find fresh food to eat. . .'.

'But isn't it a bit dull at times?' the Mole ventured to ask. 'Just you and the river, and no one to pass a word with?'

'No one else to — well, I mustn't be hard on you,' said the Rat with forbearance. 'You're new to it, and of course you don't know. The bank is so crowded nowadays that many people are moving away together. O no, it isn't what it used to be, at all. Otters, kingfishers, dabchicks, moorhens, all of them about all day long. . .'.
Kenneth Grahame, *The Wind in the Willows*, 1908

And Itchen water meadows, too, with their hatches and water-carriers, mean for me the sight of dabchicks getting their food under water. . .the bird as it swims is clothed from beak to tail in air bubbles, and darts, a sun-lighted creature through dark water, hunting for shrimps and spawn. And it is in the water meadows . . .that I have listened most often to that happy sound of days and nights in May, the landrails' stick-and-chatter as he runs through the high grasses.

Nineteenth-century description.

Where now are the water rats, otters, water-carriers, dabchicks, spawn and the smell of plum-cake (Figure 8.1)? No longer usual, but something special, to be wondered at. Except for settlements, mill flows, drained wetlands and the major waterways, drastic non-pollution effects of Man on channels are new. In major towns, of course, rivers have been tamed and altered all the time (see Figure 8.2), and mills on tiny streams and large rivers often required diversionary channels (those in medium-sized lowlands channels were often constructed to ease navigation around the mill(s) — as in the English Nene). Navigation has always required channels of reasonable depth and water stability, but not until the eighteenth century did canals and canalised rivers become important.

Since records began (and records are numerous from the thirteenth century, in England), rivers have been the subject of endless quarrels and lawsuits.

In early times the millers and the fishery owners often sided together against the watermen. Fishing was typically good below mills (see Figure 5.1), so there was a natural association of the two. Boats disturbed both mill flow regimes — wanting water at times unreasonable for milling — and fisheries. Owners of the banks had further conflicting interests to those of millers, fishers, swankeepers and watermen. In the nineteenth century the (river) drainers seriously entered the fray, and, considering Europe as a whole, won it. But they only won it because of a major change in public attitude and law. Until the twentieth century, rights reposed in numerous sources. This led to equally numerous discontents, but also to few alterations, since one man's improvement was another's harm. Unedifying, but preserving. Now almost all rights repose in the hands of the water authorities, the only organisations since records began to hold such power over rivers. The power has been used to drain land for the farmers, in the later phases irrespective of benefit. Power has been held over country-wide areas, and used to make drastic changes with extreme speed (extreme, compared to earlier times). Engineers have had power to physically to change rivers. Waste disposal people have had power to pollute, both to construct sewers to carry waste, and to put this waste into rivers. The checks and balances previously existing have diminished or gone over much of Europe. It is possible, however, for bodies, like individuals, to misuse power first gained, and gradually to acquire wisdom. While much is still getting worse, Figures 8.3 and 8.4 show early indications of improvement — the start of a widespread trend, one hopes. The extreme increase of both technology and wealth in the later twentieth century permitted, indeed encouraged, excesses (Figures 8.3, 8.5, 8.6). The old saying that if Nature is tossed out with a pitchfork she will soon come back — and may come back in a more unwanted way — was forgotten. Forgotten, because the 'soon' was delayed by the technology: not months but decades. It is only when, as now in parts of Germany, the altered rivers become more expensive to maintain, that the theory is queried. To the planners, it has been irrelevant that the countryside heritage is destroyed. It is relevant, though, that flash floods are increased, the fertile topsoil is lost, irrigation is needed, streams need expensive lining and relining, and flood protection construction is expanded. Was it really worth it? Conservation has no price tag, so it is not even considered like the other interests of angling, navigation, etc. Neat, tidy, machine-accessible channels lowering the water level ever further (and increasing the sediment accumulation) are what is produced. Swan-keeping and navigation with constant traffic are as destructive, be it said. That has contracted over the past few centuries and now affects only a few main rivers and the purpose-built canals.

a)

b)

Figure 8.1 Good-quality rivers (P.A. Wolseley)

Figure 8.2 Fifteenth-century city river, Germany (from Nuremberg Chronicle, by kind permission of the Master and Fellows of St Catharine's College, Cambridge). Note the river bank.

What is river structure?

Structure is the physical habitat, the patterning, in, on and among which live the plants and animals. For macrophytes, the structure is mainly the inorganic parts, the bed, banks, etc., made mainly by inorganic means — discharge, erosion and the like — but partly by trampling livestock, burrowing water rats, grazing swans and the macrophytes themselves, among others. For the fish, however, the macrophytes are as much part of the river structure as is the gravel on the bed. So the identity of structure depends on the organisms using it.

By an awkward change in common usage, however, 'loss of structure' is usually now considered just as loss of diversity and complexity in the animal and plant communities, not as loss of diversity and complexity of bed and banks as well.

The best river structures, like plant communities, vary with the river type. An Alpine stream derives structure from rocks and trees, not from macrophytes. Good structure is exemplified in Figure 8.1 (see also frontispiece). Animals and plants all over the place (but, note, no choking vegetation). An ecologists dream? Perhaps. An engineer's nightmare? It should not be! With surplus food production in the EEC, and set-aside uses for land already in action at the time of writing (1989), minor flooding in some parts is acceptable. In the Figure, the ground slope prevents floods spreading — supposing weirs and sluices are properly placed. Alternatively, the water level may be rather lower (banks higher) or wetland may be established by

Figure 8.3 'Never more like this for maintenance'. Sønderjyllands Amstkommune. (Redrawn from Markmann, 1988).

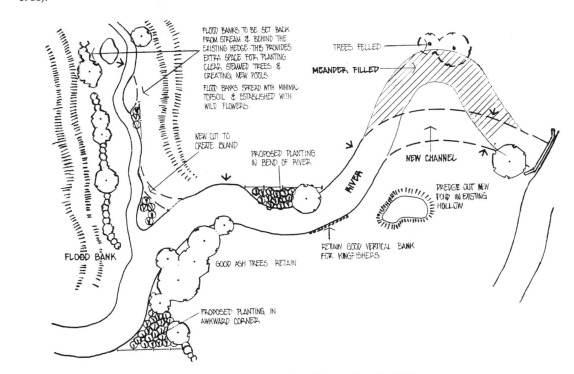

Figure 8.4 Nature Conservancy Council river plan (from Newbold *et al.*, 1983).

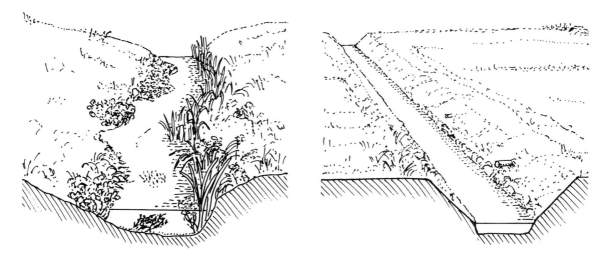

Figure 8.5 Alteration of a species-rich partly-channelled stream (left) to a species-poor much-channelled stream (right). (P.A. Wolseley in Haslam, 1987a). Note that the much-channelled stream has a flow swift enough to require stoning to protect the sides from scour, and under-drainage is present.

the river. Upstream or downstream more drainage may be needed, say for towns, but what are flow controls for? And washes for surplus water are valuable wetlands. More lowering and channelling are obviously needed in places, but not in all places.

Why is structure lost?

Some of the attitudes leading to adverse changes in structure are listed in Royal Society for the Protection of Birds (RSPB) (1977):

'I am not convinced conservation is part of my business as an engineer.'

'Here's this marvellous scheme'. 'My dear boy, it is splendid but it has no political appeal for Ministers.'

'When officials have things nicely balanced, you find a piece of elastic pulling down the scales — party political, town council, trade union, or whatever.'

The RSPB make the valid points: there is an obsession that water quality is the prime consideration. Structure is just as important. Would you take down the Tower of London to put up an office block? Then why substitute a wheatfield for a wetland and its river (given that wheat production is in surplus, adds the

writer).

Some fifteen years ago, a water authority official complained bitterly to the writer that the chairman of his authority was exerting his personal influence to keep one river satisfactory for angling. The writer replied it was better to have one good river than none. The official found this incomprehensible.

When all is said and done, *it does not, and it should not, need central government action to preserve the river bank.* Where money can be made, however, legal constraints may be needed.

It is the quantity and quality of the changes which are at issue — not the basic necessity. The watershed English publication was the prize essay by Clarke (1854–5) 'On Trunk Drainage', which pointed out the value of unimpeded movement to various east-flowing rivers. He noted the damaging floods on undrained land. Floods such as those of 1852 were perhaps extreme (1852 being an exceptionally wet year), but rather lesser ones were common. In 1852, the Vale of Gloucester, central Somerset and parts of the Thames Valley were under several feet of water. Even when flooding was minor, the constant damp led to disease. Malaria, liver complaints, scrofula, gastric fevers and rheumatism were rife. Cattle and crops were equally affected. Even as late as 1968, floods in Bristol caused, among those whose homes had been flooded, a 50 per cent increase in deaths, a

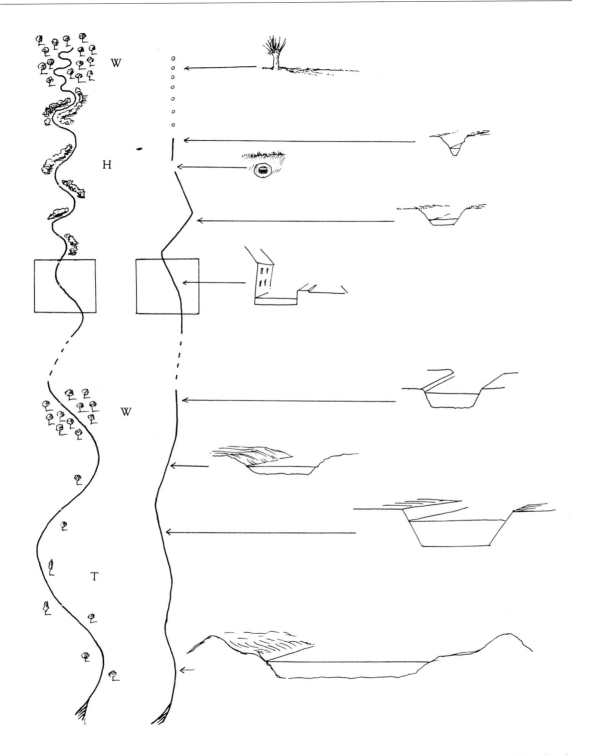

Figure 8.6 Altering of channel shape and position with intensive management. W, wood; H, hedge; T, isolated tree.

53 per cent increase in the number consulting doctors, and a doubling of hospital admissions (Bennett, 1970); and in the Fenland, £624,000 (£4 million at 1981 prices) of damage to crops (Cave, 1981).

Clarke (1854–5) realised, as over-enthusiastic followers do not, the necessity of working with the water table, and that the whole system must be considered, more land being injured by defective ditches and under-drainage than by river flooding. Improvement, to him, and to authorities since, has meant improvement of flow, not improvement of river. The excitement of drainage to give better crops and better farming comes through clearly in his essay. Reclamation is indeed exciting, see Williams (1970, Somerset Levels) and Darby (1983, Fenland) for marshland. The attitude that more is better can become an unthinking tradition.

When the rivers have gone wrong, what should be done, what should be aimed for? The full answer is discussed in Chapter 12, but two other, partial, answers are more easily available: ask a specialist (checking the expertise); and, probably more easily, look at the pictures. Those in Figure 8.7 have had the first, late-nineteenth-century round of modification for drainage. This amount of widening, deepening, removal of controls, etc., has left the river satisfactory: unlike the second round (Figures 8.1, 8.3, 8.5 and 8.6) a century later. Plenty of such pictures are available, in museums, libraries and attics. This is the desirable habitat. The only remaining question is how, technically, it can be reproduced without actual harm to Man. Reproduce the habitat, keep down disturbance and pollution and, given time, populations will recover — except for extinct species, of course (see Hynes, 1960).

Access for machinery is needed for dredging: but not access all the length on both banks. Water level must be such as to prevent destructive floods: banks need not be smooth, steep (and lined) (see Figure 8.4). Good-quality banks take up more space, but now that the EEC produces a surplus of food and is promoting schemes to take land out of farming, could not river banks be improved under a similar scheme? Good banks — and even some wetland beside? Choking vegetation — weeds — can cause floods: but a constant 20–25 per cent of the water volume apparently occupied by plants is satisfactory in most streams, and extra removal is unnecessary.

Loss and alteration of stream

A stream lost through drainage or underdrainage, or put into a pipe, represents a total loss of habitat. The old town sewers (formerly streams) and many of the tiny, and indeed small brooks have gone this way (Figures 2.2, 8.5 and 8.6).

Underground sewers, in particular, may retain a highly distorted river life, of anything from *Dreissenia* to alligators, but not including photosynthetic plants. (Underground clean streams contain less food for animals.)

Streams may be changed in position (Figure 8.8) and shape (Figure 8.1). Straightening gives more uniform channels, so decreasing micro-habitats. It frequently increases scour (by increasing slope), though channelisation may also give canal-like waterways. Neither aids structure. Changes of shape in small streams usually mean narrower channels, and for drainage these may be set deeper, giving banks able to shade the stream as well as less water surface for river life. Typically such streams are also shallower than their undrained counterparts, often too shallow for the full development of water-supported macrophytes.

Loss of structure is easy and widespread in small streams, as these are so simply altered, piped and lost. Larger streams, with greater discharges, are not so easily eliminated or piped, and the danger is more the loss of micro-habitat than the loss of total habitat: the creation of the uniform (Figure 8.2) channel.

A stream constrained into a channel still has water force. This water force acts on the banks, etc., towards returning the channel to a natural position and shape. The amount and strength of bank channel works depend on this water force (and the other habitat constraints). Where water table has been much lowered, water force will be less, for the given channel width. This allows either less construction to maintain the channel or further constraints to be made on the channel. Effects of channelisation are summarised in Table 8.4.

Another type of drastic interference is shown in Figure 8.9 in a summer-dry river in Malta, where water conservation is essential. Frequent reservoirs create deep slow silted habitats, filling rapidly in winter — and emptied, sometimes more rapidly and more than once, by abstraction for crops. Water-level fluctuations are therefore

Figure 8.7 Late nineteenth-century rivers. Lower picture is a sheep wash.

Source: Girl's Own Paper (1887)

Figure 8.8 Altering of channel. (a) Stream underground, (b) Moorland drainage, (c) Pollards marking position of stream now lost by lowering of water level, (d) Stream at side, not base of valley (here sited for mill), (e) Stream canalised early for industrial part of small town, canal disused, and downstream narrowed and stoned (also see Figure 8.10), (f) Reservoir, (g) River wharf, (h) Bed stoned in village, (i) River confluence, built-up, (j) River confluence, little-managed.

Table 8.1 Environmental implications of channel modifications in flood alleviation and land-drainage schemes and their subsequent maintenance

Factor	Physical or chemical environmental effect	Potential ecological consequences	Probable severity	Remedial or ameliorative action	Comments
1. Enlargement of channel to provide increased flow capacity	Change in physical dimensions of habitat to give (i) reduced depth under dry weather flow (ii) greater channel width (iii) change in water velocity for given discharge	Removal of biota from existing channel by reconstruction work: high, temporary, turbidity reducing plant photosynthesis, blanketing substrate downstream and affecting macroinvertebrates and possibly feeding of fish	Often very variable, but of short duration	Reinstatement by reintroduction: working from downstream to upstream helps; little can be done to avoid this but working short distances at any one time helps to reduce severity	Recolonisation occurs by drift from upstream: high turbidity and suspended solids are 'natural' phenomena associated with floods but duration may be longer during engineering operations
2. Modification of channel shape, both in profile, plan and cross-section	Reduction in habitat diversity (i) smooth profile removes variation in depth (pool:riffle configuration) with tendency towards uniform substrate material	Loss of many microhabitats and their associated flora and fauna; reduction in overall community diversity	Severe	Dig out deeper pools below level of designed profile	Fortunately, tendency for channel to return to natural configuration unless constrained by massive structures (e.g. concrete channel or piling etc.)
	(ii) trapezoidal cross-section destroys habitat diversity, especially shallower margins	Loss of habitat diversity; marginal plants unable to establish foothold; loss of some macroinvertebrates	Severe	Construct with an irregular channel cross-section and especially with marginal ledges ('berms') to allow reinstatement of marginal plants. If flood channel must be straight, encourage dry weather channel to meander within its confines	
	(iii) straight channel removes meanders having deep fast water on outside and shallow on inside of bends; modifies velocity and suspended solid carrying capacity	As above; channel length reduced and even with increased width, habitat area may be lost	Severe	Not an ideal solution but will encourage some habitat diversity	
3. Channels lost or put underground	Habitat loss	Habitat loss	Severe	Restore	
4. Channels made smaller and lower (banks steeper)	Less waterspace, shallower, over-steep banks, shading banks	Habitat change and usually loss	Moderate to severe	Widen, improve banks	
5. Bank modifications	(i) Removal of trees to provide access (mainly for machines) and reduce obstruction of flood plain	Loss of shading so that increased light reaching water encourages algal and macrophyte growth, loss of detrital input during leaf-fall and aerial insects for fish food	Variable, usually moderate	Remove only from one (preferably north) bank. Plant trees in rows parallel with flow to reduce risk of flood loss	Tree loss is slow to recover but other vegetation may return within one or two seasons
	(ii) removal of bankside vegetation by mechanical means or herbicides	As above. Herbicide spray drift may affect other plants	Moderate	Restrict control of position of bank each season	
	(iii) construction of raised or flood-banks	Increased carrying capacity of channel may modify habitat	Probably insignificant		
6. Maintenance of channels	(i) Removal of substrate and vegetation by mechanical means	Habitat modification, removal of benthic fauna, loss of flora and fauna associated with it	Variable to moderate to severe	Restrict to partial treatment in any one season	
	(ii) removal or control of vegetation by means of herbicides	Loss of plants and associated animals; invasion of more resistant species leading to community changes	Variable, usually mild to moderate		

Source: modified after Hellawell (1986)

Figure 8.9 Effect of control in upland stream. Brittany, France. Above the control, *Nuphar lutea*, *Sparganium emersum*, etc., occur in slow, deep water. Below the control, *Ranunculus* sp., etc., occur in swift, shallow water (P.A. Wolseley in Haslam (1987a).

great (up to about 3 m). Man's influence shortens the period of stream flow. Potentially lengthens that of reservoir water, increases pools, and their water level fluctuations. Since this leaves a net increase of aquatic habitat, it improves aquatic life. Where there is excess dredging in these pools, though, that lowers the quality of that aquatic life (Haslam, 1989).

On the less stable soils, bank stability is often provided by tall monocotyledons such as *Phalaris arundinacea* or *Phragmites communis*. If these are eliminated, bank erosion will follow — and subsequent walling be much more expensive as well as ecologically worse. Trampling by heavy livestock also causes damaging bank erosion. Wet grassland by rivers originally had ponds and channels, later changed to the abundant — but not ubiquitous — penned water systems like water meadows, where channels were numerous, or to less controlled drainage systems. Most have now gone. The loss, in absolute terms, of river life has been enormous.

In Britain, water table has been lowered more than in the rest of western and central Europe. Between 1930 and 1980, 35 500 km of lowland river were at least partly channelised (Brookes and Gregory, 1983). In Denmark, streams have been much straightened and often narrowed, and piped. Half the minor streams in the Stor in 1876 are now in pipes, for instance (F. Jensen, pers. comm.). Many more than this have been lost from the French Chalk.

With these complaints it must also be stated that adequate drainage for the health of Man, livestock and crops is necessary, that excess vegetation hinders flow, and that partial shade and scour are acceptable ways of preventing this excess.

Effects of loss of structure

The importance of cover and resting places for the otter is described in Chapter 6, and the same applies to other river mammals. Loss of structure here is even more important than pollution, as damaging loss has affected more rivers.

Channelisation reduced waterfowl on the Great Ouse (Campbell, 1988). River discharge capacity increased by 30 per cent, a weir was raised by 30 cm, the bottom was dredged — and, the crucial thing, the banks were smoothed. The loss of birds followed the loss in structure; the more the loss in structure, the more the decrease in birds (Table 8.1). Wider surveys, and the checks between lesser- and more-affected sections showed engineering alone was responsible. It was not the disturbance (successful nesting was seen within 50 m of the dragline), nor, to any major extent, the loss of food, but the loss of nesting habitat (95 per cent of nests in river fringes or the bushes above). Structure is vital.

Different schemes have different amounts of loss in habitat, and so of birds (Raven, 1986), which explains much of the argument about the effects of channelisation. Research not fully documented is not fully explicable.

Many birds depend on the macrophytes, and the macrophytes depend on the river characters. Put the macrophytes right, and most fauna will need no further attention. Other structures can, though, substitute to some degree for macrophytes: old bicycles or artificial vegetation, for instance. The importance of structure is seen in the industrial heart of Birmingham. Where alders can occur, mallard (and indeed where the water is clear enough to see the prey, even kingfishers) can live, together with the tree-based birds

Table 8.2 Effect of loss of river structure on waterfowl
a) The extent, as total length (m), of major vegetation features between Oakley and Bromham before (1980) and after (1983) river engineering work

	Right bank		Left bank	
	1980	1983	1980	1983
Section A (3840 m)				
Fringe: 0.5–2 m wide	549	39	341	264
Fringe: over 2 m wide	364	15	92	0
Bush	793	21	519	403
Section B (2680 m)				
Fringe: 0.5–2 m wide	1198	1203	387	110
Fringe: over 2 m wide	570	565	0	0
Bush	650	625	222	183

b) No. of territories

	Number of territories			
	1980	1981	1982	1983
Section A (3840 m)				
Coot	15	14	1	3
Moorhen	23	22	10	8
Little grebe	3	2	0	0
Reed bunting	6	7	4	3
Sedge warbler	13	8	1	4
Total	60	53	16	18
Section B (2360 m)				
Coot	18	18	13	15
Moorhen	17	19	21	15
Little grebe	4	4	1	2
Reed bunting	6	7	5	6
Sedge warbler	14	14	10	12
Total	59	62	50	50

c) The locations of the nests of Coot and Moorhen between Oakley and Bromham, 1980–1983

	Number of nests in		
	Bushes/trees, in/over water	Emergent fringe	Other
Coot			
April	16	6	1
May	10	15	0
June	3	3	0
Moorhen			
April	19	1	0
May	16	10	2
June	4	7	2

d) Summary of changes

	Change: 1983 levels as % of 1980 level	
	Section A	Section B
Habitat		
Fringe	24	87
Bush	32	93
Birds		
Coot	20	83
Moorhen	35	88
Little grebe	0	50
Reed bunting	50	100
Sedge warbler	50	86
Total territories	30	85

Source: Campbell (1988)

Table 8.3 Species distribution in relation to boat capacity in the Netherlands
Only waterways able to carry boats of 50 tonnes and more considered. Total capacities divided into three groups.
(a) Effect of boat tonnage on species diversity (figures are number of sites)

Boat capacity (million tonnes)	Species diversity				
	0	1	2	3	4
0–1	21	14	5	1	4
1–5	3	6	2		
5–10	11	2		3[a]	

[a] Plants in local shelter only.

(b) Number of species occurrences in relation to boat tonnage carried

	Boat capacity (million tonnes)		
	0–1	1–5	5–10
Phragmites communis	10	7	3
Sparganium erectum	8		
Glyceria maxima	5	1	
Rumex hydrolapathum	3		
Phalaris arundinacea	3		
Typha spp.	1		2
Other spp.	16	2	3

Source: Haslam (1987a)

Table 8.4 Frequency of occurrence of low, medium, and high-density macrophyte beds (expressed as a percentage of stations where estimates of macrophyte density were obtained) at stations in shipping and non-shipping channels (St Clair and Detroit Rivers, Canada)

Density	Shipping channels (168 stations)	Non-shipping channels (154 stations)	Probability level[a]
Low	38	19	**
Medium	50	25	**
High	12	56	**

[a] Significantly different at (**) $P \leq 0.01$.

Source: Schloesser and Manny (1989)

drinking at the water. Coot and little grebe occur in quieter waters. Less structure means fewer bird species. Moorhen, dabchick and swan occur in canalised channels, and even lined channels, with minimum diversity of structure (impossible for nesting, or walking in and out), allow drinking by sparrows and pigeons (Lovegrove and Snow, 1984).

The Po valley has seventy species of birds associated, doing well on the islands, bars, scrub etc. The middle Po is an important resting station for migrating birds, but recent engineering and other changes have made it less suitable, and the birds there are now fewer and less regular (Chiaudani & Marchetti, 1984).

Fish likewise suffer from loss of structure, from the loss of gravelly, shallow spawning grounds (for the appropriate fish), from the loss of vegetation providing shelter, spawning grounds and food (where appropriate). Trout, for instance, need some cover within the river, and are sparse in heavy shade (Hynes, 1970). Where macrophytes are removed, and fish cannot migrate, herons may eat all the fish (D.F. Westlake, pers. comm.). From riverside trees, invertebrates fall into the river, and these may be as many as or more than those living in the river, so greatly increasing fish food supply (sometimes over 90 per cent of salmonid food comes from the trees in this way). This also spreads the supply of plentiful food over more of the year, the peak supply from trees being earlier than that from streams (Mason and Macdonald, 1982). Leaves fall from

riverside trees, and may be important as food for aquatic micro-organisms and plant-eating invertebrates.

Fish themselves can also cause loss of structure. This is particularly seen with carp introduced for food (as in Wisconsin — see Haslam, 1978) or for weed control (as in the Netherlands). Excess carp remove all vegetation, leaving silty water over disturbed mud.

Substrate texture may also be considered part of river structure. Eggs and fry of salmonids, and many invertebrates, die if silt particles accumulate within the gravel. Young mussels tolerate no silting, adults tolerate some, so giving variation with both species and stages of life history (Hynes, 1970). Eleven out of thirty invertebrate species were lost in a river with sand deposition, five (including tubificids) increasing (Hellawell, 1986). Deforestation and construction are but two of the many ways of altering silting — and hence both fish and invertebrate populations. Dredging, of course, reduces the abundance and variety of fauna. Where substrate alters, invertebrates alter (Chapter 5). Consequently, altering erosion/silting regimes allows neither population to develop properly. Many invertebrates need to crawl out of water to become adult. If macrophytes are removed and the bank made unsuitable, the replacement population goes missing. (Unless, of course, suitable solid structures — say piers or rubbish — are put into the river — see Hynes, 1970.) The delayed invertebrate recovery after major channelisation is shown in Figure 8.8. Hynes (1960) records a drop in invertebrate collection from 254 specimens of twenty-three or more taxa, to 25 of mine taxa (collected in ten minutes), in the year after channelisation.

Arterial drainage on a lesser scale showed good recovery of fish and invertebrates, but a change in the species balance of the macrophytic vegetation (McCarthy 1975; 1977). The quantity and pattern remained good for the fauna. A similar but greater change in Denmark, and its significance, are discussed in Chapter 11. The effect on vegetation will depend on the original vegetation as well as on the degree of channel change: the more fragile and rare habitat types are less likely to preserve their character than are robust ones much interfered with for centuries.

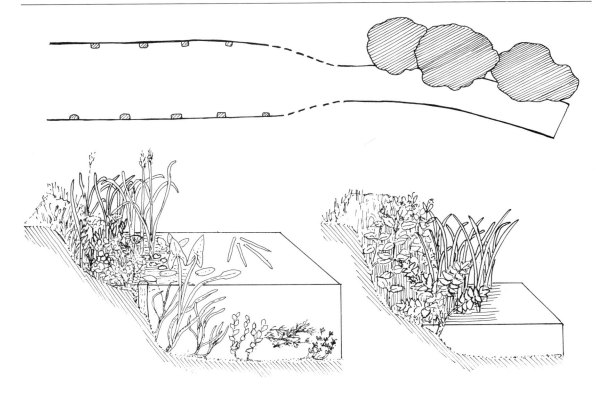

Figure 8.10 Formerly navigable river partly rechannelled. The ex-navigation part is wider, with slow flow. Because of the shape and low disturbance, the vegetation is not only species-rich and abundant, but often contains rare species. The recently channelled part has a low-quality environment and vegetation, with more shading. It is too shallow for water-supported species (P.A. Wolseley in Haslam, 1987).

Effects of flow controls on structure

Watermills were invented in Roman times, peaked in late medieval to early modern times, were superseded by steam mills in the Industrial Revolution, and became unimportant in the twentieth century. Their controls penned water upstream, and allowed it to rush downstream, often forming a pool there also. These controls might be across a river or a side channel, and if a river, a side channel might be constructed later to aid flow and navigation. So both directly and indirectly mills alter flow, primarily by creating slow silted (deep) habitats upstream which, in lowlands where mills were frequent, might occupy much of the river length. Other controls were inserted, too, for instance the hatches to control flow to water meadows. The flora and fauna would have changed with the habitat — changed, to correspond with the new habitat, rather than being eliminated (in the absence of pollution or other damage). As those controls were individually small, could be incomplete, and side-channels were plentiful, real hindrance to migrating fauna was often slight.

A typical change in habitat is shown in Figure 8.10. Where, as in lowland England, water table has been severely lowered, the subsequent shallower community below a control — or in a river with few controls — is now depauperate in vegetation, the habitat being over-shallow.

Nowadays, there are plenty of bigger and better controls, for reservoirs, hydroelectric plants, water transfers, major flow regulation, etc. These pose greater problems to migratory animals, though fish passes for salmon are standard constructions. Populations may suffer structural damage. Macrophyte populations adapt to stable conditions, but where great fluctuations in water depth occur, neither emergents nor water-supported species can develop properly on the river edges. Animals dependent on macrophytes therefore also fail (see also Table 6.3.) Chan-nelisation — whether or not (like water transfers) associated with controls — creates uniformity in channel and bank, loses many bank trees and shrubs, and may leave slow, silty, or swift shallow habitats, which may or may not be what was there before.

Channel maintenance practices affecting structure

Writing on managing chalk streams for angling, Hills (1924) was able to describe a certain village (Stockbridge) as the home of good management. The water, he says, has been cared for in every way. Good weed harbouring larvae and shrimps is encouraged, evil (*Elodea canadensis*!) eradicated. Hills made two complaints, that abstraction is decreasing both flow and, more importantly, the scouring winter floods, and that dams are being constructed which — like *Elodea canadensis* — cause mud to collect (also see the Wylye, Chapter 6). Management was an art, enhancing the vegetation understood to aid salmonids, even to pruning individual *Ranunculus* clumps separately. They were cut so that the current was thrown from side to side, giving, so it was said, maximum aeration for insets and trout (Sawyer, 1952), and in any event improving diversity and encouraging the species found desirable. Even in 1920 this was possible only where angling brought in much money, and the art, in England, is now almost extinct.

A new art for improving river structure is, however, being developed in Denmark — and Denmark is anyway the only EEC country which combines intensive farming with good-quality stream vegetation. Much intensive study of hydraulics and sediment patterns, and more geomorphological knowledge, are improving Danish streams (Neilsen, 1986; Markmann, 1988; Sønderjyllands Amstkommune, 1986, 1987; and see caption of Figure 8.3). They point out that most money goes on water quality (quite successfully as far as their rivers are concerned — see Figure 4.3, an average good stream), and most of the rest, in dredging and weed cutting, while serious research into environmentally good rivers which could maintain river structure, has been ignored. The pattern is

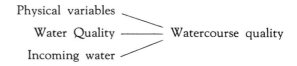

Physical variables ⎫
Water Quality ⎬— Watercourse quality
Incoming water ⎭

The natural channel is self-maintained, flooding perhaps 1 per cent of the time, but bank-full and so wetting low ground around, for longer than that. This wet ground around is unsatisfactory for crops, because of flood damage, and for livestock, because of disease. If left as wetland, however (see above) this would be ecologically satisfactory and aid river purification (see Chapter 11). Where there is but a narrow band of low ground, though, there is little problem anyway. At present, therefore, mainten-ance by Man is more needed in the flatter lowlands, than in the hills. This maintenance is to reduce waterlogging and flooding. Of course, removing silt is one thing, and removing woody plants and habitat diversity from the bank, in order to get heavy machinery in more easily, is another.

Maintenance dredging, the routine removal of deposited silt, also removes flora and fauna. Most fish swim away, invertebrates (Hellawell, 1986) tend to recover within a year, while macrophytes, not surprisingly, considering their roots and slower movement and development, usually take two or three years (varying, in fact, with the extent of the dredge, from a few months to, with severe channelisation, over ten years) (see Haslam, 1978, 1978a; Brookes, 1983). Frequency of dredg-ing depends on the amount of silt accumulated, and the damage done by floods from this silt (varying with land use, both urban and farming). British alluvial plain dykes are dredged at inter-vals of between one and five years, and lowland clay streams at intervals of two to five years. This is unusually frequent (much silt: as in lowlands with low scour; with intensive farming; and with much settlement).

Cutting is the traditional way of preventing flood hazard from macrophyte growth (apart from swans, fish, etc.), and now is by dragline, grab and boat as well as by hand. The difference between a dragline and the trimming of individual plant clumps is considerable! (see also Table 6.2 and Fig. 6.18). Structure depends, therefore, on the type of cutting — as well as its frequency, the stream type, season, and so on!

Much cutting of stream vegetation is done in Denmark, lowland England and the Netherlands (see Robson, 1974, for methods and equipment;

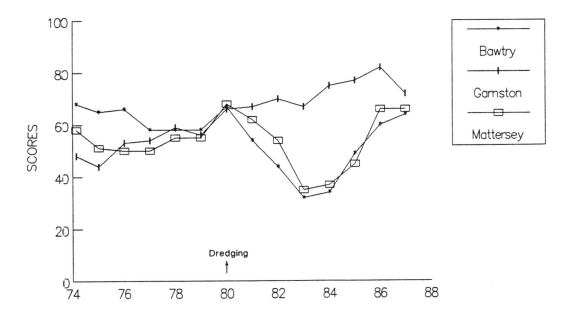

Figure 8.11 Invertebrate recovery after dredging and regarding, Idle, England. (From Severn-Trent Water, 1988). The Gamston stretch is a recently undredged control.

Source: Severn-Trent Water (1988).

and, for effects, Haslam, 1978; 1987a). Elsewhere it is more local or intermittent in lowlands, and is, of course, unnecessary in the highlands. The frequency varies from every fortnight to every few years. The cost of vegetation control in England and Wales (1988) is put at 8 million pounds on main rivers, 8 on streams, and 56 on drainage channels of various types, totalling £72m in all. three quarters is cutting and cleaning, by hand or machine, only one quarter being by aquatic herbicides (Dawson, 1989). Where cutting is done by twentieth-century tradition, or office planning, rather than on the basis of current observations, the pattern should be re-examined in the light of the Freshwater Biological Association's work (Westlake and Dawson, 1986; Dawson, 1989). They showed that cutting when 'necessary' leads to increased growth and therefore more cutting, while cutting in the autumn to match the plants' life cycle causes less future growth and so less cost. Lack of summer cutting did not cause flooding, and the

fauna were not harmed by the removal of vegetation. In fact spring flooding was also reduced. The community changed to fewer, larger, slower-growing plants (of the dominant *Ranunculus*), which impede water flow less (Dawson, 1989). Channel shape is also important in relation to cutting as is the shape cut within the channel. Leaving part uncut helps both management and conservation (Sønderjyllands Amstkommune, 1986; 1987; 1988a; 1988b; Dawson, 1989).

Loss of structure through aquatic herbicides, i.e. chemical maintenance, is also discussed in Chapter 9. Herbicide use within the water was particularly favoured in the 1960s and 1970s in channels — alluvial plains, etc. — not used for domestic water supply. The costs, compared to those of cutting, are now rising, and environmental considerations are rather more popular. Spraying wide bands of edge emergents is environmentally much more acceptable (as long as some of the vegetation is left).

Figure 8.12 Reservoir and channel alteration, Malta.

Rivers in settlements

These, necessarily, are always changed by Man. They may be wanted for navigation, power, water supply, food, waste disposal, recreation, among other things. They may cause nuisance by flooding. Most works for these purposes decrease structure (Figures 8.3 8.11, 11.3 12.3, 12.7, and landscape strips on pp. 16, 29, 40, 92, 138, 139, 158, 166, 186, 209).

Recreational and navigational disturbance to structure

Recreation has always occurred within and beside settlements; but as a serious and widespread damaging factor, it is new. It is damaging to vegetation, bed and bank, and to the stability of bed and bank. Activities include boating (Table 8.5), water-skiing, scuba-diving, paddling, bathing (less than formerly, because of more pollution and more baths in houses), angling, bird-watching, picnicking, walking — and, indeed, let us not forget the conservationists, intent on their studies (Figures 11.3, 12.5, and landscape strips on pp. 116, 157).

While some of the larger animals (such as the mallard) coexist with considerable disturbance by Man, as long as enough habitat is left, others (such as otters) are more easily sent away. Next comes the loss of macrophytes and, associated with this, some loss of physical structure, too. This means the absence of most larger animals. Finally, with severe disturbance (and unstable or lined beds in consequence) invertebrates, fish, benthic algae, etc., are also harmed.

Finally, the water

When all else is said, the river is made of water, and where water is lessened (by abstraction, drainage, or whatever) river life will be diminished, whatever other changes are associated (see, for example, Figure 6.2). Structure may also be lessened by changes to water regime increasing scour and storm flows, giving deep waters with over-steep banks for vegetation, etc. (For effects on vegetation, and recognition of these effects, see Haslam and Wolseley, 1981; as well as Haslam 1978; 1987a.)

Figure 8.13 Old industrial use of stream, Schwarzwald, Germany. Note laundry troughs on the left.

Conservationists at work

9

Within-river pollution

In any sphere of life, Man's variety and ingenuity almost boggle the imagination, and quite prohibit exact description. So it is with pollution. The division of Chapters 9, 10 and 11 into pollution from within the river, from domestic and industrial sources, and from rural land use is convenient but arbitrary. In this chapter, the pollution types described are those arising directly from the presence of the river. Most of the effects described can — to a much more limited and local extent — occur from natural causes. Excess vegetation can deplete oxygen and kill within-water animals. Putrefaction of dead buffalo is similar to that of sewage (see Chapter 1).

Gravel extraction and quarries

In mainland Italy and Sicily this is a major source of pollution and loss of structure. Elsewhere it is local. Wide gravelly channels in a hilly erodible landscape with high late-winter and spring discharges and usually low summer ones are good sources of gravel. The French ban gravel extraction within the bed of the present river, minimising direct pollution. Quarry wash may reach rivers likewise.

Italian gravel pollution causes turbidity and so low light on the river bed. The colour is distinctive, usually beige or white-grey, so easily traced to its sources. The rivers rise in hills, so have considerable water force. This, with the wet winters and dry summers of the Mediterranean, gives unstable beds: sediment is both deposited and scoured. This is bad, of course, for macrophytes, and so for all organisms dependent on them. It does not help that settlement pollution — great or little — is added. The combined damage is worse than either separately (see

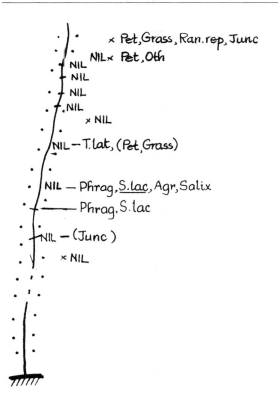

Figure 9.1 River map, Ofanto, (S.) Italy. Gravel extraction works as well as settlement pollution. Hilly Resistant rock, 1980. Notes as for Figure 4.2.

chapters 4, 6) (Figure 9.1). Good vegetation occurs only away from pollution. *Phragmites communis* grows particularly well and is tolerant in the south, and river fringes are common — even with pollution. Not, this time, *Potamogeton pectinatus*, which requires a more stable channel centre. *P. communis* has deep roots, anchoring well in finer sediment at the sides, and tolerating a regime of much annual variation in water level. (Where the water is more grey than beige on the east coast, *Scirpus lacustris* comes in, with low water force). Blanket weed is the macrophyte best adapted to summer-dry, unstable scouring stream channels. Where this occurs alone, it is most often in beige-coloured rivers of low water force. Likewise, other vegetation patterns occur with pollutions of other colours.

Invertebrate as well as macrophyte diversity can be decreased in these conditions of unstable, shifting substrate (Chapter 5; Nuttall, 1972).

Inland fisheries

Fishponds have been cultivated since ancient times. The prophet Isaiah, writing in the eighth century BC, mentions the ponds and their sluices. They were part of inland life in Christendom, partly for protein — as also outside Christendom — but also because meat was banned, but fish allowed, in Lent, on Fridays, and on certain other days. Inland fisheries were those in the river, often by mills (Figure 5.1), and ponds, constructed or enlarged, and up to several hectares in size. Inlet and outlet streams were needed. Early fishponds were in monasteries, whose inhabitants were the most debarred from meat, the greater houses, and in villages, probably typically downstream (better feeding for the fish?).

River fishing is now widespread for sport, both for salmonids and coarse fish. Salmonids are the most caught and sold commercially. Otherwise, river fishing as a livelihood is mainly confined to large rivers like the Danube — and there is doubt, to put it mildly, about the safety of some such fish as food, the Danube not being renowned for cleanliness. In sparsely populated parts of Italy, such as inland Umbria, table fish are as common as in the past.

As an activity, river fishing causes little more pollution than visitors or boats. Except for what the fishermen leave behind: which is, after all, unnecessary. Angling can be hazardous to birds and mammals. Fishing lines may trap them and eventually kill them (Lovegrove and Snow, 1984, report deaths of mallard, moorhen, dipper, wagtail, grebe and heron, among others). Nylon lines do not decay — in a reasonable time-span, that is — and even a thread line can kill many animals before it vanishes. Swans need grit, anglers drop lead weights, and the unfortunate swans cannot distinguish the two, and die of lead poisoning. (About 3000 out of a total 18 000 population in 1978 were dying annually in Britain. The number in Stratford, for example, dropped from about 50–60 in 1965 to one or two in 1978. (See Birkhead and Perrins, 1981; Lovegrove and Snow, 1984.) Lead shot is no longer used, officially, in England and swans are increasing fast. But dropped fishing lines, polybags (sticking over heads and suffocating), broken glass and jagged tins (cuts) are still found. One of any of these does but little harm to a

Figure 9.2 Sixteenth-century fish pond, fish stew, by house

whole population, whatever it may do to the individual. A hundred thousand may be a different matter. Few mind when animals die as a by-product of sport or other leisure activities. (Many anti-bloodsports groups now object when animals die as the direct result of sport, even though sporting organisations must conserve their prey to survive.)

Areas of commercial fish ponds are now local, and will be described for two regions. The Czechoslovak ponds were mainly constructed in the fourteenth to sixteenth centuries, and though

a few have inlets from or outlets to rivers, most are on smaller streams, either the natural course, or diversions. Chains occur (Figure 9.3). The top pond is termed 'sky-fed' and is the least likely to be fertilised. Organic waste will drain to the stream below, which will be added to by the lower ponds, these probably being fertilised, too. The ponds are drained every few years, and now usually in winter when obstructing vegetation is least. This scours and erodes thoroughly. In the rest of the cycle, though, these associated streams are of low water force, and usually of low flow. Waste and fertiliser pollute the water, and with the draining, polluted mud is moved into the streams too. Because the streams must carry high discharge, narrow ones have high banks, so may also be bank-shaded (Figure 9.3). Vegetation is thus pollution tolerant, scour-tolerant and may be shaded: very poor, in fact. Fish ponds draining to larger streams do less damage, other things being equal, since the pollution is more diluted, and the effect of draining discharges, less.

Fishponds are widespread, and the small streams of the region are very poor in vegetation. Most damage is attributed, as usual, to pollution from settlements, but the fishponds are an important aggravation, seldom found elsewhere.

The French fishpond area of the Dombes is up on a plateau, with many ponds, generally smaller

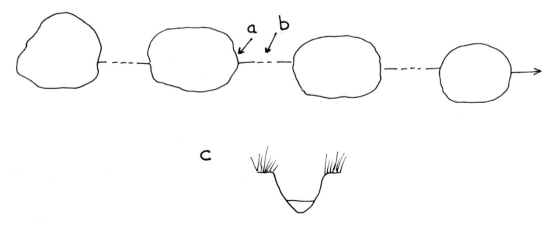

Figure 9.3 Czechoslovak fish ponds. In sequence from the 'sky-fed' pond to left. (a) Immediately below exit, still-water species may occur, washed out of the pond, e.g. *Hottonia palustris, Hydrocharis morsus-ranae, Spirodela polyrhiza.* (b) Stream from pond, typically with one or more of: *Lemna minor* agg., *Phalaris arundinacea,* Blanket weed (less common, *Agrostis stolonifera*). A very polluted assemblage. (c) Section through common pattern of streams from ponds: water too shallow for good water-supported vegetation, banks high for draining discharge, and so partly shading the centre

than the Czechoslovak (43 ha being very large). These date from early monasteries, though new ponds are still being built. They are connected by Man-made dyke-like channels or small streams, with low banks. This is a discrete area, separated from the main flowing water system in the low-lying land outside the Dombes, and so the effect on their system is minimal. This contrasts with the Czechoslovak pattern.

Fish farms are fish ponds of small size and intensive management (Figure 9.4). Control of pollution is much more possible as — theoretically — effluent standards can be set and met. Biocides from both may enter the river.

The small amount of damage from a well-run fish farm is described in Chapter 6. It is possible, even, to have negligible pollution. Many are bad, and even well-run ones, if dense upon the ground, are cumulatively polluting. And these, especially trout farms, really are spreading across the EEC.

One area where they are particularly dense is in the Italian Alps (where consecutive fish farms do cause worsening downstream pollution). Another is Denmark. In the mid–1970s a third of all inland effluents in Danish streams came from fish farms (Water Research Centre, 1977) and by the early 1980s there was much more (Heise, 1984). Organic waste pollution arises both as wastes from the fish, and from surplus food fed to the fish. They pollute the water, so solutes and suspended solids enter the river immediately, and they pollute the soil, so when the pond is flushed out (perhaps biennially) dirty mud enters the river, too. This, as usual, is smothering (see chapters 4 and 5) (Table 9.1). When a farm starts to pollute, vegetation is harmed before invertebrates, as expected with a mild organic pollution, so macrophytes should be used as the monitors, to detect potential harm to animals.

Rice paddies

These need much water. There are many in the Po plain, Italy, and fewer in the Rhône delta, France. If incoming water is contaminated by bioaccumulating toxins, these may reach levels in the rice harmful to Man (Kobayashi, 1971). Passing through the paddy — as through the fishpond — water picks up fertiliser, biocide, and polluted sediment. The Rhône water is already

Table 9.1 Macrophytes tolerant and sensitive to fish farm pollution, Denmark[a]

Tolerant	Sensitive
(Callitriche spp.)	Glyceria fluitans
Potamogeton crispus	Lemna trisulca
P. pectinatus	Ranunculus flammula
Sparganium emersum	Sparganium erectum

[a] Table incomplete

Source: partly from Rasmussen (1980)

Table 9.2 Macrophytes tolerant and sensitive to rice paddy and domestic, etc., pollution, Po plain, Italy Typical downstream patterns after pollution are:

	Rice paddy	Domestic/ industrial
Near discharge, pollution severe	Carex sp	Nil
Downstream, moderate pollution	Carex sp, Phragmites communis	Phragmites communis
Downstream, mild pollution	4 spp., e.g. Carex sp., Phragmites communis, Callitriche sp., Ranunculus sp.	4 spp., e.g. Phragmites communis, Polygonum hydropiper agg., Rorippa austriaca Blanket weed
Downstream, near-clear	6 or more spp., including	Callitriche sp. Ranunculus sp.

With the rice paddy pollution the community is reduced, with the domestic etc. it is distorted

grey with suspended solids (and pollution), and little change is seen in the water as it passes through the delta. Heurteaux (1979) is concerned about biocides, and that such polluted water should not reach the Vaccaries — nor, indeed, should it reach the clean marsh regions. The carried silt has been decreased by the Rhône dams (which encourages upstream deposition). It contains much calcium carbonate, which can carry heavy metals (Golterman, 1984). The increased Alps erosion will be increasing carried solids (see Chapter 11).

The Po paddies are a different matter. Here much of the drainage system is ancient, and it is

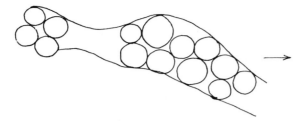

Figure 9.4 Fish tanks in former watercress beds (*Rorippa nasturtium-aquaticum*). Site could be medieval village fish pond (location downstream of village, so suitable).

complex. Fortunately for the researcher, the water colour varies consistently. Dark is for water leaving rice paddies, browns for domestic/industrial pollution sufficient to cause colour, white-grey for incoming water with gravel, etc., pollution, and clear for clean (cleaner, strictly speaking!) incoming, or downstream-purified water. Patterns can therefore be followed (Table 9.2). *Callitriche* is tolerant to the simpler pollution of the paddy, as it also is to fish farms (above) and wash-houses (below). The species quality (Table 4.2) is better, at the same diversity level, with the paddy pollution. The paddy effluent has cut diversity, the domestic/industrial one has also skewed the community: it is the worse. *Phragmites communis* is more tolerant than further north — pollution tolerance tends to be most where a species is most prolific. That is, where a species is optimal geographically, it takes more stress — of any kind — to eliminate it.

Watercress beds

Rorippa nasturtium-aquaticum, *R. microphyllum* and hybrids, = *Nasturtium officinale*, *N. microphyllum* and hybrids.

Watercress is a salad plant, which can be cut direct from (clean!) smallish streams, mainly non-mountain limestone, also sandstone, some Resistant rock, etc. It has also, for centuries, been grown in intensively managed beds, most often on limestone with a good supply of spring water. A good flow of cool water is needed for the best growth. As discussed in Chapter 6 (Wylye), a little sediment pollution is usual. Much is possible, but should not occur. The beds may be treated with biocide (for example, for molluscs) or

Table 9.3 Macrophytes tolerant and sensitive to conditions in the Danish ochre streams

	Tolerant	Sensitive
a) R. Skjern, low pH	*Agrostis stolonifera* *Glyceria fluitans* *Juncus bulbosus* *Scirpus lacustris*	*Berula erecta* *Elodea canadensis* *Glyceria maxima* *Potamogeton alpinus* *P. natans*, and many more
b) R. Vid, higher pH	*Glyceria fluitans* *Sparganium emersum* *Sp. erectum*	*Berula erecta* *Elodea canadensis* *Potamogeton natans*

pH also determines heavy metal availability, etc.

Source: a) partly from Sand-Jensen and Rasmussen (1978b) from Sønderjyllands Amstkommune (1982b).

fertiliser, which may then reach the stream. One of the very few instances where the writer has been sure that eutrophication, nutrient increase, occurred alone, and sufficiently to alter vegetation, was by a watercress bed outlet. Two species grew which should not have been there so far up a chalkstream (*Groenlandia densa* and *Zannichellia palustris*, both semi-eutrophic in habitat). A few years later, watercress bed and inappropriate species had both gone. (See also Casey, 1981, no effect on biomass.)

Danish ochre streams

It is a moot point where mine or quarry streams cease to be classified as Industrial. The Danish ochre streams, the lignite streams, are such a case. The lignite was always there, and superficial. The ancient stream vegetation is unknown. Pollution is now native to them — but lignite was mined from the First World War until recent decades, and from mines, streams drain. One of the two areas is in the Skjern (Figure 4.3, streams marked 'L' for lignite). Cadmium, chromium, copper, iron, manganese, nickel, lead, vanadium and zinc are very much present. So is the precipitated iron which provides the name, ochre. When the water level drops below the lignite level (by drought, drainage or abstraction), sulphuric acid and iron enter the stream, pH is lowered (to 4), metals become soluble and streams become toxic. Even with treatment by lime or sand and gravel, only partial improvement has resulted (but how many countries would have taken as much trouble?). Table 9.3 shows species sensitive and tolerant in these streams. Comparing species lists with other habitats shows decreased diversity, and skewing of community to the more tolerant species. Tolerant to what? To ochre? To other heavy metals? Low pH? Sulphate? Table 9.3 shows that pH is, at least, correlated (ochre is also) and communities differ between the affected streams. Presumably the pollution influence is also different.

Boats

In addition to harming river structure (Chapter 8), boats also pollute. The impact, though, is usually small compared to the impact of their physical damage. In an ideal world, there would be no boat pollution, as all is — theoretically — avoidable. It consists of:

● Petrol, oil, paint and other maintenance and repair materials, their spills and breakdown products.
● Cargo spills: industrial freight is carried on the main European waterways of the Danube, Rhine and some smaller rivers, and on the canal network. Any barge may spill.
● Domestic waste products (and domestic lost products!).

Washing

Doing the laundry in the river was formerly a standard, though far from universal, practice. As a normal activity in the EEC it is now restricted to, for instance, Madeira, which combines a pleasing climate and few modern conveniences. The soap or, more recently, also detergent enters the river, together with the dirty wastes from the clothes. There is also some disturbance (see Chapter 8 for similar). Laundry troughs beside the river are shown in Figure 8.12 (on the left).

A refinement on the river is the wash-house. Those in use are concentrated in western France, and provide but little pollution. Town and village wash-houses, on rivers, were formerly widespread (and often still remain) in France, Germany, etc. Table 9.4 shows a typical recovery pattern in a *Berula-Callitriche* stream. *Callitriche* — once more, when the pollution is not sewage/industrial — is tolerant.

Table 9.4 Macrophytes tolerant and sensitive to wash-house pollution, Brittany, France
A typical downstream pattern after mild pollution (extending over 25 m).

1. Near discharge, worst pollution (10 m length)	Nil
2. Downstream, improving	unhealthy *Callitriche* sp.
3. Downstream, where small effluent enters	very unhealthy *Callitriche* sp.
4. Downstream	healthy *Berula erecta*, *Callitriche* sp.
5. Downstream and upstream, near-clean	*Apium nodiflorum*, (much) *Berula erecta*, (much) *Callitriche* sp., *Myosotis scorpioides*, *Phalaris arundinacea*, *Rorippa nasturtium-aquaticum* agg.

Biocides

Biocides may enter the water as within-river pollution, aquatic herbicides, fish poisons, and misplaced farm biocides — those sprayed on water as well as on land, and containers dumped in the nearest ditch. No farmer would ever do the last, of course: some, however, will sadly say it is the reprehensible habit of a neighbour.

Different herbicides are cleared for aquatic use in different countries and under different circumstances. The arsenic compounds much used in America in the 1960s would, for instance, cause many Europeans to shudder. Copper sulphate was used surreptitiously in fen dykes well after it was banned: it is really most effective against algae! Most used now are the artificial herbicides, dalapon, diquat, glyphosate and the like. No one herbicide controls all unwanted species of plant, no herbicide kills only unwanted (i.e. choking) species of plant, and no herbicide is completely safe for all animals. The first of these three is better known than the second and third! Effects are shown in Figure 9.5.

Herbicides sprayed on emergents will, other things being equal, do less harm than those in the water. Indirect harm, from careless use, can be great: by removing all emergents, and also by inadvertently spraying the water, the river

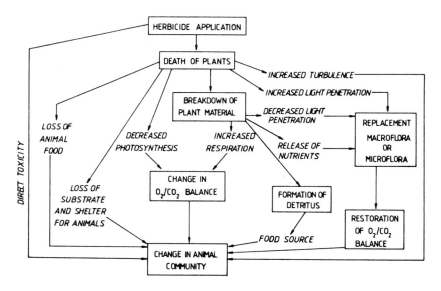

Figure 9.5 Ecological consequences of herbicide treatment (from Hellawell (1986)).

communities may be reduced and distorted (see Chapters 4, 5). Non-invasive emergents can be safely left even on the most arterial of drains, to add structure for the fauna and other flora. Non-invasive means either species not known to become choking, e.g. *Butomus umbellatus* in Britain (but not in the Netherlands), or species which, although choking in some habitats, will not be so in this one. Even *Phragmites communis*, a bane to the drainer of fens, can safely remain as small clumps for years on the banks of suitable drains, and in general it will only rarely grow into drain water over about 1 m deep.

Drainage channels in farmland on former wetland must move water if good crops are expected. Herbicides are a favourite management tool to keep such channels sufficiently clear: favourite, because easy, and, over much of recent decades, cheap. It is not the best method (Chapter 6) and so should be avoided in nature reserves. These apparently so-still dykes have moving water: water liable to move in both directions, as the wind blows, and as sluices open, and may move 'upstream' several kilometres at least. Turbid pollutions may be traced, by the curious-minded, from their source (in, say a farm) along and around the dyke system. Where one chemical can go, so can another.

Different vegetation patterns occur with different intensities and types of herbicide, modified, as usual, by other environmental factors. One example is sparse Blanket weed — *Phragmites communis* — *Lemna minor* agg., with intermittent *Potamogeton crispus*, with Casoron-G: a typical uniform, species-poor and distorted community of a polluted site. The tolerant species, of course, differ from those with other pollution types, and from those of streams. (Dyke-tolerant species are not the same as stream-tolerant ones, since wash-out is not a factor in dykes, and so species unable to survive a pollution giving poor anchorage can yet occur in dykes.) In general, physical maintenance leads to a reduced normal community; chemical maintenance, to a skewed one like the above.

Let no one doubt that the use of aquatic herbicides is pollution. It is Man chemically damaging watercourses. This is not to say that their use should be banned. Pollution of all sorts occurs, and at least this one leaves some river life, and provides an (at least short-term) benefit to Man's welfare. Which waste disposal does not: that is merely a convenience which may eventually harm Man.

It is also a thought that the dykes in which these chemicals are usually used are Man-made —

Table 9.5 Relative use of herbicides and macrophyte site diversities in farmland and roadside channels, in the Netherlands
(Herbicide use from Onderhoud Watergangen, 1976, macrophyte data 1977, 1978)

Provinces listed in order of % length of watercourse treated, lowest first	Provinces listed in order of decreasing typical (modal) diversity
Friesland	Friesland
Zuid Holland	Zuid Holland
Noord Brabant	–
Gelderland	Groningen
Groningen	Gelderland
Overijssel	Overijssel
–	Noord Brabant
Drenthe	Drenthe
Zeeland	Zeeland

The correspondence is close, except for Noord Brabant, where many channels appeared to be polluted from the rivers in the Maas plain.

Source: Haslam (1987a)

Man drained the marshes, replacing wetland, pools, meres and channels with fields and dykes. Some aquatic life is better than none — and the trend in recently 'improved' wetland is to give fewer (but larger) dykes, reducing the aquatic habitat. Studies on vegetation structure, and species rarity in the Netherlands, show persistent herbicides applied in late spring are damaging. Non-persistent ones applied from June onwards actually raise indices, because growth is suppressed (de Lange and van Zon, 1978). Not all herbicides are applied with extreme care. Table 9.5 shows that the typical dyke diversity of a province corresponds remarkably closely with the proportion of herbicide-treated dykes.

The use of aquatic herbicides is governed by the Codes of Practice of the various countries (in Britain, issued by the Ministry of Agriculture, Fisheries and Food). These should be consulted for what to use where, and how. Hellawell (1986) gives details of the action of each, and uses and applications are further described in Barrett (1981), Barrett and Logan (1982), Robson (1978), Robson and Fillenham (1976). A fuller reference list is in Haslam (1987a).

Generally speaking, within the water delicate species are killed more easily than tough ones (this does not apply to emergents). If all vegetation goes, banks may become unstable, and much Blanket weed develop. Care and judgement, properly exercised, can retain considerable conservation quality. (Deoxygenation from decomposing plant remains kills fish, etc., so should not be allowed). Protective plants are helpful on banks.

Plants in the water aid, or are needed by, some animals (see earlier), and depleted animal communities following herbicide treatment are described by Harbott and Rey (1981), Newbold (1975; 1976). A good example is a mild terbutryne treatment, which led to more tolerant macrophytes replacing more sensitive ones for one or two years or more after treatment. The invertebrate pattern depended on the vegetation — plant-associated invertebrates decreased with loss of plants, but benthic ones increased with the increased open bed. Oxygen losses occurred, which could be harmful. Again, care and knowledge are needed for ecological use (Murphy and Eaton, 1981a).

Sheep dip

As a tail-piece, and as it is shown in Figure 8.6, a former, intermittent pollution source may be mentioned: sheep dip. Former, as sheep dips are no longer allowed within streams (though unfortunately, so much of these chemicals do, eventually, reach rivers, one way or another).

10
Domestic and industrial pollution

This is the most serious and widespread type of pollution in Europe — and, no doubt, elsewhere. Necessarily: domestic pollution is found wherever there is Man. Other types depend on some particular activity of Man (such as creating sediment by land misuse, or putting biocides in drainage channels).

In this chapter, first some large-scale comparisons will be made, then various rivers are described individually.

Macrophyte variation between countries and regions, (1977–84 data only)

The Yser/IJser is an unspectacular river flowing from France to Belgium through similar and unspectacular country each side of the border. The vegetation of its small tributaries is bad in France, and even worse in Belgium (Table 10.1).

The only significant difference is the crossing of the border: pollution is greater in Belgium. Paired streams from different parts of France are shown in Table 10.2. Here there are differences in both pollution, and other reducing factors.

Variations in site diversity

Diversity is an absolute. It is not dependent on subjective thought (save that of deciding the method and month of recording and the level of taxon to be recorded). This is an advantage for study. Diversity is, though, governed by many different factors. This is a disadvantage for interpretation, because the more significant of these factors must be known if the interpretation is to be true. Table 10.3 compares streams, small and large, in different countries. Here two of these governing factors, landscape (water force) and the

Table 10.1 Comparison of Yser/Ijser tributaries in Flemish France and South Belgium
Comparable small tributaries with similar landscape, land use, etc.
Sites with water at least 10 cm deep, not shaded, recently dredged, etc. (drier sites may be choked with emergents in unshaded parts, in both countries).
Phalaris arundinacea is the main species.

	France	Belgium
Number of sites	6	6
Benthic algae prominent	0	4
Species records (excluding Benthic algae)	36	5
Sites with 3+ macrophytes	6	0
Cover at least 10%	5	1
Cover at least 50%	4	1
Records of most tolerant species	13	4
Blanket weed	4	0
Enteromorpha sp.	3	0
Phalaris arundinacea	3 (one flaccid)	3
Agrostis stolonifera	3	1
Records of not-so-tolerant species	23	1

Source: Haslam (1987a)

Mediterranean climate (summer drought and warmth), are eliminated, enabling better comparison.

Belgium (lowland) has low diversities throughout, rivers being worse than brooks. This is due to the gross pollution also throughout (see Figure 4.12).

In Germany also the brooks are in better condition than the larger rivers. Some of the species-poor large rivers are damaged by boats, but most of the low diversity is due to pollution.

The Netherlands is the third country with large channels in a worse state than small. Dutch dykes are so numerous that it would be difficult to pollute all of them. In addition, effluents go to main drains where possible. Few channels are shaded in this much farmed country. Dyke diversity is reasonably good, though reduced by herbicide and recent other maintenance. Large channel diversities are as low as in Germany, but here boats are more responsible than pollution. (N.B. The Dutch *dijk* is a bank, the English *dyke* is now usually a channel, but may also be a bank, and is used for channels in this book. Both come from the name of a defensive border embank-ment, where enemies have to cross a ditch, the defenders being in a good position on a bank above. An example is Offa's Dyke in England.)

In other countries, the small lowland streams are more depleted than large ones. They receive most shade (are most easily shaded all over), are most straightened, most affected by lowering of water table, and may have other severe reducing factors, including pollution.

In South Norway pollution is of little importance, it being shade (Figure 11.1), gorges and high water force which most reduce the vegetation. If sites can bear macrophytes at all, they are likely to bear good vegetation.

Ireland has fewer geographic difficulties, and diversity is even better.

Denmark is not far behind, in spite of the intensive farming. It has but few large rivers, and these often have other troubles, such as boat damage.

Italy is so hilly that it is difficult to restrict sites, looked at country-wide, to the lowlands, and diversities in the table are lowered by water force. Gravel extraction pollution (Chapter 9) is great in many large rivers (decreasing vegetation). Population is low, but pollution per capita is high.

France, like Norway, has numerous small streams without macrophytes. Here, though, it is more due to pollution than to other causes such as shade. The pattern shows well in Figure 10.1. The stream channel starts summer-dry (due to abstraction and drainage), then gets village pollution, from which the vegetation recovers but slowly (no vegetation, Blanket weed, then with small grasses, and further, and with dilution, before a chalkstream community can occur). Rather too many large rivers are also empty from pollution. Britain, France and Germany have similar overall pollution. Different cultural patterns of disposing of waste mean that Britain has pollution equally in small and large streams, France most in small ones, and Germany most in large ones.

Variations in site diversity of pollution-tolerant species

Classifying species as pollution-tolerant involves a judgement. Studying tens of thousands of sites shows which species occur in the Alps, which in the plains, which downstream of towns and which are found far from human habitation. Table 4.2, listing pollution-tolerant species, is an

Table 10.2 Comparisons of small tributaries in lowland landscapes in different parts of France. (Also see Table 10.1)
Comparable sites with water at least 10 cm deep, not over 2 m wide, not shaded, etc. Six sites per region. Regions selected to show contrasts.

| | Limestones | | | | Sandstones | | | Moraine | Miocene of south-west | |
| | Chalk | | Jurassic | | | | | | | |
	N.W.	S.E.	C.[a]	N.	C.	C. ('Good')	S.	N.	S.	W.	C.
Species records (excluding Benthic algae)	8	21	13	24	12	23	19	23	8	10	4
Sites with 3+ macrophytes	1	3	2	5	3	5	4	3	2	2	1
Sites with cover at least 10%	1	3	3	4	2	5	3	4	1	1	0
Sites with cover at least 50%	1	1	0	1	1	3	2	0	0	1	0

[a] Central.

The general effect of the high French pollution is shown throughout. Within this, variations occur between and within rock types. On chalk, drainage and management (and pollution) is more intense in the north than in the south. On Jurassic limestone, there is somewhat more pollution in the centre than in the north. On sandstones, there are both more- and less-intensively farmed areas, the latter having the more diverse stream vegetation. The sandstone streams further south (north of the Massif Central) have more unstable substrates than those in the north, and low vegetation. The northern sandstone streams are more stable than the southern ones, and bear moderately good vegetation even with fairly intensive farming. The mixed-rock Miocene outcrop in the south-west is very polluted, the pollution generally overriding the effects of differences in rock type. Examples cited in the table are from a very polluted and a grossly polluted region.

Source: Haslam (1987a)

Table 10.3 Macrophyte diversities in different countries
Excluding communities of mountain streams and, except in the Netherlands, dykes of alluvial plains. Size ii streams are defined as those up to 3 m wide and containing at least one water-supported macrophyte. For this series of tables, channels up to 3 m wide which are not classed as size i and which have (apparently) water deep enough for water-supported macrophytes are classed as size ii with nil diversity. Size iv streams are at least 10 m wide, usually 10–25 m. Diversities shown are: 0, no species (none obvious above ground); Low, 1–3 spp. in size ii, 1–4 spp. in size iv; Medium, 4–6 spp. in size ii, 5–8 spp in size iv; High, 7+ spp. in size ii, 9+ spp. in size iv (diversities increase with stream size, so, for example, 7 spp. is high for a size ii stream, low for a size iv one).

Country	Stream size	O	Low	Moderate	High
South Norway	ii	186	19	21	10
(hilly)	iv	5	8	7	2
Denmark	ii	46	82	105	49
(lowland)	iv	2	10	9	12
Ireland	ii	34	39	88	40
(lowland, upland)	iv	9	23	38	54
France	ii	363	193	217	93
(lowland, upland)	iv	68	177	66	36
Germany	ii	171	190	124	35
(lowland, upland)	iv	72	73	30	7
Italy	ii	146	91	75	27
(plain to low-mountain, north and centre only)	iv	63	105	22	2
Belgium, Luxembourg	ii	122	93	26	10
(lowland, upland)	iv	48	39	13	1
Netherlands	ii	46	140	118	74
(dykes)	iv	43	69	16	9

Numbers of sites recorded with diversity (column group header above O, Low, Moderate, High)

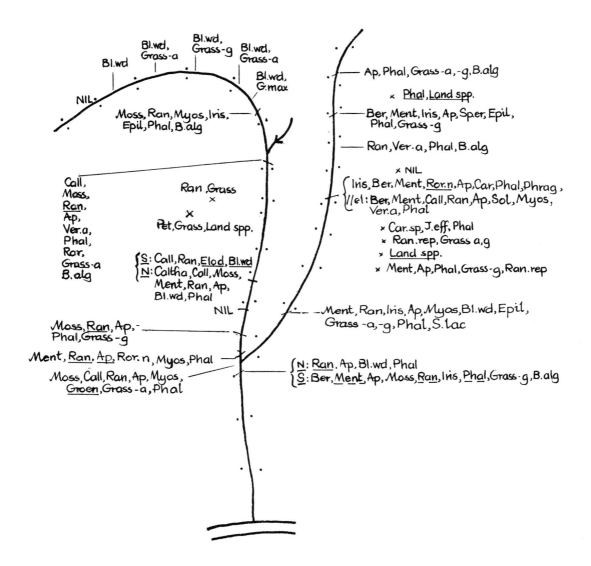

Figure 10.1 River Map, Tille, France (from Haslam, 1987a). Limestone (Jurassic), lowland. 1980. (M). Notes as for Figure 4.2. Note effect of pollution from village, upstream left, with added pollutions downstream.

Table 10.4 Comparative distribution of pollution-tolerant and other species in sites of low and high diversity in different countries.

As full-diversity sites are too few to be used for comparison, low-diversity sites for this table bear 1–2 spp. in size ii streams, and 1–3 spp. in sizes iii (4–8 m wide) and iv; and high-diversity ones, 5+ spp. in size ii streams, and 6+ spp. in sizes ii and iv streams (see text). Lowland and lowland-upland sites only. Recorded as % of occurrence of spp. tolerant to sewage and industrial pollution (and, in the dykes of the Netherlands and the Po plain, also tolerant to other damage).

	Sites of low diversity	Sites of high diversity	Mean no. of spp. in high-diversity sites
S. Norway	33	35	8.1
Denmark	78	59	7.5
Germany	76	51	6.6
Italy (Po, flow)	79	52	7.7
Netherlands	91	64	6.8
France	52	40	7.3
Italy (Po, dykes)	70	42	7.3
Belgium	80	53	5.9

overall guide. Tolerance varies with the type of pollution, the type of stream, and the geographic position of the stream (see other chapters). Any reader can, by studying local streams with proper attention and care, produce a better list for that locality.

Potamogeton pectinatus and Blanket weed can grow better in appropriate pollution than in clean watercourses. Other pollution-tolerants most often occur because they grow in that habitat if clean, and are — unlike other species — not eliminated by the pollution.

Country comparisons of pollution-tolerants are shown in Table 10.4, while Table 10.5 gives the detailed composition from one country. Pollution-tolerants are present throughout (they tolerate pollution: they are not restricted to it). Sites of high diversity have a higher number of pollution-sensitive species, so a lower proportion of pollution-tolerants. The 'high' diversities in Table 10.4 include many sites with diversity reduced by moderate pollution: hence the high 'control' proportions of pollution-tolerants. The number of full-diversity sites is too few to use for

Table 10.5 Distribution of pollution-tolerant and other species in sites of low and high diversity, Germany. Site diversity as in Table 10.4 see text for notes on Berula erecta, Callitriche spp., Ranunculus spp. and Phragmites communis

	Site diversity	
	Low	High
Total sites recorded	309	191
Pollution-tolerant spp.	291	431
Agrostis stolonifera	43	92
Phalaris arundinacea	105	120
Potamogeton pectinatus	18	19
Sparganium emersum	83	70
Sparganium erectum	33	54
Blanket weed	69	76
Semi-tolerant spp.	85	222
Carex acutiformis agg.	1	14
Glyceria maxima	12	33
Glyceria spp., short leaves	10	57
Lemna minor agg.	25	52
Nuphar lutea	28	30
(Polygonum hydropiper agg.)		2
Potamogeton crispus	8	24
Potamogeton nodusus	1	5
Enteromorpha sp.		5
Other spp.	118	615
Alisma plantago-aquatica	1	14
Apium nodiflorum		9
Berula erecta	12	53
Butomus umbellatus		11
Callitriche hamulata		5
Callitriche spp.	19	74
Carex spp.		6
Ceratophyllum demersum	1	6
Elodea canadensis	5	44
Groenlandia densa		3
Hippuris vulgaris		3
Iris pseudacorus	1	6
Mentha aquatica	1	6
Myosotis scorpioides	7	57
Myriophyllum spicatum	1	21
Phragmites communis	22	14
Polygonum hydropiper agg.		8
Potamogeton natans	4	14
Potamogeton perfoliatus		5
Ranunculus spp.	15	61
Rorippa amphibia	1	8
Rorippa nasturtium-aquaticum agg.		37
Rumex hydrolapathum		4
Sagittaria sagittifolia	2	15
Scirpus lacustris	1	4
Solanum dulcamara		11
Veronica anagallis-aquatica agg.	2	20
Veronica beccabunga	3	25
Zannichellia palustris	2	14
Mosses	12	25
Additional spp.	6	32
% pollution-tolerant spp.	59	34
% semi-tolerant spp.	17	17
% both of them	76	51
% other spp.	24	49

Source: Haslam (1987a)

this table (Table 10.3). Low-diversity sites have their vegetation much reduced by some habitat factor. This may be shade, boats, channelisation, etc. Or it may be pollution. So are the tolerant species (Table 4.2) merely damage-tolerant (as in Table 7.9, for dykes) rather than pollution-tolerant?

South Norway has a small and mainly coastal population (coastal settlements do not discharge into inland rivers). Streams suffer little from effluents. And the proportion of pollution-tolerant species is the same in depauperate and good macrophyte communities. Diversity is reduced by water regime and shade. Reduction by pollution is, overall, trivial.

Where, however, as in Denmark, species-poor sites contain 20 per cent more pollution-tolerants, then pollution is the major reducing factor. Denmark is reasonably clean, be it remembered. Impoverished sites are few compared with other Continental countries and Britain.

In Germany pollution is of overall greater importance, and the species-poor sites contain 25 per cent more pollution-tolerants than species-rich ones (in sites without the high water force of mountain streams, etc.). The difference would be more (37 per cent), except for species tolerant to other reducing factors. *Berula erecta* and *Callitriche* spp. are common in clean but shady sandstone streams. These shaded sites occur in the low-diversity column in Table 10.5. So do hilly streams with some water-force reductions in diversity, and much *Callitriche*, *Ranunculus* and moss. So also do navigable rivers with fringes of *Phragmites communis*. Pollution is the single most important reducing factor, but not the only one. Pollution skews and distorts vegetation, creating uniformity, creating a restricted assemblage of pollution-tolerants out of the diversity proper to natural rivers.

The calcareous irrigation streams of the Po plain, Italy, show a similar pattern to Germany. There is a difference of about 25 per cent in pollution-tolerant proportions between low- and high-diversity sites, and pollution is again not the only major reducing factor. *Callitriche* is tolerant to shade and the maintenance disturbance which are common in these channels.

In France there is only a 12 per cent difference between the proportion of pollution-tolerants in low- and in high-diversity sites. This is not because pollution is little, but because *Callitriche*, *Ranunculus* and mosses are all common in many streams damaged by shade, unstable substrate, etc.

Belgium has a nearly 30 per cent difference in pollution-tolerant proportions. This reflects the gross pollution there. The average diversity in 'species-rich' sites here is only 6. High-quality sites usually bear twelve and more species, so giving an average diversity of over twelve. Lowland Belgium has only half this (even south Norway, the highest, 8, has only two-thirds of what should be, the reduction being mainly due to non-pollution factors, see above.)

In dykes, the tolerants listed are tolerant to all forms of damage, not just to pollution, and so the data for the Dutch and Italian dykes are not comparable to those of streams.

Comparisons of sewerage and sewage treatment between countries were made in Chapter 1, noting pollution following roughly but not exactly the same pattern. The data presented here show the results of the waste disposal systems on the rivers. While it would be good for Denmark and Ireland further to improve their (inland) disposal systems, the importance of so doing for the rivers is less than in the other countries.

Few concerned with rivers work intensively outside their own country. They should. It is only be so doing that they can realise that there is no 'general European' river. Each country alters rivers, and therefore vegetation, differently, so river patterns are easily recognisable as, say, Danish or Irish. The Rhine Vale may be French, but the channels were patterned by a German-influenced engineer. What, to one country, is the obvious thing to do, is unthought-of or unthinkable in another. To achieve ends it is wise to see the results of what others have done: whether what one would wish to have, or the opposite.

A detailed pollution study

The Lower Don, Scotland

This mountain-rising river, on which limited surveys were done in 1969 and 1978, and full ones in 1981–9 (see Figure 10.2; Tables 10.6–10.8), has a long enough lowland stretch to take on some lowland characters. There are two long-established papermills at about 7 km and 5 km

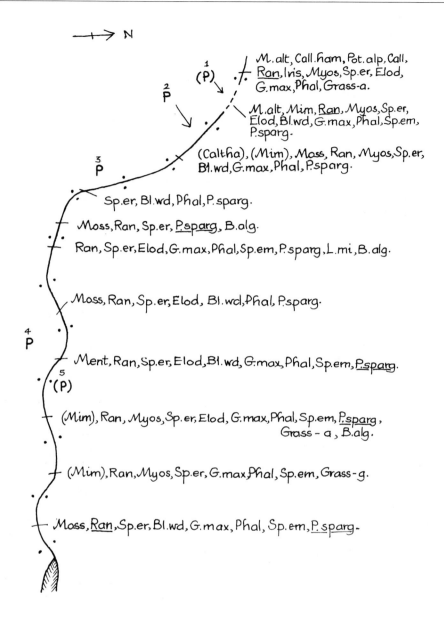

Figure 10.2 River map, Lower Don, Scotland. (Also see Figure 6.7.) P, pollution.
1. Tributary with settlement and farm pollution, formerly little, 1988 moderate.
2. 'Airport' tributaries, severe pollution within the streams, diluted in river,
increasing. 3. Upstream paper mill. 4. Downstream paper mill. 5. Sewage works,
effluent has little effect. Lowland reach of a river rising in mountains, Resistant
rock. 1984 (S). Selection of 11 sites out of 60 surveyed annually. Notes as for
Figure 4.2.

from the mouth. More recently a sewage works was added downstream of the lower (with but minor effects on the river) and very recently, an airport and consequent industry and housing developed, whose (surface, etc.) run-off enters above the mills. Paper mills deal in paper: dead plant material. The effluent is this, plus processing and other chemicals, of which those of the upper mill have the greater variety and toxicity. As late as 1969 the lower river was as if made of instant milk, white-turbid and streaky. Since then, the mills' pollution has decreased greatly.

Above all this pollution is a site of really satisfactory vegetation — though the water is carrying effluent (see below). This is the reference site for what follows. Next downstream is a site with adequate vegetation, suffering at times from too low a water level, and also from minor and increasing pollution from a tributary (more sewage fungus). Various streams — 'airport' tributaries — bring in the new and increasing run-off pollution from that development. Even in 1981 this led to visible oil, and yellowed *Glyceria maxima* above the upper paper mill. From here down, the vegetation is limited in rapid bouldery reaches by water force, in deep slow reaches by depth, and recording may be hampered by water darkness or turbidity. Lack of an *a* rating is, by itself, no evidence of pollution.

The paper mills and airport streams discharge to the south bank, the sewage works, to the north. Point-source pollution spreads diagonally downstream (Figures 10.2, 10.5), more mobile compounds moving first. So the upper mill has a simple diagonal poison-line from the discharge. The varied toxicities led, in the earlier years of the period 1981-7, to a sharp drop in macrophytes (in a physically satisfactory reach of river). The lower mill has no such obvious pattern. Flow regime changes too much at the site for a diagonal line to show, and there is little obvious change in the channel vegetation anyway. By 25 m or so below most effluents, there is cross-river mixing of (major!) pollutants toxic to macrophytes. It is different here for sewage fungus and invertebrates. In the days when sewage fungus was standard, particularly below the lower mill, it hugged its own south side until complete water mixing at a weir a kilometre or so downstream. Invertebrates followed the same pattern. Carbohydrates, suspended solids, and oxygen depletion are all important here.

When a river is grossly discoloured, and has sewage fungus reaching bloom status for over a kilometre, it is indeed polluted. Figure 10.2 and Tables 10.6-10.8 record this pollution, and its abatement.

The fish populations have not been demonstrably affected. Pigmentation was found, but the cause (1989) is still doubtful. A water chemical index (Table 10.8) shows the overall pattern: a drop with the paper mills, not properly recovered from downstream. This index uses various parameters but is biased towards oxygen regime. Most compounds in the pollutants are not measured. The BOD output of the lower paper mill was the greatest.

The invertebrate position is similar. The community is adequate upstream. It shows a small drop in the airport streams region, a greater one with the paper mills and sewage works, and there is no proper downstream recovery. The index is worse on the south of the channel, (upstream of the weir causing complete water mixing). Most of the damage is from oxygen depletion. In the past, depositing solids and smothering sewage fungus were important, and there is minor toxicity. Macrophytes indicate the position of discharges, since they respond immediately to toxicity. Invertebrates responding primarily to oxygen shortage are necessarily worst where oxygen is least, some way below the discharge. Both water quality and invertebrate indices have improved in recent years.

Sewage fungus occurs with suitable food (Chapter 4), flow and temperature. Formerly it was rare indeed above the paper mills, intermittent to the lower mill, excessive some way below that, and decreasing before the river mouth. In and about 1981, therefore, macrophyte quality decreased first, just by the upper mill, and was largely recovered by the mouth. In contrast the invertebrate and water quality indices reached their worst point further below the paper mills. These indices (both oxygen-biased) hardly improved by the mouth, where the sewage fungus had much decreased. Different groups respond to different facets of pollution.

Sewage fungus is now well established upstream of the paper mills from the erstwhile minor pollutions. Downstream of the mills the amount is small — too small to show whether or not the mill pollution is contributing to its survival: a remarkable reversal which is, of course, due to

Table 10.6 Macrophyte improvement in the Lower Don, Scotland

	1969	1978	1981	1983	1984	1986	1987
a) Damage ratings							
Upstream of paper mills, Hatton			*a*	*a*	*a*	*a*	*a*
Parkhill	*b*	*b*	*b*	*a*	*a*	*a*	*a*
Downstream of both mills. Persley Bridge	*e*	*d*	*c*	*b*	*b*	*b*	*a*
Grandholm Bridge			*c*	*a*	*a/b*	*a/b*	*a*
Seaton			*b*	*a*	*b*	*a*	*b*

	1981	1983	1984	1986	1987
b) Average site diversity–					
(No. of upstream sites 13, between sites 11, downstream sites 12, not all surveyable each year)					
Upstream of mills	9	10	10	9	10
Between mills	5	6	7	8	10
Downstream of mills	7	7	8	9	10

	1981	1983	1984	1986	1987
c) No. of sites with fringing herbs					
(Short bushy emergents, *Mimulus, Myosotis, Veronica*, etc.)					
Upstream of mills	15	12	10	7	7
Between mills	0	1	0	16	8
Downstream of mills	0	2	5	14	11

	1981	1983	1984	1986	1987
d) No. of sites with Mosses (prominent, all species)					
Upstream of mills	6	6	6	5	6
Between mills	3	4	6	3	9
Downstream of mills	2	4	1	0	3

Table 10.7 Plant deterioration in the Lower Don, Scotland

	1981	1983	1984	1986	1987	1988
a) No. of sites with yellowed *Glyceria maxima* (in water, on mud)						
Upstream of mills	4	3	1	3	2	5
Between mills	1	1	1	1	3	4
Downstream of mills	4	0	1	0	5	6

	1981	1983	1984	1986	1987	1988
b) No. of sites with sewage fungus						
Upstream of mills	0	0	0	1	1	5
Between mills	1	2	3	4	0	5
Downstream of mills	6	4	5	5	0	3
	(2much)	(1much)	(3much)			

(Sewage fungus varies greatly with temperature and flow.)
Note the upstream gain and downstream increase.

more organic pollution entering the river upstream, and less, downstream.

Sewage fungus depresses animals where they or their habitats are smothered (Chapters 4.5). For macrophytes, smothering sewage fungus cuts down light, so eliminating bottom-growing species (in the Don, *Elodea canadensis*, mosses, etc.). Tall water-supported species accumulate more sewage fungus the less the leaves move. Leaf movement depends on the overall flow, on the plant's position in the flow (more shelter, more sewage fungus, at sides) and the leaf habit. Slippery separate waving leaves collect least, fan-shaped and overlapping leaves most, and upright ones occupy an intermediate position.

The macrophyte patterns can be looked at collectively, by groups, and individually. The damage rating shows the fine upstream vegetation, the drop at the upper paper mill, and the near-recovery by the mouth. The change in the damage rating over the years is most impressive: mill pollution *e* in 1969, *d* 1978, *c* 1981, and about *a* from 1983 below the downstream mill. Remarkable.

Is it possible that some of the improvement is due to other factors? Upstream surveys show a general increase in diversity — but not nearly as much as the downstream increase. Most of that is therefore due to cleaner water. Fringing herbs (*Veronica, Mimulus, Myosotis*, etc.) are easily washed from Resistant rock streams like this one, and when their roots are smaller under pollution, wash-out is easier still. In 1981 no Fringing herbs occurred in the river below the upper mill. In 1983, the river was better, and Fringing herbs were there. Severe spates in 1985 created new good habitats for them — the peak is not due to sudden cleanliness — but the species remained afterwards; remained, and later increased, without the spates.

In a rapidly improving river with scouring spates, much alteration is to be expected in individual plants and so sites. The overall improvement is evident when the full range of sites is studied. Water-supported species show a general increase. Mosses are slightly spreading downstream. Mosses are short, anchored but not rooted, and are deterred by smothering sewage fungus and depositing solids as well as by dirty water.

Those concerned with the Lower Don may well have rejoiced over the downstream improvement. The upstream deterioration, though, casts a shadow — a shadow seen in the spread of sewage fungus, and in the return of yellowed *Glyceria*

maxima. This last appears with the airport streams. Its frequency dropped from 1981 to 1984, and the rise thereafter was small enough to seem insignificant — until 1988, when it is plain this had become a trend, and trouble was here. Whether the trouble is long-term or short-term is not yet known, but maybe a variant of Horace's axiom applies? Toss urban pollution out, and it will soon find a way back, somewhere else?

A late note in 1989: pollution is increasing. It has indeed found a way back. This time the worst is from upstream of Figure 10.2. Sources previously minor are now major. With this large input, the effect on the vegetation is hardly increased by the downstream sources, though they no doubt slow downstream recovery. The river is hardly satisfactory even near the mouth. Blanket weed is now common, and intermittently abundant, though sewage fungus is sparse. Diversity, cover, quality and index are all reduced.

Pollution ecology in various other rivers

Great Stour, England

Features: Pollution over three decades, deterioration and improvement, fish and vegetation. Recent increase in population and industry. (Figure 10.8, Table 10.9.)

Above the main town sewage works much of the Great Stour (Figure 10.3) was, and is, in bad condition: too many septic tanks, etc. This upper part is on clay. The water is therefore nutrient rich, organic-rich clay water, before any pollution is added. Below the town the rock turns to chalk — nutrient-medium and calcium-rich. The river does, rather surprisingly, take on chalkstream attributes, including a good salmonid fishery. With so much clay water, it could never become more than a fragile near-chalkstream. The upstream pollution tips the balance more. It is thus not surprising that bad sewage works effluent is devastating, even only that from a formerly small market town. The thirty-year history of the decline and improvement of the river and fish is shown in Table 10.9. This is a good example of the effects of an increasing population overloading a sewage works, and of the dramatic, though slow improvement made by — belatedly — improving the sewage works.

The vegetation improved, though not to clean,

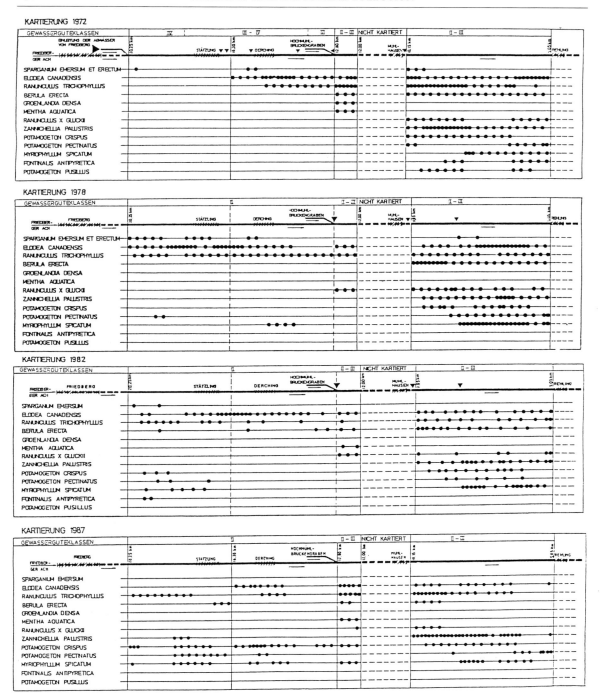

Figure 10.3 Distribution of macrophytes in the Friedberger, Germany (from Wamek and Kohler (1988)). The dots indicate species presence (on full horizontal lines: the broken lines indicate unmapped zone). The arrows show direction of flow and the tributary input, and the cross-hatching indicates settlement.

Table 10.8 Comparison of indices on the Lower Don, Scotland

	Assessment			
	Macrophyte damage rating	Scottish Chemical Water Quality Index (mean)	Scottish Invertebrate Index (NESI), Average Score per taxon (mean)	
	1981	1982	1980	1982
Upstream (Kinkill or Hatton)	*a*	87		6
Parkhill	*b*	85	5.6	5.6
Between mills	*c*	–	5.3	–
Downstream of mills	*c*	79	3.0	–
Persley Bridge				
Grandholm Bridge	*c*	73	3.3	3.8
Seaton	*b*	73	2.9	3.4

Table 10.9 River and fish pollution over 30 years in the Great Stour, England

Date	History	Fish	Invertebrates	Vegetation
1951	Crystal clear, clean gravel, little sign of pollution	Numerous trout	Fly hatches profuse	Good
1958	Detergent foam increasing	(decrease)	No mayfly. Other ephemeroptera decline	Blanket weed increases spreading downstream. Other macrophytes decline
early 1960s	Poor	Trout decrease	No ephemeroptera in Blanket weed reaches	Blanket weed stretches downstream of sewage works
1965–6	No gravel seen	Upstream, only a few sickly trout; downstream trout rising	Upstream, no fly hatches	Upstream, Blanket weed; downstream, Blanket weed patches
1966 and later	New sewage works, bare gravel increases	Trout improve	Fly hatches return	Blanket weed decreases, other macrophytes increase
1970s to early 1980s	Improves	Good	Mayfly hatches return and spread, less than 1950s, but good in part	Improves
mid-1980s	Pollution incidents	80% loss trout in part of river (1985)		Decline (see Figure 10.8)

Source: D.S. Martin, pers. comm.

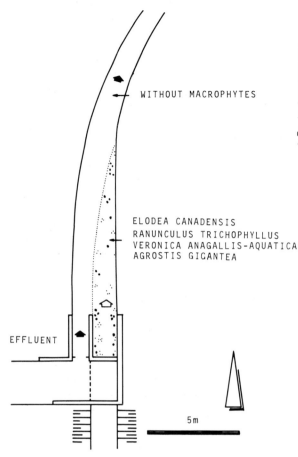

WITHOUT MACROPHYTES

ELODEA CANADENSIS
RANUNCULUS TRICHOPHYLLUS
VERONICA ANAGALLIS-AQUATICA
AGROSTIS GIGANTEA

EFFLUENT

5m

Figure 10.4 Macrophyte pattern below an effluent source in the Friedberger, Germany. Channel structures are marked. (The triangle indicates north.)

Source: redrawn from Köhler *et al.* (1974)

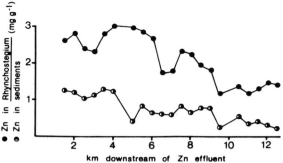

Figure 10.5 Zinc in moss and sediments, Team, England (from Wehr *et al.* (1981)). Zinc (mg g^{-1}) in the aquatic/moss *Rhyncostegium riparioides* and sediments (less than 210 μm size fraction) plotted against distance below the first major metal effluent.

Friedberger and Moosach, Germany

Features: Pollution over a decade. Fragile waters. Individual pollutants. Channellisation. Vegetation changes (surveyed in 1972, 1978, 1982 and 1988, see Figures 10.3 and 10.4; see also Kohler & Zeltner, 1981; Kohler & Schoen, 1984; Kohler *et al.*, 1974).

Over the decade, uniformity has increased in the Friedberger. A clean-up has improved the dirtiest parts. The cleanest-water species (like *Ranunculus trichophyllus*) have, however, retreated further. The clean-up measurably reduced tenside, phenol, ammonia (from 5–6 to 0.35 mg l^{-1}), phosphate (from 2 to 0.2 mg l^{-1}) carbon dioxide, hydrogen sulphide and turbidity.

The correlation between ammonia and macrophytes in the Moosach (Figure 4.21) is quite as much as expected with any single nutrient. Vegetation declined between 1971 and 1979, perhaps because of channelisation; measured pollution did not increase, and structure changes are equally important to macrophytes.

Both rivers show a downstream progression from the fragile *Potamogeton coloratus* community, lost with only trivial pollution, through a *Ranunculus-Berula* community, found in near-clean lower waters or slightly polluted upper ones, to a depauperate *Sparganium emersum* assemblage in greater and downstream pollution. This becomes more depauperate, and is eventually lost, where pollution increases more. The *P. coloratus* community is in less than 0.02 mg l^{-1} phosphate-phosphorous. The more nutrient-rich communities have higher concentrations in the water.

but started to decline again in 1986. Diversity and cover are good, but quality, inadequate. *Ranunculus* is too little, and there is too much *Potamogeton pectinatus*, Blanket weed and — even considering the upstream clay — there is too much tolerant clay flora (*Nuphar lutea*, *Scirpus lacustris*, *Sparganium emersum*). Significantly, yellowed *Glyceria maxima* remains throughout.

What of the future? The Channel Tunnel will have a terminal here, increasing traffic, industry and probably housing. Without a spectacular change in waste disposal, what will happen to the river?

Nitrate-nitrogen levels in the *P. coloratus* community vary from less than 0.003 to 50 mg l^{-1}. The most eutrophic of the species-rich communities is associated with over 0.3 mg l^{-1} phosphate-phosphorous and 20–30 mg l^{-1} nitrogen (nitrate plus ammonium).

Amal, Israel (Agami, 1989)

Features: Southern differences. Importance of emergents. Pollution. Modification. Vegetation changes.

This far south, a trend first seen in south Italy and Sicily continues further. Under pollution, water-supported species are more sensitive than emergents, and disappear first. Also as in south Italy, *Phragmites communis* is particularly tolerant to pollution. Because of the dry summers, land/marsh species grow at the river's edge, so the pollution-tolerant species appear odd, like *Juncus maritimus* and *Cladium mariscus* (see also Figure 4.11). Odd, not of course for the Mediterranean, but for those used to more northerly rivers. In the north *Cladium mariscus* is a marsh, not a river species, and has a very specialised habitat within the marsh; in Britain, lime-rich, nutrient-poor, wet but not stagnant, woody plants kept out: not exactly a tolerant species! The Amal is a valuable example of how principles of species behaviour are world-wide, but application varies with habitat.

Ceratophyllum demersum is the only tolerant water-supported species. Probably because of variation in the clay substrate, it grows better in moderate than low pollution (i.e. the cleaner parts have unsatisfactory substrate). Further north, *C. demersum* is not very pollution-tolerant, though occurring in eutrophic was well as mesotrophic places.

Transplant tests confirmed the pollution tolerance of the relevant species.

Small-scale vegetation patterns (Figure 10.10)

Variation between nearby small streams is useful for wider interpretations, as the governing factors are usually easy to see. The Danish Fiskbaek has channels both sides of a road. That with Blanket weed carries sewage effluent. Where it enters the main beck, the dominant *Sparganium emersum* is eliminated in the pollution pathway, a total removal of vegetation spreads diagonally downstream (see also Figure 10.5). As pollution is slight, *Sp. emersum* returns at about 10 m below the discharge, and soon thereafter is again dominant.

The Italian pattern is from the Po plain calcareous streams. The most polluted irrigation channel is higher up, and spills to the level below, making a not-clean channel moderately polluted — *Potamogeton pectinatus* and Blanket weed appearing, and cleaner species decreasing. Cleaner water is then added, diluting the pollution — but the vegetation still deteriorates. Why should this be? The cause needs other observations. Flow slows here, and so mud accumulates. Mud contains the most pollutants. Although the water is cleaner, the soil is dirtier: the vegetation deteriorates.

The French patterns show combined effects of pollution and shade.

Sicily is a badly polluted land. Near the source of the Ciano is one of the few good sites recorded: eight species no less (and no more — eight is far from good). This was protected by its holiness, being anciently a sacred spring. History may be necessary to full interpretation.

Cole, England

Features: Pollution. Oxygen. Invertebrate patterns. Figure 10.9 (Davies and Hawkes, 1981; Hawkes and Davies, 1971; Hellawell, 1986).

The Cole, tributary of the Tame (Figure 4.6), is on the outskirts of industrial Birmingham, suffering accordingly. An organic pollution was studied. Clean-water species dropped sharply below, with only minor downstream recovery in the surveyed length. A second group of species peaked by the discharge, where conditions are at their worst, and a third did so in the recovery zone further down.

The leech, *Helobdella stagnalis*, followed its prey, *Asellus aquaticus*. *Asellus* is genuinely related to the oxygen levels (see Figure 10.6). *Helobdella* is so only through the *Asellus*. Seasonal oxygen can be important. Near the discharge, oxygen is very low in summer. The summer generation of chironomid (*Eukiefferiella hospitus*) is lost, and the spring and autumn generations are small. Further downstream, as conditions improve, the spring and autumn generations are properly present, and downstream again, all three do well. Adults can fly up to the middle zones, compensating for the lost summer population there.

The distribution of the (clean-water) *Gammarus* was related to oxygen: but not to the maximum,

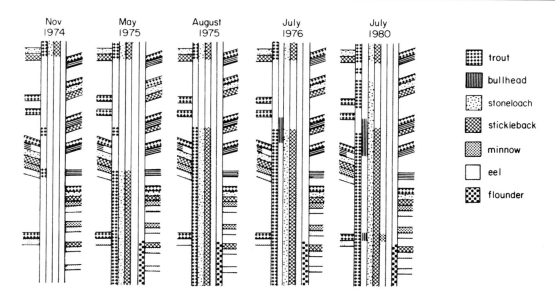

Figure 10.6 Fish recolonisation between 1974 and 1980 in the Ebbw, Wales (from Edwards *et al.* (1984)). Fish present in tributaries and upstream of the town in 1974 and gradually colonising nearer to the town, from downstream.

Table 10.10 Zinc pollution and invertebrates in the Allen, England
Summary of experiments to determine the toxicity of dissolved zinc to invertebrates

Species	96 hour median lethal concentration, (mg l^{-1})	Median survival time at mean ambient Zn conc. at site 7, (hours)	Maximum conc. tested at which no deaths occurred within 2 weeks (mg l^{-1})
Gammarus pulex	2.1	140	–
Baetis rhodani (May)	28	>336	–
(Aug)	1	90	–
Rhithrogena semicolorata	>100	>130	Test terminated at 130 hours – no deaths up to 100 mg l^{-1}
Leuctra sp.	37	>130	Test terminated at 130 hours
Chloroperla torrentium	>362	>336	362
Limnephilus sp.	>100	>336	100

Source: Abel and Green (1981)

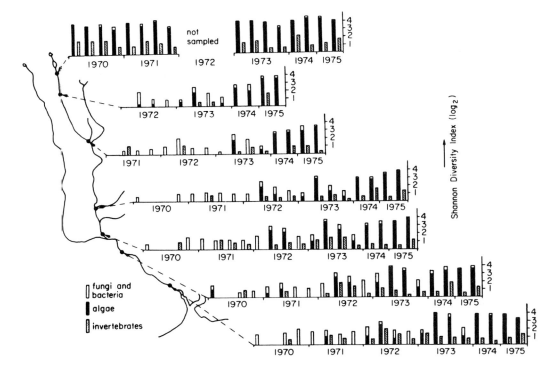

Figure 10.7 Annual and seasonal changes in bacteria, algae and invertebrates in the Ebbw, Wales, between 1970 and 1975, measured by the Shannon Diversity Index.

Source: Edwards *et al.* (1984).

minimum or mean oxygen values. *Gammarus* depended on having enough oxygen at night. Its distribution was governed by the length of time the oxygen was low, in the night (see also Grant and Hawkes, 1982).

Eleven species of chironomids are common. *Chironomus riparius* is the most pollution-tolerant and extends its range in summer when pollution is most — and other species suppressed. High oxygen is, in fact, toxic to *C. riparius*. *Polypedium arundeti* is the most numerous, and is also pollution-tolerant, but it occurs, as well, not infrequently downstream.

The other species are typical of cleaner waters, and peak in different parts of the recovery zone (*Prodiamesa olivacea* at the least-polluted, downstream site, *Cricotopus sylvestris* and *C. bicinctus* further up).

Lake District stream, England

Features: Small change in pollution, large change in invertebrate community.

Mains water came to a village and, as usual when water is available on tap, much more was used. The septic tanks became overloaded, and sewage entered the stream. Several invertebrates increased, but one, *Polycelis felina*, went on increasing. This carnivore traps its prey with mucus strings laid down while it moves over the top of stones. Mayflies living on stop tops decreased or vanished, and since mayflies and stoneflies living *under* stones stayed the same, *P. felina* seemed responsible. A small chemical change, not enough to harm any of the invertebrates present, altered the invertebrate community indirectly, through favouring one species. How little is known about life in rivers! Such changes commercially presumably occur

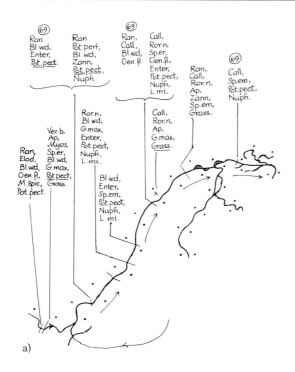

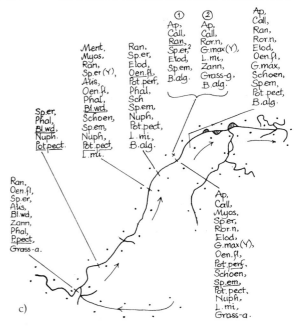

c)

Figure 10.8 River-map, Great Stour, England. (a) 1969 and 1974, (b) 1979, (c) 1986. Three intermediate surveys confirm the trends shown. The river rises on clay, and flows on to chalk just downstream of the marked sewage treatment works (Ashford). Lesser effluents enter elsewhere. (M). Notes as for Figure 4.2.

often, with habitat changes which appear small.

Ebbw, Wales

Features: Coal and heavy industrial pollution. Decrease and clean-up. Fish, invertebrates, algae, bacteria. (Figures 10.6, 10.7.)

The coal valleys of south Wales have been notoriously polluted for over a century (Chapter 2). Heavy industry followed the coal. Clean-up measures were late, and were in fact largely overtaken by the industrial blight of the 1970s: rivers becoming cleaner because factories closed, more than because their effluents became cleaner. However, effluents were improved, and the results of cleaning steelworks effluent are seen in Figures 10.7 and 10.8. There are dramatic increases in fish, invertebrates and algae, and a satisfactory decline in bacteria. Clean tributaries had good fish populations all along. This gave sources from which fish could spread into the main river once the conditions there allowed it. The fish gradually

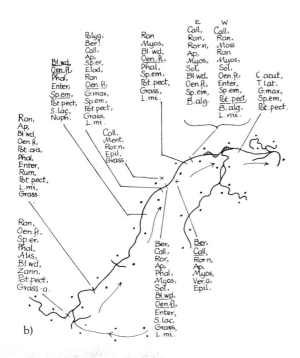

b)

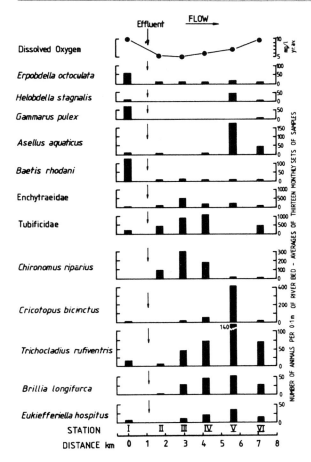

Figure 10.9 Changes in the composition of the benthic invertebrate community in the Cole, England, below an organic discharge (from Hellawell (1986)).

moved upstream, getting closer to the effluent discharge, as the river becomes cleaner.

The industrial base in England and Wales has partly moved, in recent decades, from the hills of the north and west, polluting rivers such as the Ebbw, to the south and east, polluting such as the Great Stour (above).

Tyne tributaries, England

Features: Metal and other pollution. Inverte-brates. Bryophytes. (Figure 10.5, Table 10.10 East and West Allen, Abel and Green 1981; Team, Wehr et al., 1981)

These are metal-polluted streams, though also with domestic and (for the Team) other industrial waste. The West Allen carries much zinc, possibly

with associated cadmium, lead, etc. It has but few invertebrates. The East Allen has less metal, and seventy-two invertebrate species. Transplants show that cleaner-river animals either died with more pollution (for example, *Gammarus pulex*) or suffered in other ways (for example, larval feeding rate slowed in *Limnephilus*). In the Team, metal concentrations in Bryophytes followed those in the sediments, so were a useful indicator.

Syr, Luxembourg

Features: Some detail on a moderately-polluted inland Europe rivers. Vegetation. (Figures 10.11, Table 10.11.)

The Syr (Figure 10.11) rises on a lowland plateau, flowing down through a wooded gorge to the Moselle. In the lowland, there is mainly grass near the stream, with cereals increasing elsewhere, and maize and fruit trees among the minor crops. Villages are frequent, factories sparse and mostly near the Moselle. In villages, streams may be walled.

The lowland part is a typical managed West European stream, formerly used for water and power, latterly (past century or so) for effluent disposal and lowering water table (drainage). It used to be winding with low banks (high water level) and slow flow. It is now straighter, banks typically 1 m above water, and with a frequently scouring flow. The recent management, including removal of trees, has enhanced bank instability. In the gorge, the tree-anchored banks create a variety missing in the dredged farmland stream. The water is, of course, swift in the gorge, though water force remains low (little run-off from hill slopes). The tributaries are the muddiest, the lowland stream the sandiest and the gorge stretch the stoniest.

Pollution is from untreated and undertreated domestic, farm (and factory) wastes, fertilisers and biocides, as usual. The field run-off is currently the least harmful (Haslam and Molitor, 1988; and see Chapter 11). There is also run-off from the airport, and from roads.

The river is polluted from near the source. Vegetation drops further in the first village (where partial shade and garden cultivation aggravate pollution) then recovers well (the mill stream now carrying most of the water, the other not having enough for assessment).

Apart from places with known other reducing factors, such as shade, walling and the gorge, the

Table 10.11 Changes in the vegetation of the Syr, Luxembourg 1980–85
a) Changes in frequency.
The numbers related to the number of sites in which the relevant species were recorded. Since more sites were surveyed in 1985, the proportions must be considered as well as the actual numbers.

	1980	1985
No. of sites recorded	16	46
EMERGENT SPECIES		
Phalaris arundinacea	9	25
Other tall monocotyledons	2	7
Agrostis stolonifera	8	27
Other small grasses	11	4
Fringing herbs	1	15
Other short dicotyledons	11	4
		(inc. 2 *Ranunculus repens*; typical of disturbance)
Tall dicotyledons	11	3
WATER SUPPORTED SPECIES		
Blanket weed	7	26
Mosses	8	11
Callitriche spp.	5	6
Elodea canadensis, Groenlandia densa, and *Zannichellia palustris*	7	7
Potamogeton pectinatus	3	3
Other species	1	2

b) Changes in cover.
Replicate sites (including one nearly replicate, and shaded non-replicate but similar sites). Cover values are approximate, and small differences are not reliable. 'Main species' indicate the species providing most of the cover.

Site	1980 Cover (%)	Main species	1985 Cover (%)	Main species
Moutfort	30	(Angiosperms)	30	Blanket weed
Munsbach	60	*Callitriche*	80	Blanket weed
Übersyren	90	*Elodea canadensis*	25	Blanket weed *Elodea canadensis*
Downstream of Übersyren	< 10		70	Blanket weed
Übersyren				
Roodt	< 10		50	Blanket weed
Downstream of Roodt	< 10		60	Blanket weed
Betzdorf	< 10		70	Blanket weed
Hagelsdorf	15 +		50	Blanket weed *Potamogeton pectinatus*
Wecker	< 10		20	Blanket weed
Downstream of Wecker	20	*Zannichellia palustris*	< 10	
Manternach	< 10		10	
Downstream of Manternach	< 10		30	Blanket weed
Shaded sites	< 10		< 10	
Upstream of Mertet	< 10		< 10	
Downstream of Mertet	50	*Zannichellia palustris*	25	Blanket weed

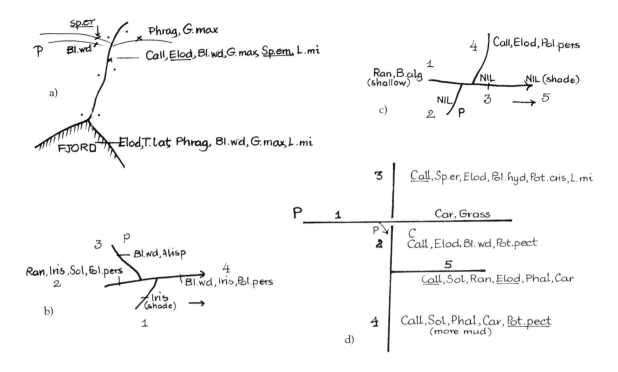

Figure 10.10 Polluted stream portions (from Haslam (1987a)). (a) Fiskbaek, Denmark, 1977; (b) near Perpignan, France, 1981; (c) Gers, France, 1981; (d) Po plain, Italy, 1977. P, pollution. See text for description.

main drops in rating are from villages — from one village after another. Here is the main pollution. The airport tributary has poor vegetation, and yellowed bank species. Yellowing also occurs further down, by a sewage works. (Similar pollutions have similar effects: see the Don yellowing from the airport pollution, above.) Diversity rises in one village — where the channel bed is lined. Lining may encourage anchorage by providing stable substrates. It may also prevent good rooting, as there may be too little sediment, as seen on a tributary site. *Ranunculus repens* occurs, typical of disturbance. Here the disturbance is slumping banks.

Potamogeton pectinatus enters as soon as the stream is large enough. Blanket weed requires but little water, and is present from near the source. Most of the plant cover (in 1985) is from Blanket weed. *Enteromorpha* has been spreading in polluted British rivers for some decades, but is rare on the Continent. Its appearance in the Syr

is therefore unexpected — though consistent with pollution (and the airport?).

In the gorge, of course, the rating drops: the moss and *Petasites* show the habitat. (Rating is the simplest indicator. Species lists are more informative.)

Overall, how much pollution is there? Over 60 per cent of river species records are of pollution-tolerants (see Figure 4.2 and Table 10.4), and there are no *a* ratings. This means consistent pollution. There is little overall change in rating between 1980 and 1985, but the species composition giving rise to this rating has changes (Table 10.11). Pollution was stable, but changes were:

- Increase in short emergents, particularly Fringing herbs (*Mentha*, etc.).
- Channel species alter from angiosperm cover, with some diversity, to Blanket weed, and little diversity.

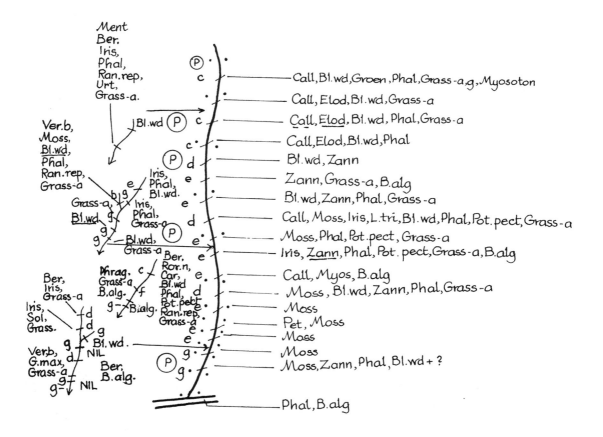

Figure 10.11 River map, Syr, Luxembourg. Sandstone etc., lowland above, gorge below. (S), 1980 with 1985 tributary vegetation. P, pollution, Y, yellowed. Damage ratings marked. Pollutions at villages and from airport

These are typical of shallowing. The decrease in *Potamogeton natans*, not an especially shallow-water species, supports the deduction. Whether the reason is climate, drainage or abstraction is for the researcher to discover.

This tail-piece river is included to show how easily anyone with some interest and no previous knowledge can quickly learn to interpret rivers.

The plant is always right. Naturally no beginner should rush to water authority or Member of Parliament about dreadful pollution just discovered! Wait, and carefully check a variety of rivers against their known pollution and other habitat influences first. To report recent dredging, swans, boats, bathers, shade or Alpine force rivers as pollution helps no one!

11

Pollution from rural land use

Man alters land use — greatly in densely populated areas such as Europe, and slightly even in Antarctica, the least-affected area. The influence may be direct, such as growing wheat, or indirect — cliff ledge habitats rarely if ever touched by Man may be much altered through human pressures on birds. All may influence streams.

Rain, falling on the land, runs to the stream. In so doing, it picks up dissolved substances and particles. The nature of these is governed by land use. Hence land use influences river environment. The land use wanted near the stream also influences drainage, with all the effects on structure described in Chapter 9. Rain penetrates through porous rock or subsoil. As it passes through the soil, downwards and laterally, the original particles and some of the dissolved substances are filtered out or absorbed. More are collected from the soil, subsoil and rock. It used

to be thought it was quite safe to use fertilisers above aquifers, since they could not possibly reach and contaminate ground water. Unfortunately this is now known to be false; when enough of any substance is put on the land, it will — years or decades later — reach the ground water. Dumped poisonous waste, used in landfill, etc., may also reach ground water unless carefully sealed — and if need be, resealed. Ground water then feeds rivers: and domestic supplies.

Land use affects the quantity, as well as the quality, of water reaching streams. Other things being equal, more rain will sink into porous soil covered with thick vegetation than into bare soil, compacted soil — or tarmac. The latter encourage flash floods — rapid run-off, scouring and removing soil after storms. This produces unstable river flows and more loose silt. Run-off is further speeded by nice and smooth, so efficiently-draining, landscapes, rather than criss-crossing small hedged fields; and by ditches placed for drainage rather than to conserve water. The faster the surface flow, the more the sediment picked up. This sediment has, of course, a chemical as well as a physical influence on the river (for physical effects, see Table 3.1). The underlying rock sets the scene: water will be absorbed into chalk in a way it cannot be into clay or indeed hard Resistant rock, and into lowland as it cannot be into Alpine landscape. It is Man who determines what will happen within these preset limits.

Flow regime and sedimentation both govern physical structure, therefore affecting habitat also through these. The type of alteration produced by land use may be very obviously pollution, as when land biocides are sprayed on the river. It may also be subtle, recognisable only by those knowing what to look for. For the change may be to a stream type proper somewhere else. In Jutland, Denmark, the best-quality small streams are those where *Ranunculus*, *Callitriche* and Fringing herbs grow. These are good winding streams in little-drained grassland (Figure 8.1b). Commonly however, good-quality small streams — and they are good, too, though less fragile than the other — have abundant *Sparganium emersum-Berula erecta* instead. This community is characteristic of channelled and more managed watercourses. It is a more eutrophic community, occurring in more managed farmland. Is this second type polluted? It all depends on definitions. For rating, a stream is compared to the best available vegetation. Yes, but — available where? Ordinarily available means the *Sparganium emersum* types (see, for example, Figure 4.2). And what if the rare *Ranunculus* streams were not found because the survey was inadequate? What if there were none left at all? Indeed, how many stream types have been lost, Europe-wide, by intensive land management?

Again, originally most of Europe was forest, with trees shading and forming bank habitat. Forest removal much altered the streams. That we can be sure of. What their communities were before is much less certain. In North American woods there is full river vegetation in open glades. Plantations, tree-lines and hedges are regular; this means glades and therefore abundant vegetation are rare (Figure 6.1). Waterside birds and mammals are less dependent on shading, and more on the amount of tree habitat, and the habitat created by the trees. Invertebrates are influenced by leaf fall and vegetation, and fish by the structure created by the other components.

Forest removal led to grassland and small-scale arable, in which nutrients were largely retained and recycled. We now, though, have output (waste) of nutrients from the land, from fertilisers and increased stocking, giving eutrophication of both lowland and — more newly — hill streams. And we also have acidification of other streams, by peat disturbance, conifer afforestation and acid rain.

Total impact

Cover-Diversity Number

The intensity of Man's use of the land (here including all aspects of pollution), and so the intensity of pressure on river vegetation, can often be shown by the *Cover-Diversity number*. This is a very simple index, using only cover and diversity, with no weighting for species quality. Table 11.1 shows a range of communities in a range of countries. A number of points arise:

- In small (size ii) streams, Britain has low values. The water table has been lowered more here, and small streams reflect such drainage the most (a drop of, say, 10 cm in water level is 20–40 per cent of total depth in small streams).
- In Italy, the Latina plain has the lowest values and most intensive farming of those plains surveyed.
- South Norway has large values on Resistant rock even in fairly high hills: human pressures are light. More managed land has lower values.
- Hill limestone has very low values in Britain. This is quite likely due to shallowing and peat removal.
- French sandstone values are very low, even in lowlands: they have unusually unstable sand.
- Corsica and the Pyrenées are both Alpine regions, Corsica being the steeper (of the surveyed parts). Corsica, though, has much better stream vegetation. This is because there is less pressure on the land.
- Sardinia, with the lowest pressure on land, has exceptionally high values.
- Denmark has better vegetation than Germany.

Table 11.1 Cover-diversity assessment for different countries and different stream types
The numbers have been obtained from the data in Table 7.4 and Haslam and Wolseley (1981) where the minimum community diversity and covered expected in undamaged sites are listed. Each species, and each 10% of cover, score 1. The surveyed communities, though incomplete, are considered to be representative. N., north; S., south; E., east; W., west; C., central. Stream sizes (i–iv) are as in Figure 1.3

Rock type	Landscape/vegetation type	Stream size		
		ii	iii	iv
Belgium				
Alluvium	Dykes	17	17	19?
Clays	Lowland	13	15	15
Miscellaneous	Lowland	14	15	16
Acid sands	Lowland	14	16	15?
Sandstone	Lowland/upland	15	16	16
Resistant	Upland	10	15,8	14,8
Britain				
Clay	Lowland	4,11	13	15
Clay	Upland	6	6	–
Sandstone	Lowland	6	12	–
Sandstone (acid)	Lowland	8,2	11,13	–
Sandstone	Upland	8	8	7
Sandstone	Mountain	0	0,2	0,7
Sandstone	Alpine	0	0	–
Sandstone (Caithness)	Moor	5	6	9
Limestone (chalk)	Lowland	15	14	17
Limestone	Upland	7	9(7)	9
Limestone	Mountain	0	0	1
Resistant	Bog	8	9	6
Resistant	Moor	6	10	8
Resistant	Lowland farmland	2	5	4
Resistant	Upland	5	3	1+
Resistant	Upland fertile	7	8	9
Resistant	Mountain	1	1,5	0,6
Resistant	Alpine	0	0	0
Coal Measures	Lowland/upland	4	4,10	–
Coal Measures	Mountain	0		
Corsica				
Resistant, etc.	Lowland	14	15	10
Resistant	Mountain	14	4	2
Resistant	Alpine	8	3	2
Denmark				
Clay	Lowland	15	16	16
Moraine sand	Lowland	14	16	16
Fluvial sand	Lowland	14	16	–
Alluvium	Lowland	14	16	17
France				
Limestone (chalk)	Lowland	15	14	17
Limestone (Jurassic)	Lowland	15	17	18
Limestone	Upland	15	16	17
Limestone	Mountain	7	9	5
Limestone	Alpine	0–2	0–4	0–3
Muschelkalk	Lowland	12	13	–

Table 11.1 contd.

Rock type	Landscape/vegetation type	Stream size		
		ii	iii	iv
Sandstone (main)	Lowland	14	16	14
Sandstone (Jura)	Lowland	15	16	17
Sandstone (acid)	Lowland	14	15	14?
Sandstone (Girondes)	Lowland	14	15	16
Sandstone (main)	Upland	11	12	13?
Sandstone (Brittany)	Upland	9	10	10
Sandstone	Mountain	0–5	0–5	0–4
Clay	Lowland	15	16	17
Clay	Hilly	2–4	7+?	7+?
Alluvium/moraine (N.C.)	Lowland	15	16	–
Alluvium/moraine (E.C.)	Lowland	14	15	16
Alluvium/moraine (Rhine)	Lowland	12	13	–
Alluvium/moraine (Rhône)	Lowland	14	14	–
Alluvium/moraine (S.)	Lowland	14	13	–
Alluvium/moraine (Outwash)	Lowland	0–5	0–9	0–5
Alluvium/moraine (W.)	Lowland	15–17	16	17
Alluvium/moraine (Gironde)	Lowland	15	15	–
Resistant (Brittany)	Lowland/upland	9–13	9–14	11,15
Resistant (E.C.)	Lowland/upland	6–11	6–12	7,13
Resistant (S.)	Lowland/upland	5	6	7
Resistant (Massif)	Upland	4	5	7
Resistant (Massif)	Mountain	6,10	6,8	5,7
Resistant (Alp)	Alpine	0–1	0–1	0–1
Resistant (Alsace)	Alpine	0–3	0–4	–
Resistant (Pyrenées)	Alpine	0–1	0–2	0–2
Germany				
Sandstone (acid)	Lowland	11–12	13	14
Sandstone (main)	Lowland	11	13	13
Sandstone (sand/clay)	Lowland	12–13	13	14
Sandstone	Upland	6–10	7–11	13
Sandstone	Mountain	0–6	0–7	1–8
Resistant	Lowland	10	11	13
Resistant	Lowland/upland	6–12	–	–
Resistant	Upland	0–11	0–6	0–11
Resistant	Mountain	0–12	0–7	0–7
Resistant	Alpine	0–5	0–3	0–5
Limestone	Lowland	14	15	16
Calcareous moraine	Lowland	13	14	13
Muschelkalk	Lowland	11	13	14
Muschelkalk	Mountain	0–9	0–10	0–12
Moraine (neutral)	Lowland	13	14	14
Moraine (eutrophic)	Lowland	11	12	13
Chalk + alluvium	Lowland	13	13	14
Loess, etc.	Lowland	12	15	13
Alluvium (dykes)	Plain	14	15	15
Ireland				
Limestone	Lowland	15	15	16

Table 11.1 contd.

Rock type	Landscape/vegetation type	Stream size ii	iii	iv
Limestone	Hill	0–6	0–5	0–6
Resistant	Lowland	13	13	14
Resistant (acid)	Lowland	11	8	–
Resistant	Hill	0–4	0–4	0–3
Sandstone	Lowland	13	15	14
Sandstone	Hill	0–5	0–5	0–5
Coal measures	Hill	7	8	9
Italy				
Calcareous alluvium	Plain	15	15	15
Alluvium (main)	Plain	16	18	19
Alluvium (Latina)	Plain	8–13	7–19	4–9
Clay	Plain	3	3	–
Limestone (Alp edge)	Hill	9–13	3?	–
Hard clay (N. Appenines)	Hill	0–1	0–6	0–13
Hard clay (Piedmont)	Hill	0–12	0–6	0–13
Resistant (Alps)	Alpine	0	0	0
Hard clay (N.W. coast)	Hill	0–10	0–12	0–5
Hard clay (W. coast)	Hill	0–17	0–15	0–10?
Limestone (Carrara)	Hill	11–17	9–16	0–17
Hard clay (S. Appenines)	Hill	0–7	0–2(–12)	0–1
Salt pans (Apulia)	Plain	0–13	0–3+	0–3+
Limestone (Garigue)	Lowland	10–13	10–13	–
Hard clay (E. coast)	Hill	0–17	0–18	0–5+
Resistant (S.)	Hill	0–19	0–19	0–19
Limestone (S.)	Hill	0–14	0–14	0–14
Hard clay (S.)	Hill	0–14	0–14	0–14
Sicily	Hill/mountain	Up to 16,18[a]	Up to 10,13[a]	Up to 10,13[a]
Sardinia: Resistant	Hill/mountain	Up to 17,21[a]	Up to 17,21[a]	Up to 17,21[a]
Sardinia: Limestone	Hill	Up to 13,17[a]	Up to 13,17[a]	Up to 13,17[a]
Luxembourg				
Sandstone	Lowland	11	14	16
Resistant	Hill	11–13	1,12	1,11
Netherlands				
Alluvium (high diversity)	Plain	17	18	18
Alluvium (low diversity)	Plain	14	15	16
South Norway				
Resistant (Ostfold)	Hill	13(0–17)	12(0–17)	12(0–15)
Resistant (Vestfold)	Hill	11(0–17)	13(0–17)	12(0–18)

[a] Numbers very variable.

Source: adapted from Haslam (1987a)

There is a greater change from mainland Germany to the Isthmus of Schleswig-Holstein (comparing like with like) than there is between Schleswig-Holstein and Denmark. The earlier farming system was laid down while Schleswig-Holstein was Danish (before 1864): present ways are, still, more Danish than German. (By their 'rivers' ye shall know them?)

Cover-Diversity number also reflects water force, rock type etc; it is not solely man's influence! It is, though, a very useful first tool of general assessment. If the number is high, there is no need to look further. If it is low, an explanation is required: often, intensity of land use.

Nutrient-poor influences

Nutrient-poor communities develop in nutrient-poor habitats, of course. The most basic cause of such habitats is rock type. Most Resistant rocks and some sands are nutrient-deficient, and nutrient-deficient their streams will be, too, unless a nutrient-rich influence is also present. Another cause are the variable and changeable acid influences of the catchment above the rock (though some of these are consequent on the type of rock). The most widespread are acid peats (bog peat, moor and so on), and acid-leaved forests.

The effect of forests is not just physical (as described above), it is also chemical, through leaf fall. Conifer and birch (*Betula*) leaves, like acid peat, have an acid, oligotrophic influence on the streams. Not just acid, either: plenty of organic chemicals are involved. The influence of leaves was termed toxic by Huet (1951), when describing the trout loss from the spawning grounds in the Ardennes hills after conifer planting.

Bog peat has been drained, lost by oxidation or physically removed from part of its former distribution in Britain, Ireland and, to a lesser extent, Scandinavia. Acid heaths have been claimed for farming and made more fertile, particularly in Denmark, Germany and the Netherlands (or in part built on, as in Britain also). Conifer and birch woods have been removed, and, particularly in Belgium, Britain, France and Germany, conifer woods have been

added. In Germany, nutrient-poor stream sites are perhaps commonest on birchwood heath, in Britain, Ireland and Norway on acid peat, in France on Resistant rock, and in Denmark (and Belgium) on heath. (Acid peat may occur over limestone in Britain, giving peculiar results on streams subjected to two, almost opposed, influences.)

In northern and western Europe acid peat rivers tend to rise in the peat of the hills, and may flow through to farmland below. This means much downstream eutrophication (see, for example, Figure 6.7 (Don)), the nutrient-poor influence fading out, and that of farming being added. Further east, in Czechoslovakia, acid peat may instead be found in forests along the middle reaches (see, for example, Figure 6.15, Luznice). This gives intermittent downstream acidification, opposing the general downstream eutrophication. The stream and land-use pattern of one's own land is 'normal' only for oneself, not for others.

The effect of peat removal — and channel alteration — is well illustrated in South Norway (Figure 11.1). Farming and forest occur on the same topography. Forest streams, when relatively untouched, are oligotrophic. Equivalent streams in grassland, with the woods removed, the grassland added and the stream dug out, are mesotrophic. (Vegetation is usually absent: shallow scouring flow from channel alterations, and shade from now steep banks.) A similar pattern is seen in South Czechoslovakia, in quite a different climate. Small peat streams (just in farmland but rising in forest peat) bear oligotrophic species like *Juncus bulbosus*. In a village, though, the absence of peat allows mesotrophic species such as *Veronica beccabunga* to occur, and the presence of pollution brings in Blanket weed.

Of recent decades, there has been large-scale removal of peat in Ireland. This is for electricity generation and also for sale to gardeners. There has also been the usual drainage and drying of farmland. Milled peat is harvested from 4000 ha in the Suck catchment (Caffrey, 1986a). Deep drains criss-cross the flat bog, carrying eroded peat silt to the slow-flowing Suck, which is ill adapted to carrying this extra sediment. The peat reached 3 m thick near the main outfall, and was noticeable for 0.5 km. Alkalinity, conductivity and pH dropped, because of the acid peat, while turbidity and suspended solids rose, decreasing river light. Upstream, *Potamogeton lucens*

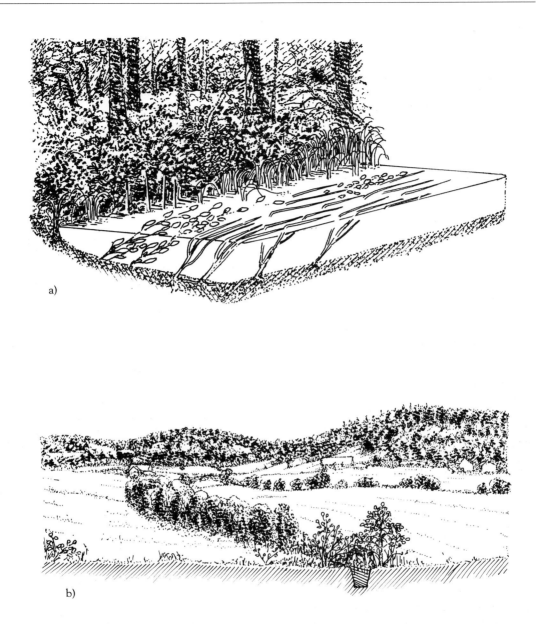

Figure 11.1 Managed and little-managed streams, S. Norway (P.A. Wolseley, from Haslam (1987a). (a) Nutrient-poor forest stream. Deep enough for water-supported species. Silty, good vegetation. Species include *Potamogeton polygonifolius* *Glyceria fluitans* (long-leaved forms), *Sparganium emersum*, *Alisma plantago-aquatica*, *Sparganium erectum*, *Ranunculus flammula*, *Luzula sylvatica*. (b) Managed, nutrient-medium, grassland stream. Narrow, dry or wet, scouring flow, gravel, shaded. Where suitable (which is rare), species include *Agrostis stolonifera*, *Phalaris arundinacea*, *Alisma plantago-aquatica*, *Lythrum salicaria*, *Lysimachia thyrsiflora*.

Table 11.2 Characteristic nutrient status bands of streams under different types of land use
Nutrient status bands, in order from low to high, are: Brown, Orange, Yellow, (Green), Blue (Turquoise, Pink), (Mauve), Purple, Red (see Table 7.5 for details, and Table 4.1 for typical species occurring in each band).
  The named bands are those for the mean of the species present.
  Stream sizes as in Figure 1.3. Downstream eutrophication shown (see also Table 11.4).

Land use	Stream size		
	ii	iii	iv
Much blanket bog	Brown (Brown–Orange)	Brown, Brown–Orange	Orange
Moorland	Orange, Yellow	Yellow (Orange–Blue)	Orange–Yellow
Heath, conifers	Yellow, Blue	Blue	Blue–Purple
Poor-quality farmland	Yellow, Blue	Blue, Blue–Purple	Blue, Purple
Good-quality farmland on Resistant rock	(Yellow–)Blue	Blue	Blue–Purple
Good-quality farmland on richer rocks	Blue (Blue–Purple)	Blue, Purple	Blue–Purple, Purple, Purple–Red

Source: Haslam (1987a)

dominated the river; downstream, such vegetation could not tolerate the unstable substrate, smothering peat silt and lack of light. All macrophytes were absent, for over 0.25 km, then tall monocotyledons and a little *Nuphar lutea* appeared. Tall monocotyledons live on the edges where they need firm anchoring, but since they are strong emergents they are little affected by turbidity, etc. *Nuphar lutea* has deep rhizomes and tough floating leaves. This was the first species to occur, and occurred first at the sides, where there was least of the unstable peat silt. An efficient settling pond produced near-recovery in a year or two. The deposit was reduced to a thin layer on leaves and soil.

This is an extreme case, but peatland drainage is increasing peat pollution generally. Peat drainage near sources will mean peat and peat influence extending further into the downstream farmland region, altering vegetation, etc. Overgrazing causes erosion and more river peat, and so do drainage (for farming) and fires (unless very carefully controlled). (Recreation, more intensive farming, etc., also contribute; see below.) Drying the peat upstream also leads to browner water because more humic substances are washed in (Yorkshire Water and Water Research Centre, 1987). The drying can be from drought, drainage, burning or change of climate. It comes, therefore, not just from *whether* Man interferes, but from *how* he interferes. Humic

substances react and interact in river processes in ways not fully understood — but whose significance is easily understood. Any major change in river chemistry will influence the river's reaction to pollution.

Other acidifying influences of Man include mines. This pollution is local and only near those mines, and in Britain is decreasing from purification and mine disuse. Acid rain (see Chapters 2, 4, 5) is increasingly important, the influence being most where there is both pollution arriving and the land already base-poor (as on Resistant rock and in boggy areas like parts of Canada and Scandinavia). Exacerbating factors are acid-leaf forests and high rainfall (which dilutes neutralising substances picked up by the water from the land). (See Gorham and Gordon, 1963, for loss of macrophyte diversity caused by precipitated sulphate, and a list of Canadian species tolerant and sensitive to this pollution.)

Nutrient-rich influences (Tables 11.2, 11.3)

Apart from the powerful and underlying influence of rock type (see Table 11.3), the most important of these influences are the obverse of the nutrient-poor ones — the removal of acid peat and acid leaf-fall forests. (While peat is being removed, the rivers are acidified by the extra incoming peat. After the peat is gone, however, the rivers become more nutrient-rich.)

Table 11.3 Macrophyte nutrient status bands in relation to rock type landscape and stream size
Cation numbers are a measure of calcium, magnesium, sodium and potassium regime: hardness ratio plus a grading of the total present.
The communities as defined by country, rock type and landscape are arranged in order of increasing nutrient status as defined by nutrient status band (see Table 7.5). The bands are based on adequate records, the Cation numbers on only about 1200 samples in all, so their selection may not be truly representative of the habitat as a whole: the nutrient status band is considered the more reliable. Nutrient status bands from low to high (Br, Brown not included). Or, Orange; Y, Yellow; G, Green; B, Blue; T, Turquoise; Pi, Pink; M, Mauve; P, Purple; R, Red.

Country	Rock type	Landscape	Stream size ii	iii	iv	Cation number	Mean of group
Ireland	Resistant	Hill and lowland	OrY–YB	Y–YB	Or–B	2.25	
Germany	Resistant	Hill	YT(T)	YT(T)	YT,T	3	2.8
Germany	Acid sand	Lowland	Y–T	TM	M	2.5	
Britain	Resistant	Hill	Or–B	OrB,B	YB,B	3.5	
France	Acid sand	Lowland	Y,YG	(Y)G–T	T,TP	2.25	
Ireland	Sandstone	Hill	(Y)B	(Y)B	(Or),B,(BP)	2.25	
France	Resistant	Hill	G,T	(G)T	(G)T,TP	2.5	3.4
Belgium	Acid sand	Lowland	T	(YT),T,(TM)	T(TM)	4.5	
Denmark	Sand (fluvial)	Lowland	OrB(B)	BM,M		4.25	
Germany	Sandstone	Hill	Or–T	T,TM	TM,M	4.75	
Belgium	Acid sand	Lowland	(YT,),T	T,(TM)	M,MP	4.5	
Britain	Coal measure	Hill	(YB,B,(BP)	BP	P?	5.6	
France	Sandstone	Hill	T	T,TP	TP	4	
Denmark	Sand (marine)	Lowland	B	BM,M	M	5	4.9
Italy	Lime (Po)	Lowland	B	B,BM	B,BM	5	
Ireland	Sandstone	Hill	B	B	BP,P	4.25	
France	Sandstone	Hill	T	T,TM	M	6.25	
Ireland	Clay	Lowland	B	B,P		4.75	
France	Limestone	Hill	GT,(T)	(GT),T,(TM)	T,TP	8	
Germany	Calc. moraine	Lowland	T	T,TM	TM,M	7.5	
Britain	Limestone	Hill	B	B(BP)	B,BP	7.75	
Germany	Limestone	Hill	T	T,M	TM(M)	8	5.6
France	Limestone	Lowland	T	T(TP)	T,TP	8	
Denmark	Clay	Lowland	BM	M	M,MP	5.5	
Britain	Sandstone	Lowland	B(BP)	BP	BP	6.5	
Germany	Muschelkalk	Lowland	T	(TM)M	MP	8	
Germany	Eutrophic moraine	Lowland	TM	M	MP?	6.5	
Britain	Limestone	Lowland	B	B,BP	BP,P	7	
Italy	Limestone (Carrara)	Hill and lowland	B(BM)	M		7	
Belgium	Clay	Lowland	TM	(TM)M	MP	7	7.3
Italy	Clay (Po)	Lowland	M	MP		7	
Britain	Sandstone	Hill	B	BP	BP,P	8	
Germany	Sandstone	Lowland	T(TM)	TM	M	8.5	
Britain	Clay	Lowland	B,BP,P	P(PR)	PR,R	8.5	8.5

Source: Haslam (1987a), q.v. for details.

Table 11.4 Downstream eutrophication in macrophytic vegetation
Downstream shift in nutrient status band.

a) All species. Occurrence of species of the two most nutrient-rich (Purple and Red) bands compared with the occurrence of all other species. The table shows the probability of random distribution as estimated by χ^2 analysis between stream sizes i, ii, iii and iv.

	Belgium and Luxembourg[a]	France[a]	Germany[a]
Lowland landscape	0.001	<0.001	<<0.001
Hill landscapes	0.001	<<0.001	<0.001

b) Emergent species. Occurrence of species of the three most nutrient-rich (Mauve, Purple and Red) bands compared with the occurrence of all other species. Notes as for (a)

	Belgium and Luxembourg	France
Denmark (lowlands)[a]	(lowlands and hills)[a]	(lowlands and hills)[a]
<0.001	<<0.001	<0.001

[a] The non-random distribution found is due to species of the Purple and Red bands occurring in downstream parts of rivers.

c) Species in – quite short – acid sands streams.
   These typically show the most change, in short streams.
   Species restricted to upstream reaches include: *Callitriche hamulata, Comarum palustre, Juncus bulbosus, Caltha palustris, Glyceria fluitans* (long-leaved), *Ranunculus omiophyllus.*
   Species typical of middle reaches, often extending up and down: *Berula erecta, Mentha aquatica, Ranunculus* spp. (Batrachian), *Veronica beccabunga, Callitriche* spp., *Rorippa nasturtium-aquaticum* agg., *Potamogeton natans.*
   Species restricted to lower reaches include: *Rorippa amphibia, Myriophyllum spicatum, Potamogeton crispus, Sagittaria sagittifolia, Butomus umbellatus, Nuphar lutea.*

d) Detailed breakdown (France)
   Species distribution in relation to rock type, divided into those showing much ($\chi^2 < 0.1$), medium (χ^2 0.1–0.4) and negligible (χ^2 0.5–0.99) variation.
   Chalk has loess overlay and is more drained, with more river pollution etc., than Jurassic limestone: hence the difference.

χ^2 probability less than 0.1	χ^2 probability 0.1–0.4	χ^2 probability 0.5–0.99
Chalk		
Callitriche spp.	*Berula erecta*	*Apium nodiflorum*
Iris pseudacorus	*Ceratophyllum demersum*	*Caltha palustris*
Mentha aquatica	*Epilobium hirsutum*	*Elodea canadensis*
Myosotis scorpioides	*Polygonum amphibium*	*Glyceria maxima*
Myriophyllum spicatum	*Potamogeton crispus*	*Groenlandia densa*
Nuphar lutea	*Rorippa nasturtium-aquaticum* agg.	*Lemna minor* agg.
Phalaris arundinacea	*Sagittaria sagittifolia*	*Ranunculus* spp.
Phragmites communis	*Sparganium emersum*	*Rorippa amphibia*
Polygonum hydropiper agg.	*Typha* spp.	*Veronica beccabunga*
Potamogeton pectinatus	*Zannichellia palustris*	Mosses
Rumex hydrolapathum		
Scirpus lacustris		
Sparganium erectum		
Small grasses		
Blanket weed		

Table 11.4 contd.

χ^2 probability less than 0.1	χ^2 probability 0.1–0.4	χ^2 probability 0.5–0.99
Jurassic limestone *Callitriche* spp. *Polygonum hydropiper* agg. *Rorippa nasturtium-aquaticum* agg. *Scirpus lacustris* *Sparganium emersum* Small grasses Mosses	*Berula erecta* *Butomus umbellatus* *Elodea canadensis* *Glyceria maxima* *Groenlandia densa* *Iris pseudacorus* *Mentha aquatica* *Myosotis scorpioides* *Nuphar lutea* *Phalaris arundinacea* *Potamogeton pectinatus* *Rorippa amphibia* *Sagittaria sagittifolia* *Typha* spp. *Zannichellia palustris* Blanket weed	*Apium nodiflorum* *Epilobium hirsutum* *Lemna minor* agg. *Myriophyllum spicatum* *Phragmites communis* *Polygonum amphibium* *Potamogeton crispus* *Ranunculus* spp. *Rumex hydrolapathum* *Sparganium erectum* *Veronica anagallis-aquatica* agg.
Sandstone *Callitriche* spp. *Nuphar lutea* *Polygonum hydropiper* agg. *Potamogeton crispus* *Potamogeton natans* *Potamogeton pectinatus* *Ranunculus* spp. *Rumex hydrolapathum* *Sparganium emersum* *Sparganium erectum* *Typha* spp. *Veronica beccabunga* Small grasses Mosses	*Alisma plantago-aquatica* *Apium nodiflorum* *Butomus umbellatus* *Caltha palustris* *Ceratophyllum demersum* *Eleocharis palustris* *Elodea canadensis* *Glyceria maxima* *Iris pseudacorus* *Lemna minor* agg. *Myriophyllum spicatum* *Petasites hybridus* *Phragmites communis* *Rorippa nasturtium-aquaticum* agg. *Sagittaria sagittifolia* *Scirpus lacustris* Blanket weed	*Berula erecta* *Callitriche hamulata* *Epilobium hirsutum* *Groenlandia densa* *Mentha aquatica* *Myosotis scorpioides* *Phalaris arundinacea* *Rorippa amphibia* *Veronica anagallis-aquatica* agg. *Zannichellia palustris*
Acid sands *Berula erecta* *Lemna minor* agg. *Nuphar lutea* *Elodea canadensis* *Myriophyllum spicatum* *Phragmites communis* *Polygonum hydropiper* agg. *Rumex hydrolapathum* *Typha* spp. *Veronica beccabunga*	*Apium nodiflorum* *Callitriche* spp. *Iris pseudacorus* *Myosotis scorpioides* *Butomus umbellatus* *Callitriche* spp. *Myosotis scorpioides* *Nuphar lutea* *Polygonum amphibium* *Potamogeton natans* *Ranunculus* spp. *Rorippa amphibia* *Sparganium emersum* *Veronica anagallis-aquatica* agg. Blanket weed	*Apium nodiflorum* *Caltha palustris* *Ceratophyllum demersum* *Elodea canadensis* *Epilobium hirsutum* *Iris pseudacorus* *Lemna minor* agg. *Mentha aquatica* *Phalaris arundinacea* *Potamogeton pectinatus* *Rorippa nasturtium-aquaticum* agg. *Sparganium erectum* Small grasses Mosses

Table 11.4 contd.

χ^2 probability less than 0.1	χ^2 probability 0.1–0.4	χ^2 probability 0.5–0.99
Clay		
Berula erecta	*Apium nodiflorum*	*Ceratophyllum demersum*
Lemna minor agg.	*Callitriche* spp.	*Elodea canadensis*
Nuphar lutea	*Iris pseudacorus*	*Mentha aquatica*
Polygonum amphibium	*Myosotis scorpioides*	*Phalaris arundinacea*
Sparganium emersum	*Rorippa nasturtium-aquaticum*	*Ranunculus* spp.
Sparganium erectum	*Rumex hydrolapathum*	*Rorippa amphibia*
Veronica beccabunga		*Sagittaria sagittifolia*
Mosses		*Scirpus lacustris*
		Veronica anagallis-aquatica agg.
		Small grasses
Resistant rock		
Callitriche hamulata	*Butomus umbellatus*	*Apium nodiflorum*
Callitriche spp.	*Caltha palustris*	*Berula erecta*
Lemna minor agg.	*Epilobium hirsutum*	*Groenlandia densa*
Nuphar lutea	*Glyceria maxima*	*Iris pseudacorus*
Potamogeton natans	*Mentha aquatica*	*Myosotis scorpioides*
Ranunculus spp.	*Phalaris arundinacea*	*Polygonum hydropiper* agg.
Rorippa nasturtium-aquaticum agg.	*Polygonum amphibium*	*Rumex hydrolapathum*
Sagittaria sagittifolia	*Potamogeton crispus*	
Scirpus lacustris	*Rorippa amphibia*	
Sparganium emersum	*Typha* spp.	
Sparganium erectum	Mosses	
Small grasses	Blanket weed	

Source: Haslam (1987a)

Downstream eutrophication (Table 11.4)

This is the shift to more nutrient-rich habitats and vegetation downstream. It is a widespread, natural process, now usually enhanced by Man's activities. It has a number of causes. Silt is nutrient-rich, so increased silting in downstream parts — as is common — will increase the nutrient status there. Going downstream, the nutrient levels within the silt may increase. They may increase even within a grassland river, but do so much more if there is an upstream nutrient-poor habitat which is lost downstream. This may be acid peat or acid leaf-fall woodland, or a less fertile rock or subsoil type (e.g. Resistant rock above, alluvium, clay or sandstone below). Other eutrophic influences are incoming tributaries from more fertile catchments, and, of course, added nutrients from pollution by effluents, fertilisers, slurry, etc.

There is a slight complication. Water space increases downstream (streams get larger), and various water-supported species of more nutrient-rich habitats increase solely because of that. Table 11.4b shows, from emergents alone, that there is a shift in species downstream. The increased water space over-enhances this effect.

The greatest downstream changes in nutrient status are perhaps those on acid sands with acid leaf-fall forests upstream. Near the source the community is influenced by the nutrient-poor sand and dead leaves. Downstream, there is instead the influence of the nutrient-rich silt which occurs in sandstone. The smallest downstream changes are perhaps in Resistant rock and limestone rivers with but a little downstream change in landscape. Farming enhances downstream eutrophication by increasing both silting, and downstream (solute) nutrients.

Within this overall pattern, species behaviour

differs. Table 11.4c shows species with more, and with less downstream change on different rock types. Drainage and other aspects of channel management lead to more downstream variation. The lowland French limestones are a good example. In landscape and rock chemistry the Jurassic limestone and chalk are similar. The chalk is more intensively managed, particularly in the north. There are, in Table 11.4c, more species showing upstream–downstream variation on the chalk. *Iris pseudacorus* and *Rumex hydrolopathum*, for instance, can grow better along the damper banks of the (Jurassic) limestone streams, and are more restricted to downstream reaches on the most drained chalk.

Human interference, here as elsewhere, has crystallised species assemblages, restricting what is present at a site to those species which are most tolerant of its specific conditions.

Farming

In general, the more intensive the farming, the greater the eutrophication (Table 10.2). Unfortunately many components are involved, non-nutrients as well as nutrient:

a) Removal of acid humus and peat (improved grassland, etc., management).
b) Adding fertilisers, whether to grassland or arable. If ash from burnt stubble is washed into the river, this also adds nutrient.
c) Where livestock are gathered together, such as sheep for lambing, the nutrient output may be sufficient to pollute streams, particularly if there is also surplus, rotting feed. This can happen in spite of an overall stocking rate of about two dairy cows or ten sheep per hectare. This livestock pollution happens much more easily when low-nutrient moors and heaths are converted to fertilised, drained, livestock pasture. The solute content — buffering capacity — of streams in such places is low, and damaging pollution is all too simple. Animals dying in streams likewise pollute.
d) Ploughing. Ploughing increases silt run-off, and the nutrient content of the water. First ploughing (or first for some years) releases much nitrogen and other nutrients into the soil and so also into the river. Deep-ploughing

releases more nutrients to the river than shallow ploughing, and as bare soil is more easily eroded, the duration of bare soil is also important.
e) Removing buffer zones for nutrients. Grassland, particularly if 'unimproved' (low fertiliser and drainage), or wetland by the river acts as a 'sink' for chemicals from the land beyond. Cross-ditches (rather than speed-the-flow ditches), hedges and all other hindrances to the free downstream movement of water decrease eutrophication of rivers, by decreasing the loss of nutrients and sediment.
f) Discharge of silage liquor, slurry, etc., (considered as effluent pollution).
g) Adding biocides.
h) Drainage/irrigation (more by drainage, in most of Europe), involving stream alterations and loss (see Chapter 8).

The farming influence is therefore complex, and it is not easy to separate its different components and assess the effect of fertiliser alone. The task is made more difficult by the presence of the other factors influencing nutrient status: downstream eutrophication, rock type changes, effluents, etc. Vegetation type varies with nutrient balance, particularly cation balance (Haslam, 1987a).

Due to fertiliser and sewage, nitrate and phosphate levels in chalk water are now far above levels of plant uptake (Casey and Westlake, 1974). The vegetation has changed only little, and not in either composition or biomass.

With more intensive farming, river — and lake — macrophytes have declined, and, in parts, declined drastically. It is often difficult to assess the comparative importance of the above factors and only the final loss is known. In northwest France, Mériaux (1982a; 1982b) notes, between 1900 and 1970–80 the loss of, for example, *Potamogeton actutifolius* and *P. gramineus*, and the decline of, for example, *Ranunculus hederaceus* and *Stratiotes aloides*. And Aymonin (1982), also for France, describes regression in over 100 species, most of them nutrient-poor species. For Belgium, Delvosalle and Vanhecks (1982) show alarming decreases, between 1960 and 1980 of, for example, *Potamogeton lucens*, *P. perfoliatus* and *Sium latifolium*, and serious losses in fourteen other species. For British wetlands, aquatic macrophyte distribution was reduced by 60 per cent between 1930 and 1980 largely by loss of habitat (Newbold, 1981).

In the Scottish moorlands, however, where farming has become more intensive, stream quality, including vegetation quality, has dropped appallingly. Moorland streams are, of course, fragile (Chapter 3), so can be changed by much less alteration in land use than is required for solute-rich rivers. Two factors are responsible for the decline. Using high stocking rates and adding fertilisers over much more of the land; and discharging silage liquor and other farm effluents direct to the rivers. Even if there are only two farms in a small catchment, if both are managed intensively, the stream can be ruined. The conversion to grassland is, by itself, not responsible: that would just increase nutrient status, changing some of the macrophyte species, while leaving diversity and cover high (Chapter 10). Instead, the overall change in land use is destroying the life in these Scottish streams (P.A. Wolseley, pers. comm.).

Alluvial plains

Studies in these former wetlands often show clearer patterns of alteration and loss since they are separate, identifiable areas, with more rapid and dramatic change, and so are more likely to attract researchers. Even here, though, it is seldom possible to separate the different effects of farming (plains have more channel alteration and water table lowering, associated with drainage).

Habitat loss in the plains has come from altering wet grassland to arable. Wet grassland dykes have low, gentle banks and a potentially excellent flora of water-supported and emergent macrophytes, with all their associated fauna. Arable dykes are high-banked (Figure 11.2) and often very infrequent. These are less suitable for emergents. In particular, they are less suitable for a diverse emergent community, as the habitat is too uniform, and they may be unsuited to a good water-supported community through being either too shallow (Figure 11.2) or too deep (given that these dykes are usually turbid with algae and eroded soil, in deeper water lack of light reduces submerged macrophytes). Habitat loss has also come from the partial replacement of cutting and dredging by aquatic herbicides.

Modern farming removes many dykes, so

decreasing aquatic habitat. A third of ditches were lost by 1979 in the Ouse Washes (Grose and Allen, 1978), and the same between 1973 and 1981 in part of Broadland (Driscoll, 1983) (both England).

The effects on vegetation are shown in Tables 11.5, 11.6 and 11.7. Dyke diversity and quality increase with the duration of stable management, and arableisation causes eutrophication and decline. The effects are not restricted to macrophytes, though. Waterfowl are reduced by replacing many and diverse channels with a few featureless straight uniform ones (Baldock, 1984). However, to look on the bright side, Palmer (1984) found only minor decreases in invertebrates, in some arable ditches (Pevensey Levels, England). (See Chapter 6 for the Somerset Levels research, identifying the effects of different facets of management.)

Purifying influences

Stream macrophytes purify water (Chapter 3). Most research has been done on emergents (particularly *Phragmites communis* and *Scirpus lacustris*), but, to a greater or lesser extent, all stream vegetation purifies. Purification is surprisingly rapid when rivers run through little-managed grasslands and peat (for example, River Luznice, see Chapter 6). The Moors River, England, shows much quicker recovery in vegetation (diversity) when meandering through unimproved pasture than in more-used land. Purification here is also aided by cutting vegetation (weed control) from the middle of the river. Polluted silt from upstream can therefore be washed through, rather than accumulate (P.A. Wolseley, pers. comm.). (Pollutants accumulate in silt, see Chapter 4.)

An important cause of the recent general increase in river pollution is that natural purification is hindered by reducing stream vegetation itself (drainage, channelisation, weed control), and by adding farm pollution from much-managed land along the streams — as well as effluents. The streams are increasingly overloaded with pollutants.

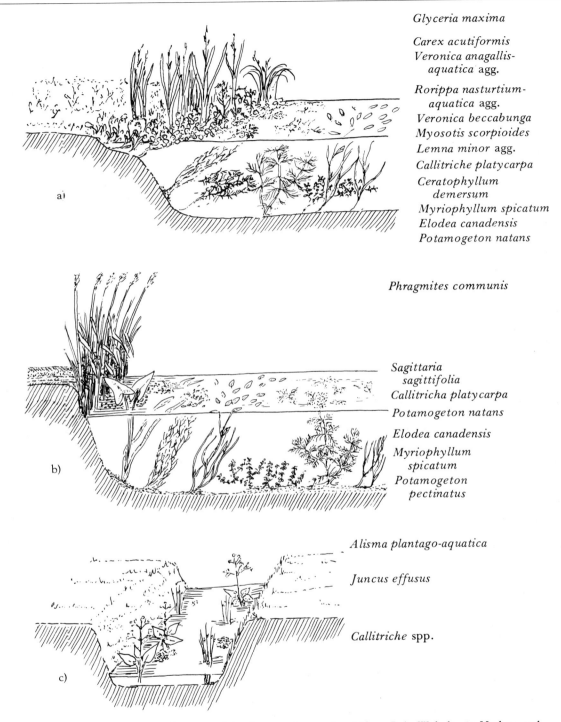

Glyceria maxima

Carex acutiformis
Veronica anagallis-
    aquatica agg.

Rorippa nasturtium-
    aquatica agg.
Veronica beccabunga
Myosotis scorpioides
Lemna minor agg.
Callitriche platycarpa
Ceratophyllum
    demersum
Myriophyllum spicatum
Elodea canadensis
Potamogeton natans

Phragmites communis

Sagittaria
    sagittifolia
Callitricha platycarpa
Potamogeton natans

Elodea canadensis
Myriophyllum
    spicatum
Potamogeton
    pectinatus

Alisma plantago-aquatica

Juncus effusus

Callitriche spp.

Figure 11.2 Dyke vegetation in relation to dyke shape and water level (from P.A. Wolseley in Haslam and Wolseley, 1981). (a) Gentle banks, adequate water, good vegetation. (b) Over-steep banks, adequate water, good water-supported vegetation, poor emergents. (c) Too little water, over-steep banks, sparse emergents.

Table 11.5 Flora of drainage channels in the
IJsselmeer polders in the Netherlands
Based on surveys of about 60 sites. a, abundant; d,
dominant; f, frequent; o, occasional; r, rare.

Species	Oost-flevoland	Noordoost polder	N.E. of Noordoost polder
Agrostis stolonifera	f	o	o
Berula erecta	—	f	—
Butomus umbellatus	—	—	o
Callitriche spp.	o	—	o
Ceratophyllum	—	o	a
Elodea canadensis	—	o	o
Elodea occidentalis/ nuttallii	f	—	o
Glyceria fluitans/ plicata	o	o	o
Glyceria maxima	—	—	a
Hydrocharis morsus-ranae	—	—	o–f
Lemna minor	—	f	a
Lemna trisulca	—	—	o
Nuphar lutea	—	—	o–f
Phragmites australis	d	d	o–f
Potamogeton crispus	o	—	—
Potamogeton natans	—	o	o
Potamogeton pectinatus	a	o	f
Potamogeton perfoliatus	o	—	—
Sagittaria sagittifolia	—	f	f
Scirpus maritimus	f	—	—
Sparganium erectum	—	f	f
Spirodela polyrrhiza	—	f	f
Typha latifolia	o	o	r
Wolffia arrhiza	—	—	o
Mean (and max.) no of species per site	3(6)	4(6)	6(9)
Main drainage	1957	1942	earlier

Source: Haslam (1987a)

Sediment erosion

(Incoming sediment from the land contains
chemicals: for this effect see acidification,
eutrophication, and purification above, and
poisonous run-off, below.)

Sediment always has, and always will, erode
from the land by natural processes, and be
moved by rivers down to its resting place in the
sea. Pollution by sediment is therefore accen-
tuating an existing process, not creating a new
one. But the accentuation may be dramatic. The

Table 11.6 Effect of arablisation on dyke vegetation
Pevensey Levels, England

	Old pasture 17 ditches	2 drains	Arable land 18 ditches
No. of macrophytes in 20 m length	20.5 (S.D.2.9)	20	13.7 (S.D.3.7)
Vegetation cover in 20 m length (%)			
Submerged	11 (S.D.17)	65	34 (S.D.28)
Floating	52 (S.D.7)	13	39 (S.D.36)
Emerged	38 (S.D.22)	20	41 (S.D.24)

Source: Palmer (1986)

removal of trees, for instance in Nepal for fuel
for the increasing population, causes horrific
floods and silting downstream in Bangladesh.
Europeans know such disasters principally
through television. A bad situation is, however,
developing in the Alps.

For centuries, the Middle Alps have been
cleared and settled. Good stable grassland and
enduring small settlements have developed. No
problem. The High Alps, though, had minor
summer pasture, and the occasional climber and
skier. In the nineteenth century hardy moun-
taineers came for enjoyment, but still in small
numbers. Use was little, and kept within safe
limits.

Alas now for the High Alps (Figure 11.3), no
longer inaccessible to mass population movement!
The habitat is fragile. The avalanches — a sign of
instability — were well known, but the general
fragility was not. Indeed it still is not. It should
have been, since all highly stressed systems are
fragile, and the altitude plus the steepness of the
landscape create such stress. Erosion results —
from drainage, and from disturbance: both for
recreation. Once vegetation cover is broken,
erosion soars.

Change of land use has meant swifter, and so
higher-force, run-off. Drainage channels have
increased. This means that water is more able to
pick up sediment, and that, as erosion progresses,
there is increasingly more bare soil to be eroded.
Flash floods progressively increase, and
progressively increase erosion. Affected streams
may be characteristically turbid grey (France)

Table 11.7 Arablisation, conductivity and dyke macrophytes, Pevensey Levels, England

Habitat	Conductivity (approx.) (μmhos)	Primary species
1. Arterial drains, intensive land management	680	*Lemna minor, L. polyrrhiza* (*Ceratophyllum submersum, Potamogeton trichoides*), and others
2. Arable, widened channels, etc.	2400	*Alisma plantago-aquatica, Sparganium erectum* (no *L. polyrrhiza, Hydrocharis*) and others
3. Pumped area. Deep, narrow. Species-rich channels.	600	*L. trisulca, L. minor, Hydrocharis morsus-ranae, A. plantago aquatica* and others
4. Gravity drained channels, wider than (3) above. Species-rich	680	*Potamogeton natans, H. morsus-ranae, Sp. erectum*, (*P. acutiformis*) and others
5. Gravity drained, minor channels, farmer-cleaned. Most species-rich	800	*H. morsus-ranae, L. minor, A. plantago-aquatica, Sp. erectum, Bidens* and others
6. Medium. Few open-water species	915	*Sp. erectum, Glyceria maxima, G.* spp, *Phragmites communis* and others
7. Narrow, neglected	1000	*Phragmites communis, L. minor*, etc.
8. Narrow, dry, or shaded, summer-dry	–	*Solanum dulcamara, Oenanthe aquatica* (*Agrostis stolonifera, G.* spp.)

Source: Glading (1986)

(Figure 11.4), deficient even in the little vegetation which, in similar but better conditions (for example, in Corsica; Table 11.1) would be there. Frequent hydroelectric dams hold up water — but do so much too far downstream to stop erosion. Similar erosion occurs elsewhere from recreation as in the Scottish Cairngorms, from skiing.

In areas liable to sudden severe storms, such as Malta and Madeira, large quantities of soil are washed to sea (Figure 11.5). The loss of the topsoil, the most fertile soil, in a country short of soil, as well as of water, is regrettable. In Malta (annual rainfall about 80 cm) dams are frequent, up to every 50 m or so (Figure 8.12). This conserves water — and also silt, but silt dredged from the river is unfortunately not soil on the land (see Haslam, 1989).

Civil engineering works can create much sediment, usually only temporarily (for ploughing, see above and see Table 3.1 (base)). Small differences in land use may make immense differences in sediment loss to the river. This is especially so when the fine line which separates stable land from that which is unstable and eroding, is crossed.

Polluted ground water

Pollutants may and do enter the ground water and so pervade the whole freshwater system. They are slow to get in, and even slower to get out. Nitrates are there already, in EEC farmland aquifers. In absolute concentrations, more nitrates go on the land than do other nutrients or biocides, so it is not surprising, that nitrates were reported first and most. Nitrates are less held and less broken down in the soil than other fertilisers. As they are readily released, they are also the most affected by ploughing. They are also measured often, so long-term records are frequently available.

Contaminated ground water has been much studied in the Netherlands, for example, with

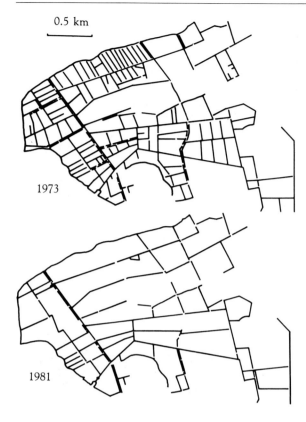

0.5 km

1973

1981

Figure 11.3 Loss of dykes on arablisation, Norfolk, England. 1973 on grazed marshland, 1981 on cereal land (from Driscoll (1983)).

nitrogen, phosphate, suspected biocides (Steenvoorden, 1976), solid wastes (Hoeks, 1976), heavy metals (Hoecks, 1977), and sewage (de Haan, 1976). The Maas and Rijn pollute their lower plains. Table 9.5 shows decreased dyke macrophyte diversity in this plain, possibly due to this pollution. Salt water can contaminate ground water near the sea (by incursions, or by uprising after excess abstraction), altering both invertebrate and macrophyte populations (*Potamogeton pectinatus*, *Phragmites communis*, *Scirpus* spp. and *Myriophyllum spicatum* are encouraged).

In the Rhine flood plain in Alsace, polluted water from the Rhine moves into the ground water. This movement and pattern can be monitored by the macrophytes in surface channels (Carbiener and Ortscheit, 1987) (Table 11.8).

The clean water community is fragile, lime-rich but solute-deficient. It is therefore easily altered by pollution. The 'worst' community is still species-rich but its characteristic species are ones tolerant of high solutes. The communities can be linked to ammonia and phosphate levels, each community being separated by some 20 ppb. The spread of the polluted communities, and the increase of measured contaminants, have been watched and measured. There is no doubt about this: contaminated water is there, spreading, and altering macrophytes. Krause *et al.* (1987) note the decline of plants, such as *Hippuris vulgaris*, *Chara hispida* and *Hildenbrandia rivularis*, in channels in the east of the Rhine vale.

In the Rhine Vale the Rhine river itself is very polluted, and the water flowing from it can be followed. Elsewhere it may be difficult to separate the effects of pollution in the ground water from those of the intensive farming with which it is often associated. The Rhône is cleaner, and its vale has decreasing diversities and rare species in the channels: from which or both causes?

Poisonous run-off

Silt, nutrients, etc., pass from land to river anyway, human interference merely altering quantity and balance. Other substances reaching the rivers would not do so at all without Man's intervention.

Motorways, for instance, have as much and as bad run-off as comparable urban surfaces and much worse than villages — petrol rubber, heavy metals, salt, spills, etc. (see Chapter 6). These run through the country, forming part of the rural run-off: they are not localised like the towns. And their run-off does not go into the sewage treatment works, as does urban drainage water in the best-purified towns (though settling ponds may be present in recent constructions).

Figure 11.6 shows the effect: the ditch by a minor road is a fragile nutrient-poor habitats. It is not, therefore, the presence of a road which does the damage. It is the presence of a busy motorway. Here, its run-off collects in a settling lagoon. This, like the polluted Rhine Vale channels described above, bears a good diversity of macrophytes: but the species are those tolerant of general pollution (e.g. *Sparganium erectum*, Blanket weed) or encouraged by salt (e.g. *Scirpus*

Figure 11.4 Erosion in the High Alps, France. Ski and summer tourist damage. Below, Alp stream, grey–turbid with suspended particles, water level — discharge — fluctuates considerably.

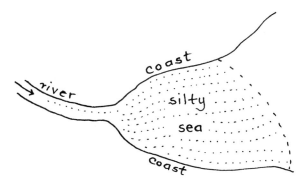

Figure 11.5 Silt pollution of sea, Malta, after heavy rain.

maritimus). When this solute-rich water discharges into the solute-poor stream, vegetation is sparser than expected, though it is at least still mesotrophic in nutrient regime. The next stream along shows what the vegetation should be: mesotrophic and abundant — in the absence of the motorway run-off.

Biocides applied to the land reach the river (see chapters 4, 5, 6). One must suppose they are slowly, but ever more surely, penetrating down to the ground water like the fertilisers. They are more usually tested for breakdown and adsorption than for penetration — which is in very low concentrations. What do they do in the rivers? This

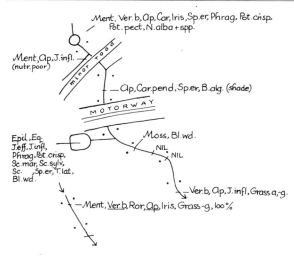

Ment, Ver.b, Ap, Car, Iris, Sp.er, Phrag, Pot.crisp.
Pot. pect, N.alba + spp.

Ment, Ap, J.infl.
(nutr.poor)

MINOR ROAD

Ap, Car.pend, Sp.er, B.alg. (shade)

MOTORWAY

Epil, Eq.
J.eff, J.infl,
Phrag, Pot.crisp,
Sc.mar, Sc.sylv,
Sc.   Sp.er, T.lat,
Bl.wd.

Moss, Bl.wd.
NIL
        NIL

Ver.b, Ap, J.infl, Grass a,-g.

Ment, Ver.b, Ror, Ap, Iris, Grass-g, 100%

Figure 11.6 Pollution by motorway, Seabrook, England.
The drainage ditch of the minor road bears middle-
nutrient species with colouring/contraction indicating
nutrient deficiency. Middle-nutrient to motorway.
Tolerant species in lagoon with motorway run-off.
Recovery slow and stream to left (Saltwood) shows
good vegetation, in habitat similar except for pollution.

is less easy to answer. Chronic low-level effects
may include such things as inhibiting germina-
tion. Major losses of river vegetation are
probable. It is widely reported that in the
American Midwest excess herbicides and excess
silt reach rivers, and that the lack of
macrophytes is due to both. In Michigan and
Ontario, the writer noted that in intensively
farmed areas, streams were apt to bear mainly or
only tough edge emergents. Where there was a
buffer strip of wetland by the stream, though,
delicate species were more likely to be present,
and diversity was higher. This strip would reduce
chemical input to the river. (Species possibly
resistant to biocides included: *Alisma plantago-
aquatica*, *Nuphar advena*, *Peltandra virginia*,
Pontederia cordata, *Typha* spp. Species possibly
sensitive to biocide included: *Callitriche* sp, *Chara*
spp, *Heteranthera dubia*, *Potamogeton foliosus*, *Wolf-
fia arrhiza*.) (Also see Purifying Influences, above.)

At the other extreme, Luxembourg rivers were
surveyed to see whether the recent spread of
maize was damaging river vegetation. Not the

Figure 11.7 Damage from recreation.

Table 11.8 Effect of Rhine ground water pollution on macrophytes
Alsace. Water reserve in gravel of floodplain. Ground water rivers. Exchange of polluted Rhine water and ground water. Macrophytes detect and assess contamination.

Communites:
A. Clean, oligotrophic
    Potamogeton coloratus, Juncus subnodulosus, Mentha aquatica, Berula erecta, Chara hispida, Agrostis stolonifera, and others

B. *Mentha aquatica, Berula erecta, Agrostis stolonifera, Phalaris arundinacea* and others

C. *Berula erecta, Callitriche obtusangula, Veronica anagallis-aquatica, Potamogeton friesii, Sparganium emersum* and others

D. Eutrophic
    Berula erecta, Callitriche obtusangula, Veronica anagallis-aquatica, Elodea nuttallii, Nuphar lutea, Zannichellia palustris, and others

E. Most polluted and eutrophic. Impoverished *Ranunculus fluitans* community.
    Berula erecta, Callitriche obtusangula, Potamogeton friesii, Sparganium emersum, Potamogeton crispus, Hippuris vulgaris, Ranunculus fluitans.

Groundwater near Rhine is grade D. The Ill floodplain is cleaner and A, B or B/C. Clean-up is needed.

Source: Carbiener and Ortschiet (1987)

maize itself, of course! (It was merely replacing other arable crops.) Atrazine was used on the maize, and this was feared dangerous, especially as much is applied near the river. The surveys were both before (1978–80) and after maize increased (1985) (Haslam and Molitor, 1988). No evidence of herbicide pollution emerged. Many sewage and industrial effluents reach the rivers. These alone are enough to account for the polluted vegetation found: the species and the cover are consistent with this interpretation.

There was also no effective change in river vegetation with the introduction of atrazine. Given the degree of existing pollution, any minor effects of atrazine would be masked. They might show if the effluent pollution were reduced. Alternatively, after many years atrazine could penetrate through soil to the streams themselves in sufficient quantity to affect the vegetation. At the time, no effects of increased atrazine could be detected in the vegetation.

Finally, to illustrated unexpected dangers of biocides, an instance in Canada will be described (Doarks, 1980). *Myriophyllum spicatum* arrived, thrived, and harmed recreation, fishing, etc. Tourism declined. What was to be done? 2, 4-D was used to kill the plant. This was followed by cries of mutagenesis, teratogenesis, carcinogenesis — and tourism declined further. Genuine effects were fish kills, and decline of local farming of crops grown for tourists.

Recreation

Recreation is a rapidly-expanding activity, with the increasing wealth and leisure in Europe. It is a great pity that so few wish to spend their leisure improving the environment. There are far too few to counteract the hordes that seek to destroy, directly or indirectly (Figure 11.7). Firstly, there are those who throw pollutants — anything from old sandwiches to petrol — into, or near, the water. This direct pollution ranges from trivial to severe. Recreation away from the river causes increases in waste disposal, land erosion and river silting; increased dirty and poisonous run-off from built-over surfaces (roads, recreation areas, car parks); and perhaps flash floods.

Immediately beside the river, recreation increases disturbance (bad for birds, mammals, etc., generally), bank erosion (loss of bank structure, plants and animals), and bank lining (when erosion has gone too far. This is ecologically even worse than erosion!)

Within the river, there are water sports, ranging from water-skiing to bathing. Disturbance, bank erosion, perhaps piling, and minor pollution result, disturbance here harming plants and animals within the water, as well as those beside it. Paddling and bathing denude the stream, fishing can leave damaging impedimenta behind

(Chapter 9) and cause disturbance. (There are usually agreements intended to prevent fish removal being enough to damage populations; see also Chapter 6.)

It is always possible for a river in a derelict state to be improved as part of a development for recreation. But too often one kind of disaster is merely exchanged for another. Figure 12.3 is not ugly to the human eye — but a good-quality river it is *not*!

12
Conclusions

We can command Nature only by obeying her.
F. Bacon

Be not deceived, God is not mocked: for whatsoever a man soweth, that also shall he reap.
Galations 6:7

If you drive Nature out with a pitchfork, she will soon find a way back. Horace

The fathers have eaten sour grapes, and the children's teeth are set on edge. Ezekiel 18:2

In his finest image, man cannot use wastefully the resources of the earth or destroy their beauty.
Warren (1971)

Man cannot, at the present rapid rate of change, depend on chance to lead to desirable circumstances. Warren (1971)

(Nature among the mountains is too fierce, too strong for man. He cannot conquer her, and she awes him.)
The lowlander ... is stronger than Nature:

and right tyrannously ... he lords it over her, clearing, delving, dyking, building, without fear or shame. He knows of no natural force greater than himself, save an occasional thunderstorm ... Why should he reverence Nature? Let him use her and live by her ... With the awe of Nature, the awe of the unseen ... the sense of the beautiful dies out in him more and more ... He may ... remain for generations gifted with the strength and industry of the ox, and with the courage of the lion, but, alas! with the intellect of the former and the self-restraint of the latter.
Charles Kingsley

Rivers grow and develop, like living systems — and many of their component parts are animate. Like living systems, they change and respond to each influence upon them.

These are the themes of this book, and indeed the emphasis is needed only because the 'soon' of Horace's axiom has, by modern technology, been extended to decades or even centuries, so it has been forgotten or ignored: until the children's teeth are set on edge, by rivers unfit for bathing, by fish unfit to eat. Intentions are irrelevant. All know a dearly-loved puppy treated with ignorance can suffer as much as one treated with sadistic cruelty. Intent does not affect results: actions do that. The earth was so created and developed that, if the balance is upset (by such as channelisation or dieldrin application) a train of events has been set in motion. This, if left, continues in the sequence preset by the physical, chemical and biological attributes of the river. If the perturbation is sufficient to overcome the natural resilience, the system will be damaged, as has been happening, and as is, to a large extent, unnecessary. Man and the horse formed a good partnership, from which Man benefited. The horse flourished in population and diversity. The horse, though, unlike the river, is entirely animate and died if maltreated. While the horse was so necessary to Man, the horse's well-being was a prime (if sometimes inadequate) consideration. Such a partnership should occur between man and the river.

Pictures and descriptions mark the changing relationship of man and river. In the, say, fifteenth century, the Hall, its cottages and crops, formed isolated pockets of settlement and land use. The charm of the river here is contrasted with the hostile world beyond, with the unharnessed river into which the adventurous mariner perils his life. Later, Man extended his range and worked with the river. Eighteenth-century rivers are portrayed with Man central — Man overlooking, Man-altered, and so on. Here we have local domination. Cowper's eighteenth-century 'I am monarch of all I survey, my right there is none to dispute' was written for a man on a desert island. In England, the small-scale pattern meant there were usually plenty to dispute. In the late twentieth century the river is no longer tamed, as a horse is tamed — it is caged, as a tiger is caged. That a caged river may be the more dangerous is just becoming visible: heavy metals, loss of topsoil, flash floods. There is no partnership with the river. It is Man's: like his car. The more enlightened set up nature reserves. Nature no longer inhabits the earth, she is confined to reserves. Schoolchildren do not play by and with streams (in the villages where disturbance is great anyway), they are taken to

experience a stream (and they probably destroy part of it, by for instance removing some invertebrates). Finding a good — a moderately good — stream may become, as in lowland Belgium or central Czechoslovakia, a matter for a specialist. Thus is heritage lost. The article is too rare to be part of ordinary life, so its complete loss is not generally noticed.

Pollution has now spread to affect practically all lowland streams, and most highland ones. This is a twentieth-century — nay, late twentieth-century — development. Earlier, although rivers by towns, mines, mills and factories could be dreadful, those in more remote places were mostly clean. With no piped water, much less water was used, and therefore there was less waste to be disposed of, there were few consumer goods, constant fires, and a need for manure, all factors decreasing waste in the river. Outside the main towns, few houses used the river for waste disposal anyway. The few small roads produced negligible dirty run-off compared to the numberless hectares of the present tarmac and concreted, etc., areas. Farming pollution now affects nearly all the lowlands and much of the highlands. And the earlier waste was much less dangerous anyway. Have readers ever counted the number of solids and liquid chemicals disposed of in their houses and places of work, other than food remains, water and sewage? The total is staggering, in both number and quantity. And even the water from houses may not be safe for all river life, since the chlorination to render it bacteria-free for Man may end with derivatives which may also harm the natural river bacteria, etc. Drainage damage is also near-ubiquitous. Sediment pollution has vastly increased.

In Britain, in 1989, there is for the first time some acceptance that the parts of the earth are related: that the loss of the ozone layer may cause cancer, that the loss of tropical rain forests may submerge coastal cities. In the past, when population was small and technology man-powered, such knowledge was not essential. Damage was local and did not interact. The North Sea was enormous and clean. It was inconceivable that it could be polluted.

Effort alone does not create good conservation — or indeed good anything. It must be rightly directed. We will put many thousands of pounds into conservation. 'Haven't we done well?' Probably not. Add 'and we will employ an

Figure 12.1 The good river (P.A. Wolseley).

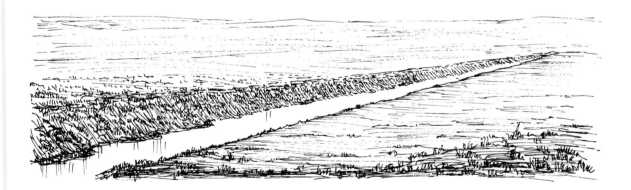

Figure 12.2 The good river as the drainage engineer sees it?

Figure 12.3 The good river as the planner sees it?

Figure 12.4 The good river as the farmer sees it?

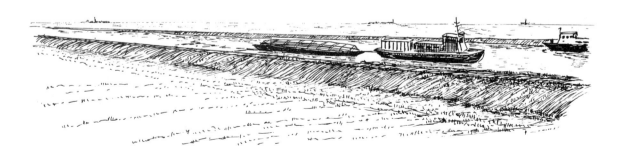

Figure 12.5 The good river as the navigator sees it?

Figure 12.6 The good river as the conservationist sees it?

Figure 12.7 Nobody's river. Compare this lifeless scene with Figures 2.1–2.4. Which gives the best life and is most likely to nurture excellence?

Figure 12.8 The good river as the environmental manager sees it?

engineer and an ecologist. Won't that be superb?'
Not necessarily. As one star differs from another
in glory, so all know one garage differs another
in efficiency. What sort of environment does the
environmental manager manage towards? that of
Figure 12.8? It is, after all, a well-managed
environment. Few make the link that the title
'conservation officer' does not, by itself, convey
omniscience — or even the ability to conserve.
Most plumbers can stop a leak for the time
being. Is this particular conservation officer doing
more? However skilful and well intentioned, how
can expert conservationists re-create a habitat of
whose existence they are unaware? They can only
try to create the second- or third-rate that is the
best they have seen.

Few make another link. That the man who
washes his dirty and salty car, particularly if
using detergent, and does not ensure that the
water goes to a sewage works, has no right to
complain of the effluent from the local factory.
First do that which one has the power to do.
(One car-wash is little — but 10 000 every week?)

When damage creates gaps in the ecosystem,
nature, which abhors a vacuum, fills it — unless
it is at the time, too toxic, in which case nature,
however slowly, restores it. Get rid of acceptable
vegetation, nuisance plants will grow. Get rid of
these, and the sterile poisonous water will do
harm somewhere else.

There is now a slight realisation that we can
command nature only by obeying her. That if
farming puts high biocides into rivers, people will
fall sick, that if daffodils are grown with dieldrin,
Cornish trout are not fit to eat. What is still
needed is a full realisation that — irrespective of
whether Cornish trout appear on the table —
trout should not be poisoned for Man's convenience.

So what should be done? Most of the responsibility lies with the town hall and water
authority — but the underlying responsibility is
with those who put the officials there, and do
not disagree with their actions. In the town hall
and water authority, there are planners and
engineers by the score: biologists by the unit.
Inevitably the ecological expertise and ecological
influence are trivial compared to those of the
other disciplines. What happens? Put in a fish
ladder, for the salmon to pass. Put a gentler
slope, to allow local emergent vegetation. These
are the sticking plaster, the First Aid for rivers.

Desirable, but of little use where a wrong needs
to be righted. And where more is done, what
actually happens? In Figure 12.3 the recreation
river — all neatly designed by planners — will be
familiar to most readers. Is it not pretty? The get-
out-of-my-way river, nobody's river (Figure 12.7)
is obviously a planner's dream, or it would not
be spreading so fast across Europe. The navigation river (Figure 12.5) — well, what else could be
expected? The engineer's river (Figure 12.2) —
who has not seen plenty? The farmer's river
(Figure 12.4) here has cows. Elsewhere it may
have ducks and geese, with less damage to banks,
though often equal damage to flora and fauna.
The conservationists' river is better. But how far
it is from the good river (Figure 12.1, part Frontispiece)!

Why should this be? Why are good rivers not
even aimed for? Basically, because research has
been directed to First Aid (First Aid is cheaper!).
That river is a mess. Find out why. Then clear
it up. But clear it up to what? To be like the
river over the hill which, on any absolute standard is deplorable, but is better than the dreadful
one? By sitting in an office and thinking — with
no data base of good rivers? How many conservationists and river researchers have even seen, let
alone studied, a river in healthy condition? Even
the writer, with a data bank of some 33 000 site
records, has seen only a few lowland ones, in
Europe, and has done intensive study on none.

Good rivers do still exist, rivers with good flora
and fauna. They need investigating in relation to
their damaged counterparts. Study of Corsican
rivers would be helpful in recommending
improvements to those of the Alps or Pyrenees,
but would hardly benefit the streams of lowland
England or Germany. The best relevant should
be found, and studied. Somehow, even in
Europe, the occasional good or fairly good
lowland river has survived — a few in Denmark,
Scotland and Ireland come to mind. In North
America, good rivers abound — but the researchers do not, so the position is equally bad.

What is a good river like? The physical shape,
the texture and pattern? The stability? The water
regime, day to day, season to season, and year to
year? What are the microclimates? How do they
vary? What is the chemical regime of water and
substrate, and how does this vary? How does it
differ from other 'clean' rivers? How much extra
can it take, of what, before change occurs? What

species are present and in what habitat niches? In what balance — for all species from *Ranunculus* and diatoms to perch, *Gammarus* and rotifers? How do they behave and change seasonally? What affects reproduction and development — what is required to maintain breeding populations, in terms of river structure, food, lack of disturbance, lack of pollution, lack of predators? What checks and balances occur? Over what time-scale do larger fluctuations occur? It is probable that if the groups larger in size, and higher up the food chain, are satisfactory, then the others are also. What land use is required beside the river? Little-managed along much of the length, undoubtedly. But how little management? Extending how far from the stream? Over what proportion of the length? This is what should be studied. Freshwater biologists and institutes please note!

The infrastructure and suprastructure of good rivers are unknown. How, then, can these be planned for in other rivers? Remember, please, that it is as easy to produce the recreation river as the conservation river. Both take equally more time than the engineers' river. Why should something closer to a good river take any more time or money? (Dethioux, 1989a,b; lists suitable trees and bank fringe species for planting, with the planting positions.)

Healthy ecosystems are buffered against considerable perturbation. It is those already stressed which are most vulnerable. The closer a river is to a natural river, the more it can take, without suffering. Even the conservationists' river here is fragile — not for the main macrophytes or the coarse fish — but for the totality of the rare species and the variability which make up the natural river.

It is the will, not the technology, which keeps our rivers bad. Laziness in the individual: how easy to spray the garden and let the biocide leach to the ditch: how difficult to weed. And laziness in the authority. The inertia described in Chapter 2 in dealing with rivers is still with us, and likely to remain. To pollute, and to make uniform, is easy and cheap. To provide clean water, and good river structure, require imaginative thought, energy, perhaps sacrifice, and certainly money. Who would pay an extra tax for the rivers? (The Icelanders set a splendid example by their voluntary tax to plant trees. Who can follow it?) Despite enormous EEC food surpluses, farmers are still draining wetlands, further draining rivers, and

putting chemicals into the rivers. Despite previously unthinkable wealth, EEC industrialists are putting vast amounts of waste into the rivers — previously unheard-of chemicals are entering, and previously known ones are entering over wider ares. The waterworks have not changed with the change in pollutants. But then, why should they? Waste products are what Man has used, or by-products of what he intends to use. Why should this be 'waste' at all? It should be recycled, reused, and not be allowed into the rivers at all.

What an impossible idea! Impossible, impractical, and astronomically expensive.

Except that an equally big upheaval has already been done, and what Man has done once, he can do twice. A century and a half ago there was little understanding of water-borne disease, or of sewage treatment. Who could have foreseen that bacteria-free water, of reasonable taste and colour, could and would be supplied as the standard in Europe, that every inland town, and most villages, would treat their sewage, pipes entering houses to bring clean water and remove dirty? That was equally unthinkable and impossible. Perhaps even more impossible, as no such thing had been done before, on such a scale. It is easier, the second time around. The nineteenth-century clean-up of drinking water had to remove solids, unpleasant tastes, and bacteria. Now there is a new and vast range of toxic compounds. Microbiological pollution is negligible, chemical is now important. But need the chemicals be put there in the first place?

To improve river ecology — and this book has shown how much room there is for improvement! — ecologists must know what to aim for, how and why. This comes from the study of good rivers. Even within the existing framework, much more could be done. And this framework should be changed. Water quality standards, as part of EEC law, are improving: when will all EEC rivers meet these standards? When they do, they will not be clean, merely better. And the standards now apply to only a handful of added chemicals. When will others be controlled? There are no standards for river structure. These, of course, should not be necessary, as but few organisations are involved — those managing banks and channels can carry out what they propose, in a way which does not apply to water quality. Also, the chemical standards are applied equally to all watercourses. Downstream reaches of clay streams can carry pollution to EEC standards without major harm, while such water could perhaps destroy French *Potamogeton coloratus* communities. This is not yet recognised in such laws. Ecologists' ignorance, often even wilful ignorance, must take partial responsibility. When Man can predict reliably he can plan his activities intelligently (Warren, 1971).

There is hope. What the public sees as wrong, will be changed, however slowly. Public opinion, over the past two decades, has increasingly favoured rivers. This favour is, however, negative: on what a river should not be — it

should not be without trees, it should not be grossly polluted, it should not be without fish, etc. Rivers need more than this to be good. Opinion must be guided to the positive aspects of good rivers (Figure 12.1 part Frontispiece). If the ecologists are not ready, the planners have shown they are not just ready, but willing and able — and a river like those in Figures 12.3 or 12.8 could become the accepted one! And how long will it be before the ecological results of hundreds of kilometres of that set the children's teeth on edge?

Appendix 1
Early river vegetation

The British river vegetation studied by Butcher (1927; 1933) in the 1920s and 1930s was of better quality than that found now. It seems likely that the same applies to rivers in other populated, non-mountainous lands in the EEC. In well-populated areas it seems probable that during the first half of the twentieth century stream vegetation was of better quality than occurred during the previous thousand years. Use for power, transport and livestock had declined, use for intensive drainage and modern pollution had not started.

Of necessity, settlements were based on fresh water. Realistic river scenes are portrayed in medieval illuminations and woodcuts, in landscape paintings from the seventeenth century, and in etchings, engravings and early photographs, etc. Details are filled in by old records and river descriptions.

Early lowland rivers were much more winding and less even-sided (less straightened) than rivers today. They were bank-full of water (valleys undrained), with a rounded grazed bank or, in swifter or more populous parts, a very steep edge (neither resembling the typical dredged bank outline found today). Small brooks could bear much vegetation, tall monocotyledons being shown the most often. Streams over c.3 m wide,

however, typically had little or no vegetation, the commonest macrophytes being occasional patches of emergents. The detailed pictures of populated places often show these confined to sheltered indentations, or occurring in a 'damage form' near wharves — showing the pictures were reliable. Obviously, in regions little affected by Man, plentiful vegetation must have occurred, it is just that these have seldom been portrayed.

River straightening increased considerably from the seventeenth century and drainage from the nineteenth century. During the nineteenth century, river vegetation increased. *Glyceria maxima* is the most palatable of the common tall monocotyledons, and, interestingly, it is the only one not portrayed frequently until the late nineteenth century in the paintings, photographs, etc.

In early centuries, most people worked on the land, and the numbers working on, or in association with, the water were much higher than now (e.g. a river scene which might now have at most one farm worker would earlier have perhaps up to a dozen people).

The lack of vegetation was due to both short- and long-distance uses. Either could alone prevent vegetation. For instance, at least from the seventeenth century, streams of 3–6 m wide could have plentiful macrophytes in openings in woods (where stream use, such as by transit horses, was light) but none where there were farming and village uses of the streams. Short-distance uses included grazing of edges, watering of people and domestic animals, geese flocks, ducks, laundry, bathing, fords (bases laid down, sometimes bends inserted), ferries, waterfronts and harbours, small boats for within-farm and within-village transport (even in water c.20 cm deep), cattle being moved between fields, etc. and waste disposal. In general, there was markedly less pollution than now occurs. Where population was low or (as by water meadows) easy access was less, damage was also less, and vegetation more. Village density must have been very important. Another source of damage in shallow streams (or those intermittently shallow near mills) was the passage of carts and ridden horses, where the stream bed formed a better surface than the road.

Mills, swans and freight navigation affected rivers over long distances (mills often occurring every few miles). Mills, especially those handling wool or leather, and those on small brooks, caused much pollution. They also caused damaging fluctuations in water regime (where intermittent controls were placed across the main stream). Above the mill the water level varied. Below the flash-lock both level and water force varied, the water force being intermittently great. From the seventeenth century the improvements (e.g. the spread of pound locks) gradually increased the stability of flow. Freight transport was difficult, whether by land or by water. By land it was also expensive, so water transport was preferred. Even in fully navigable rivers in the nineteenth century barges were being towed in water that was intermittently 20 cm deep, so in former centuries, with no railways, worse roads and less management for navigation, boats would have passed in streams now considered barely suitable for canoes. Much damage to vegetation occurred from the scrapings of hard beds and the disturbance of silt on soft ones, from wash to the banks, from tow ropes, from gangs of men or horses towing, and from mooring (more intermittent and on the opposite side of the towpath, frequently) (Haslam (1987a).

Appendix 2

Recommendations for maintaining dykes (ditches) in alluvial plains, for conservation*

Cleaning

1. Channels should never all be cleaned at the same time throughout an area, as this destroys much of the potential for colonisation as well as the diversity of seral stages. Some overgrown non-arterial channels should be retained to provide diversity of habitat for other species.

2. Use cleaning methods that favour good ditch communities. Machine cleaning is not necessarily inimical to conservation requirements. If the channel bottom is not disturbed and the banks are not scraped, good aquatic communities can develop. However, at present, hand cleaning always provides less disturbance to aquatic communities while much machine cleaning is destructive because the bottom is removed and the banks are scraped. Given this situation it is worth encouraging hand cleaning where possible.

3. Avoid using any sort of chemical weed control, as this decreases diversity, particularly of the more sensitive species, and it does not appear significantly to reduce vegetation cover of weed species in the following season and

may provide problems of removing dead plant material.

4. Cleaning intervals should be as varied as possible throughout an area as some species flourish in regularly cleaned channels and other (perennial) species favour less disturbance. However, high species diversity in all layers is most often associated with field channels that are managed on a 3–5-year interval.

5. Where arterial and through flow drainage channels are of exceptional wildlife interest, collaboration with bodies concerned in the management of these channels is essential. In order to do this effectively, monitoring of the aquatic flora of these channels, throughout the cleaning cycle, would provide the basic information for future maintenance of diversity. Although dredging of the channels temporarily destroys the habitat, if this is done in sections, colonisation can occur from upstream sites.

Water Levels and Quality

6. Ensure that penned summer levels are maintained at a high level in channels of conservation interest, and that winter levels do not

* adapted from Wolseley et al., 1984

allow ditches with a good submerged and floating community to become dry for long periods, especially in cold weather.

7. As yet we have little evidence concerning the particular effects of changes in management, but the combined effect of improved pasture with its attendant increased fertiliser application, bank fencing, and ditch deepening create a species-poor habitat with low structural diversity.

8. Most arterial channels carry water from other areas, and this can quickly downgrade good ditches. Therefore all channels of conservation interest must have sluices on them to prevent this water entering the channel. This has become a major source of damage to good ditches.

9. Man-made channels were created either for arterial drainage or as field boundaries. In the latter situation they were created with the major constraint of using as little land space as possible while providing an efficient stock-proof boundary. In order to avoid the expense of regular maintenance that is essential for the requirements of a good submerged aquatic community, conservation bodies should consider increasing the width of some field channels on one side (so as not to destroy both marginal habitats) to create a larger area for the development of a submerged community, and to slow down the rate at which the channel will become shaded by tall monocot vegetation.

Appendix 3
Biological methods of assessment, survey and analysis

Methods for the examination of waters and associated materials

Standard methods, not usually for beginners, on all aspects of water quality. Biological papers listed here. Date in title may be earlier than date of publication. Published by Her Majesty's Stationery Office, London.

Quantitative sampling of benthic macroinvertebrates of shallow flowing waters 1980.

Acute toxicity testing with aquatic organisms 1981.

Bacteriological examination of drinking water supplies 1982.

Methods of biological sampling. Sampling of benthic macroinvertebrates in deep rivers 1983.

Methods of biological sampling. Colonization samplers for collection of macroinvertebrates/ indicators of water quality of lowland rivers 1983.

Sampling of non-planktonic algae (Benthic algae or periphyton) 1982.

Biological sampling. Sampling of macroinvertebrates in Water supply systems 1983.

Sampling fish in shallow rivers 1983.

Determination of biomass of aquatic macrophytes, *and the direct measurement of underwater light 1985.*

Methods for the use of aquatic macrophytes for assessing water quality in rivers and lakes 1986.

Five-day Biochemical Oxygen Demand (BOD$_5$). Second Edition. 1988. (With dissolved oxygen in waters. Amendments. 1988.).

The enumeration of algae, estimation of cell volume and their use in bioassays 1988. In preparation.

Sampling for zooplankton. In preparation.

Methods for the use of epilithic diatoms for assessing river water quality 1989. In preparation.

Use of plants to monitor heavy metals in freshwater. In preparation.

Other sources of methods, some for beginners

Freshwater Biological Association, Ferry House, Far Sawrey, Windermere, Cumbria.

Field Studies Council, Leonard Wills Field Centre, Nettlecombe, Taunton, Somerset.

Also see Chapter 7 of this book and Hellawell (1986) for invertebrate indices.

Appendix 4

Directives of the Commission of the European Communities governing the required water quality in the Community

Relevant directives (some of which have been altered since their original date) in 1989 include:

Water protection and management

75/440/EEC: Council Directive of 16 June 1975 concerning the quality required of surface water intended for the abstraction of drinking water in the member states

76/160/EEC: Council Directive of 8 December 1975 concerning the quality of bathing water

76/464/EEC: Council Directive of 4 May 1976 on pollution caused by certain dangerous substances discharged into the aquatic environment of the Community

77/795/EEC: Council Decision of 12 December 1977 establishing a common procedure for the exchange of information on the quality of surface freshwater in the Community

79/869/EEC: Council Directive of 9 October 1979 concerning the methods of measurement and frequencies of sampling and analysis of surface water intended for the abstraction of drinking water in the member states

79/923/EEC: Council Directive of 30 October 1979 on the quality required of shellfish waters

80/68/EEC: Council Directive of 17 December 1979 on the protection of groundwater against pollution caused by certain dangerous substances

80/778/EEC: Council Directive of 15 July 1980 relating to the quality of water intended for human consumption

82/176/EEC: Council Directive of 22 March 1982 on limit values and quality objectives for mercury discharges by the chlor-alkali electrolysis industry

83/513/EEC: Council Directive of 26 September 1983 on limit values and quality objectives for cadmium discharges

84/491/EEC: Council Directive of 9 October 1984 on limit values and quality objectives for discharges of hexachlorocyclohexane

86/280/EEC: Council Directive of 12 June 1986 on limit values and quality objectives for discharges of certain dangerous substances included in List I of the Annex to Directive 76/464/EEC

Chemicals, industrial risk and biotechnology

73/404/EEC: Council Directive of 22 November 1973 on the approximation of the laws of the member states relating to detergents

73/405/EEC: Council Directive of 22 November 1973 on the approximation of the laws of the member states relating to methods of testing the biodegradability of anionic surfactants

76/769/EEC: Council Directive of 27 July 1976 on the approximation of the laws, regulations and administrative provisions of the member states relating to restrictions on the marketing and use of certain dangerous substances and preparations

78/618/EEC: Commission Decision of 28 June 1978 setting up a Scientific Advisory Committee to examine the toxicity and ecotoxicity of chemical compounds

Conservation of wild fauna and flora

78/659/EEC: Council Directive of 18 July 1978 on the quality of fresh waters needing protection or improvement in order to support fish life

Waste management and clean technology

75/439/EEC: Council Directive of 16 June 1975 on the disposal of waste oils

75/442/EEC: Council Directive of 15 July 1975 on waste

76/403/EEC: Council Directive of 6 April 1976 on the disposal of polychlorinated biphenyls (PCBs) and polychlorinated terphenyls

78/176/EEC: Council Directive of 20 February 1978 on waste from the titanium dioxide industry

78/319/EEC: Council Directive of 20 March 1978 on toxic and dangerous waste

86/278/EEC: Council Directive of 12 June 1986 on the protection of the environment, and in particular of the soil, when sewage sludge is used in agriculture

International co-operation

Convention concerning the International Commission for the Rhine (Berne Convention)

Convention on the protection of the Rhine against chemical pollution (Bonn Convention)

Proposal from the International Commission for the Protection of the Rhine against Pollution to supplement Annex IV to the Convention on the protection of the Rhine against chemical pollution, signed in Bonn on 3 December 1976

77/586/EEC: Council Decision of 25 July 1977 concluding the Convention for the protection of the Rhine against chemical pollution and an Additional Agreement to the Agreement, signed in Berne on 29 April 1963, concerning the International Commission for the Protection of the Rhine against Pollution

82/460/EEC: Council Decision of 24 June 1982 on a supplement to Annex IV to the Convention on the protection of the Rhine against chemical pollution

Appendix 5
EEC 'black' and 'grey' list substances*

EEC 'black' and 'grey' list substances

List 1 ('Black List')
The substances on this list were selected mainly for their toxicity, persistence or bioaccumulation
1. Organohalogen compounds and substances which may form such compounds in the aquatic environment
2. Organophosphorus compounds
3. Organotin compounds
4. Substances, the carcinogenic activity of which is exhibited in or by the aquatic environment (substances in List 2 which are carcinogenic are included here)
5. Mercury and its compounds
6. Cadmium and its compounds
7. Persistent mineral oils and hydrocarbons of petroleum
8. Persistent synthetic substances.

List 2 ('Grey List')
These substances are regarded as less dangerous than those of List 1 and the impact of which may be local.
1. The following metalloids/metals and their compounds:

1. Zinc	6. Selenium	11. Tin	16. Vanadium
2. Copper	7. Arsenic	12. Barium	17. Cobalt
3. Nickel	8. Antimony	13. Beryllium	18. Thalium
4. Chromium	9. Molybdendum	14. Boron	19. Tellurium
5. Lead	10. Titanium	15. Uranium	20. Silver

2. Biocides and their derivatives not appearing in List 1
3. Substances which have a deleterious effect on the taste and/or smell of products for human consumption derived from the aquatic environment and compounds liable to give rise to such substances in water
4. Toxic or persistent organic compounds of silicon and substances which may give rise to such compounds in water, excluding those which are biologically harmless or are rapidly converted in water into harmless substances
5. Inorganic compounds of phosphorus and elemental phosphorus
6. Non-persistent mineral oils and hydrocarbons of petroleum origin
7. Cyanides, fluorides
8. Certain substances which may have an adverse effect on the oxygen balance, particularly ammonia and nitrites

* from Hellawell, 1986

Appendix 6
List of principal insecticides

List of principal insecticides, grouped according to the three main types (organochlorine, organophosphorus, carbamate) and herbicides, with examples of proprietary names, together with their full chemical names from Hellawell (1986)

	Common name	Proprietary name(s)	Chemical name
Insecticides			
Organochlorine	Aldrin (HHDN)		(1R,4S,5S,8R)-1,2,3,4,10,10-hexachloro-1,4,4a,5,8,8a-hexahydro-1,4:5,8-dimethanoaphthalene
	BHC (see HCH)		
	Camphechlor	Toxaphene	mixture of chlorinated camphenes
	Chlordane		1,2,3,4,5,6,7,8,8-octachloro-2,3,3a,4,7,7a-hexahydro-4,7-methanoindene
	Chlordecone	Kepone	dedecachloro octahydro-1,3,4-methano-2H-cyclobutal[cd] pentalen-2-one
	DDD (see TDE)		
	DDE		1,1-dichloro-2,2-bis (4-chlorophenyl) ethylene, formerly dichlorodiphenyl-dichloroethylene
	DDT		1,1,1-trichloro-2-2-bis (4-chlorophenyl) ethane, formerly (dichlorodiphenyl-trichloroethane)
	Dieldrin		(1R,4S,5S,8R)-1,2,3,4,10,10-hexachloro-1,4,4a,5,6,7,8,8a-octahydro-6,7-expoxy-1,4:5,8-dimethano-naphthalene
	Endosulphan	Thiodan	1,4,5,6,7,7-hexachloro-8,9,10-trinorborn-5-en-2,3-ylene dimethyl sulphite

	Common name	Proprietary name(s)	Chemical name
	Endrin		(1R,4S,5R,8S)-1,2,3,4,10,10-hexachloro-1,4,4a,5,6,7,8,8a-octahydro-6,7-epoxy-1,4:5,8-dimethanonaphthalene
	HCH	Lindane	1,2,3,4,5,6-hexachlorocyclohexane
	(gamma-HCH)		γ-1,2,3,4,5,6-hexachlorocyclohexane
	Heptachlor		1,4,5,6,7,8,8-heptachloro-3a,4,7,7a-tetrahydro-4,7-methanoindene
	Lindane (see HCH)		
	Methoxychlor		1,1,1-trichloro-2,2-bis (4-methoxyphenyl) ethane
	Mirex		dedecachloro octahydro-1,3,4-methano-2H-cyclobuta[cd] pentalene
	TDE (DDD)		1,1-dichloro-2,2-bis (4-chlorophenyl) ethane
	Thiodan (see Endosulphan)		
	Toxaphene (see Camphechlor)		
Organophosphorus	Azinphos methyl	Guthion	S-(3,4-dihydro-4-oxobenzo[d]-[1,2,3]-triazin-3-yl methyl O,O-dimethyl phosphorodithioate
	Chlorthion	Bayer 22/190	O-(3-chloro-4-nitrophenyl)O,O-dimethyl phosphorothioate
	Chlorpyrifos	Dursban	O,O-diethyl O-(3,5,6-trichloro-2-pyridyl) phosphorothioate
	Coumaphos	Asuntol, Co-Ral, Muscatox, Resitox	O-3-chloro-4-methylcoumarin-7-yl O,O-diethyl phosphorothioate
	Demeton	Systox	O,O-diethyl O-2-ethylthioethyl phosphorothioate and O,O-diethyl S-2-ethylthioethyl phospohorothioate
	Diazinon	Basudin	O,O-diethyl-O-2-isopropyl-6-methyl-pyrimidin-4-yl phosphorothioate
	Dichlorvos (DDVP)	Vapona, Dedevap	2,2-dichlorovinyl dimethyl phosphate
	Dimethoate	Rogor	O,O-dimethyl S-methylcarbamoyl-methyl phosphorodithioate
	Dioxathion	Delnav	S,S'-(1,4-dioxane-2,3-diyl) O,O,O',O'-tetraethyl di(phosphorodithioate)
	Disulfoton	Bayer 19639 Di-syston Di-thiosystox	O,O-diethyl S-2-ethylthioethyl phosphorodithioate
	Ethion	Embathion	O,O,O'-tetraethyl S,S'-methylene di(phosphorodithioate)
	Fenitrothion	Folithion, Sumithion	O,O-dimethyl O-4-nitro-m-tolyl phosphorothioate
	Fenthion	Bayer 29493, Baytex	O,O-dimethyl O-4-methylthio-m-tolyl phosphorothioate
	Malathion		S-1,2-bis (ethoxycarbonyl) ethyl O,O-dimethyl phosphorodithioate
	Methyl Parathion		O,O-dimethyl O-4 nitrophenyl phosphorothioate
	Mevinphos	Phosdrin	2-methoxycarbonyl-1-methylvinyl dimethyl phosphate
	Naled	Dibrom	1,2-dibromo-2,2-dichlorethyl dimethyl phosphate
	Parathion		O,O-diethyl-O-4-nitrophenyl phosphorothioate

	Common name	Proprietary name(s)	Chemical name
	Phorate	Thimet	O,O-diethyl S-ethyl thiomethyl phosphorodithioate
	Phosmet	Imidan	O,O-dimethyl S-phthalimidomethyl phosphorodithioate
	Phosphamidon	Dimecron	2-chloro-2-diethylcarbomoyl-1-methylvinyl dimethyl phosphate
	Temephos	Abate	O,O,O',O'-tetramethyl O,O'-thiodi-p-phenylene bis (phosphorothioate)
	Trichlorophon	Dipterex Dylox	dimethyl 2,2,2-trichloro-1-hydroxyethyl phosphonate
Carbamates	Aminocarb	Metacil	4-dimethylamino-m-tolyl methyl carbamate
	Carbaryl	Sevin	1-napthyl methylcarbamate
	Carbofuran		2,3-dihydro-2,2-dimethyl-benzofuran-7-yl methylcarbamate
	Methiocarb (Mercaptodimethur)	Mesurol	4-methylthio-3,5-xylyl methylcarbamate
	Mexacarbate	Zectran	4-dimethylamino-3,5-xylyl methylcarbamate
	Propoxur	Baygon	2-isopropor oxyphenyl methyl carbamate
Herbicides	Asulam	Asulox	methyl 4-amino-phenylsulphonylcarbamate
	Atrazine	Weedex, Residox	2-chloro-4-ethylamino-6-isopropyl-amino-1,3,5-triazene
	Chlorthiamid	Prefix	2,6-dichlorothio-benzamide
	2,4-D	Dormane, Fernimine	2,4-dichlorophenoxyacetic acid, esters or salts especially amine salts
	2,4,-DB	Embutox	4-(2,4-dichlorophenoxy) butyric acid, esters or salts
	Dalapon	Dowpon, Radapon	2,2-dichloropropionic acid
	Dichlobenil	Casoron	2,6-dichlorobenzonitrile
	Dichlone	Phygon	2,3-dichloro-1,4-naphthoquinone
	Diquat	Reglone, Midstream	1,1'-ethylene-2,2'-bipyridyldiylium dibromide
	Diuron	Karmex	3-(3,4-dichlorophenyl)-1,1-dimethylurea
	Glyphosate	Round-up, Spasor	N-(phosphoromethyl) glycine (iso-propylamine salt)
	IPC (see Propham)		
	Molinate	Ordram	S-ethyl N,N-hexamethylenethio-carbamate
	Paraquat	Gramaxone [Weedol]	1,1'-dimethyl-4,4'-bipyridyldiylium dichloride
	Picloram		4-amino,-3 5,6-trichloropyridine-2-carboxylic acid
	Propanil (DCPA)		3',4'-dichloroproprionanilide
	Propham (IPC)		isopropyl phenylcarbamate
	Simazine		2,chloro-4,6-bis (ethylamino)-1,3,5-triazine
	2,4,5-T		(2,4,5-trichlorophenoxy) acetic acid
	Terbutryne	Clarosan	2-tert-butylamino-4-ethylamino-6-methylthio-1,3,5-triazine
	Trifluralin	Treflan	α,α,α-trifluoro-2,6-dinitro-N,N-dipropyl-p-toluidine
	Vernolate		S-propyl dipropylthiolcarbamate

Appendix 7
A key to the commoner species of rivers

From Haslam *et al.* (1982). The further a European country is from lowland Britain (latitude being more important than distance), the more likely it is that other species will also be common.

Specimens should be carefully compared with good illustrations as rare species may be locally abundant.

1. Leaves (or leaf-like shoots) simple, over 4 mm wide, strap-like, grass-like or cylindrical   2
    Not as above   17

2. Mainly emergent   3

    Mainly in or on water   9

3. Plant dark green, cylindrical   *Scirpus lacustris*
    Plant with flat leaves   4

4. Stem bearing grass-like leaves well above water level   5

    Leaves arising from base of plant, usually under water   7

5. Leaves parallel sided,
bright or yellowish green — *Glyceria maxima*

Leaves wider in centre,
tapering, bluish-green — 6

6. Hairs present where leaf blade joins the sheath — *Phragmites communis*

A membranous ligule present where the leaf
blade joins the sheath — *Phalaris arundinacea*

7. Leaves almost triangular in cross section — *Sparganium erectum*

Leaves flat or *V*-shaped in section — 8

8. Leaves iris-like, *V*-shaped below — *Iris pseudacorus*

Leaves flat throughout, not *V*-shaped
(leaves less than 7 mm wide *T. angustifolia*, and
T. latifolia has leaves over 7 mm wide) — *Typha*

9. Leaves grass-like — 10

Leaves strap-like — 15

10. Tall erect reed-like plant with wide leaves (often more than 10 mm wide)   11
Stems trailing in water often with short erect branch-shoots, not reed-like, leaves usually less
than 10 mm wide, somewhat blue-green   13

11. Ligule (at base of blade, on upper surface) a dense
     eyelash-like fringe of hairs, leaves bluish-green

                                                                    Phragmites communis

     Ligule membranous, conspicuous   12

12. Leaves flat, tapering above, blue-green, leaf sheaths
     without keel, ligule over 6 mm long   *Phalaris arundinacea*

     Leaves channelled, parallel-sided for much of their length,
     bright green, leaf sheaths keeled on the back,
     ligule less than 6 mm long   *Glyceria maxima*

13. Youngest leaf rolled in shoot,
     which is thus cylindrical   *Agrostis stolonifera*

     Youngest leaf folded in shoot,
     which is thus laterally compressed   14

14. Leaf blades tapering for much of length, up to 8 mm
     wide (submerged leaves may be parallel-sided)   *Catabrosa aquatica*

     Leaf blades parallel-sided nearly to tip
     usually above 5 mm wide   *Glyceria fluitans*

15. Leaves long-tapering, bending in current from junction of blade and sheath
                                                                     Scirpus lacustris
     Leaves parallel-sided almost to the tip, bending in current from the base
                                                                               16

16. Leaves usually over 1 cm wide, separated.
     Veins at the leaf-tip few, well separated   *Sagittaria sagittifolia*

     Leaves usually under 1 cm wide, grouped into
     shoots. Veins at the leaf tip many, crowded

                                                                    Sparganium emersum

17. Plant 2–3 mm wide, floating   *Lemna minor*

    Plant composed of stems and leaves, rooted   13

18. Plant submerged and leaves thread-like   19
    Leaves not submerged, or leaves not thread-like   23

19. Leaves opposite   or alternate   15

    Leaves whorled   22

                                                                  Ranunculus spp
                                                                  (*Batrachian*)

20. Leaves finely divided

    Leaves entire   16

21. Leaves usually opposite   *Zannichellia palustris*

    Leaves alternate

                                                         Potamogeton pectinatus

22. Leaves pinnately divided

                                                         Myrophyllum spicatum

    Leaves forked   *Ceratophyllum demersum*

23. Leaves over 12 cm long and wide,
    submerged or floating   *Nuphar lutea*

    Not as above   24

24. Leaves divided into leaflets   25

     Leaves simple   28

25. Leaves over 20 cm long,
     much-divided
     segments not ovate

Oenanthe fluviatilis

     Leaves less than 20 cm long,
     once pinnate,
     segments ovate   26

26. Terminal leaflet larger than
     the lateral ones

Rorippa nasturtium-aquaticum agg.

     Terminal leaflet not larger than the lateral ones   27

27. When the leaf is held up to the light, a dark
     line (septum) is visible across the stalk,
     below the lowest pair of leaflets

Berula erecta

     No septum are visible below the lowest pair of
     leaflets

Apium nodiflorum

28. Leaves in whorls of three   *Elodea canadensis*

     Leaves alternate or opposite   29

29. Leaves opposite, pale green; floating leaves   *Callitriche*   30
     (if present) less than 2 cm long, submerged
     leaves notched at the tip

     Leaves alternate, floating leaves (if present)
     longer than 2 cm, submerged leaves not as above.   34

Callitriche simple identification
(Keying out more specimens as *C. platycarpa* than belong to this species)

30. Leaves translucent   *C. hermaphroditica*
     (Living) leaves not translucent, though sometimes pale and thin   31

31. At least some of the parallel-sided lower leaves with a
     spanner-like tip. Usually in acid water   *C. hamulata*

     Lower leaves with a notched but not spanner-like tip   32

32. Floating leaves with leaf outline angular, leaves ribbed above,
     often blue-green. Lower leaves often parallel-sided

                                                            C. obtusangula

     Floating leaves with rounded outline, leaves smooth above,
     usually mid-green   33

33. Floating broad leaves nearly circular, smooth above.
     Lower leaves broad or long and narrow,
     widest towards tip, seldom
     parallel-sided. Shoots sometimes bronzed. Common   *C. stagnalis*

     Floating broad leaves narrower.
     Lower leaves usually parallel-sided, widest in mid-leaf. Infrequent

                                                            C. platycarpa

34. Leaves less than 1.5 cm long, translucent   Various mosses
    Leaves more than 1.5 cm long   35

35. Ovate floating leaves present   36

    Only submerged leaves present,
    sometimes at surface of water but never ovate   38

36. Floating leaves with pinnate side veins

                                                        Polygonum amphibium

    Floating leaves with sub-parallel veins   28

37. Submerged leaves linear, opaque.
    Widespread   *Potamogeton natans*

    Submerged leaves ovate, translucent   38

    (*P.x. sparganifolius* is abundant in some Scottish rivers, has submerged leaves narrow, strap-shaped and translucent)

38. Leaves more or less oblong, margins with small teeth,
    leaves often curly (crisped)

                                                        Potamogeton crispus

    Leaves ovate, margins entire, sometimes wavy but not curled   39

39. Leaf base
    clasping stem   *Potamogeton perfoliatus*

    Leaf base not clasping stem   40

40. Leaves (when free of mud) pale
    shining green, tapering at base.
    Mainly southern

Potamogeton lucens

Leaves dark green, less tapering.
Mainly northern

Potamogeton alpinus

Bibliography

Water quality directives of the Commission of the European Communities (British) Methods for the examination of waters and associated materials, and similar sources are listed in Appendices 3 and 4, and are not repeated here. Publications of official organisations with no author named are placed after the references for individual authors. Dutch names are listed under the first name with a capital letter.

Abel, P.D. and Green, D.W.J. (1981). Ecological and toxicological studies on invertebrate fauna of two rivers in the northern Pennine orefield. In P.J. Say and B.A. Whitton (eds), *Heavy metals in Northern England. Environmental and Biological Aspects.* Department of Botany, University of Durham, 109–22.

Abo-Rady, M.D.K. (1980). Aquatic macrophytes as indicators for heavy metal pollution in the River Leine (West Germany). *Arch. Hydrobiol.* **89**, 367–404.

Agami, M. (1989). Effects of water pollution on plant species composition along the Amal River, Israel. *Arch. Hydrobiol.* **100**, 445–54.

Anderson, G., Blindao, I., Hargeby, A. and Johanssen, S. (1986). The importance of submerged vegetation for waterfowl. *Proc. EWRS/AAB 7th Symposium on Aquatic Weeds, 1986*, 29–30.

Anderson, L.W.J. (1986). Recent developments and future trends in aquatic weed management. *Proc. EWRS/AAB 7th Symposium on Aquatic Weeds, 1986*, 9–16.

Alabaster, J.S., Garland, J.H.N., Hart, I.C., and Solbé, J.F. de L.G. (1972). An approach to the problem of pollution and fisheries. *Symp. Zool. Soc. Land* **27**, 87–114.

Aymonin, G.G. (1982). Phénomènes de déséquilibre et appauvrissements floristiques dans les végétations hygrophiles en France. In J.J. Symoens, S.S. Hooper and B. Compère (eds), *Studies on Aquatic Vascular Plants.* Brussels: Royal Botanic Society of Belgium, 377–89.

Bache, B.W. (1984). Soil–water interactions. *Phil. Trans. R. Soc. Lond. B* **305**, 393–407.

Baldock, D. (1984). *Wetland Drainage in Europe.* Institute for European Environmental Policy/The International Institute for Environment and Development.

Ball, R.C. and Bahr, T.G. (1970). Intensive Survey: Red Cedar River, Michigan. In B.A. Whitton (ed.), *River Ecology.* Oxford: Blackwell Scientific Publications, 431–60.

Balocca-Castella, C. (1988). Les macrophytes aquatiques des milieux abandonnés par le Haut-Rhône et l'Ain. Diagnostic phytoecologie sur l'évolution et le fonctionnement de ces ecosystèmes. Doctoral thesis, Université Claude Bernard, Lyon, France.

Baredo, R. (1985). Transfer of trace elements along the aquatic food chain. *Mem. Ist. Ital. Idrobiol.* **43**, 281–309.

Barrett, P.R.F. (1981). Aquatic herbicides in Great Britain, recent changes and possible future development. *Proc. Aquatic Weeds and their Control, 1981*, 95–103.

Barrett, P.R.F. and Logan, P. (1982). The localised control of submerged aquatic weeds in lakes with diquat alginate. *Proc. EWRS 6th Symposium on Aquatic Weeds, 1982*, 193–9.

Barrett, P.R.F. and Murphy, K.J. (1982). The use of diquatalginate for weed control in flowing waters. *Proc. EWRS 6th Symposium on Aquatic Weeds, 1982*, 200–8.

Barrett, P.R.F. and Robson, T.O. (1974). Further studies on the seasonal changes in the susceptibility of some emergent plants to dalapon. *Proc. 12th Brit. Weed Control Conf, 1974*, 249–53.

Battarbee, R.W. (1984). Diatom analysis and the acidification of Lakes. *Phil. Trans. R. Soc. Lond. B* **305**, 451–77.

Baudo, R. (1985) Transfer of trace elements along the aquatic food chain. *Mem. Ist. Ital. Idrobiol.* **43**, 281–309.

Baudo, R., Canzian, E., Galanti, G., Giulizzoni, P. and Rapetti, G. (1985). Relationships between heavy metals and aquatic organisms in Lake Mezzola hydrographic system (Northern Italy) 6. Metal concentrations in two species emergent of macrophytes. *Mem. Ist. Ital. Idrobiol.* **43**, 161–80.

Baudo, R., Galanti, G. and Guilizzoni, P. (1979). Heavy metal distribution and the effect of photosynthesic rates of two submerged macrophytes of Lake Mezzola (N. Italy). Unpublished report, Freshwater Contact Group, Commission of European Communities.

Baudo, R., Galanti, G., Guilizzoni, P., Marengo, G., Muntau, H., Schramd, P. (1985). *Heavy Metals in the Environment.* Athens, 337–9.

Baudo, R., Galanti, G., Guilizzoni, P., Merlini, L. and Varini, P.G. (1981). Relationships between heavy metals and aquatic organisms in Lake Mezzola hydrographic system (Northern Italy) 5. Net photosynthesis of the submersed macrophytes *Potamogeton crispus* L. and *Potamogeton perfoliatus* L. *Mem. Ist. Ital. Idrobiol.* **39**, 227–42.

Baudo, R., Galanti, G., Guilizzoni, P. and Varini, P.G. (1981). Relationships between heavy metals and aquatic organism in Lake Mezzola hydrographic system (Northern Italy). 4. Metal concentrations in six submersed aquatic macrophytes. *Mem. Ist. Ital. Idrobiol.* **39**, 203–25.

Baudo, R. and Varini, P.G. (1976). Copper, manganese and chromium concentrations in five macrophytes from the delta of River Toce (Northern Italy). *Mem. Ist. Ital. Idrobiol.* **33**, 305–24.

Bennett, G. (1970). Bristol floods 1968. Controlled survey of effect on health of local community disaster. *Brit. Med. Journ.* **3**, 454–8.

Beresford, M.W. and St. Joseph, J.K.S. (1979). *Mediaeval England: An Aerial Survey*. Cambridge: Cambridge University Press.

Best, E.P.H., van der Zweerde, W. and Zweitering, F.W. (1986). A management approach to the Netherlands' water courses using models on macrophytic growth, aquatic weed control and hydrology. *Proc. EWRS/AAB 7th Symposium on Aquatic Weeds*, 1986, 37–42.

Birkhead, M. and Perrins, C. (1981). Decline of the royal swans. *New Scientist* **91**, 75–7.

Bonham, A.J. (1980). *Bank Protection Using Emergent Plants against Boat Wash in Rivers and Canals*. Report IT **206**. Wallingford: Hydraulics Research Station.

Borum, J. (1983). The quantitative role of macrophytes, epiphytes and phytoplankton under different nutrient conditions in Roskilde Fjord, Denmark. *Proc. Internat. Symposium on Aquatic Macrophytes*, 35–45. Faculty of Science, Department of Aquatic Ecology, Nijmegen.

Boudou, A., Delarche, A., Ribeyre, F. and Marty, R. (1979). Bioaccumulation and bioamplification of mercury compounds in a second level consumer, *Gambasia affenis* — temperature effects. *Bull. Environm. Contam. Toxicol.* **22**, 813–18.

Bovey, R.W., Burnett, E., Richardson, C., Merkle, M.G., Baur, J.R. and Knisel, W.G. (1974). Occurrence of 2,4,5,–T and Picloram in surface runoff water in the Blacklands of Texas. *J. Environ. Qual.* **3**, 61–4.

Brock, T.D. and Yoder (1970). Thermal pollution of a small river by a large university: bacteriological studies. *Proc. Indiana Acad. Sci.* **80**, 153–81.

Bronwel, J. (1976). *Incoltarisortie van plezier vaartnigen haar ligplants situatie* **1**, 1–76, Rijswijk, Ministerie van Cultuur, Recreatie en Maatscheppelijk Wet.

Brookes, A. (1986). Response of aquatic vegetation to sedimentation downstream from river channelisation works in England and Wales. *Biol. Cons.* **38**, 351–67.

Brookes, A. (1987). Recovery and adjustment of aquatic vegetation within channelization works in England and Wales. *J. Environ. Mgmt* **24**, 365–82.

Brookes, A. (1988). *Channelized Rivers*. Chichester: Wiley.

Brookes, A. and Gregory, K.J. (1983). An assessment of river channelization in England and Wales. *The Science of the Total Environment* **27**, 97–111.

Brown, V.M. (1975). fishes. In B.A. Whitton (ed.), *River Ecology*. Oxford: Blackwell Scientific Publications, 199–229.

Bryan, G.W. (1976). Some aspects of heavy metal tolerance in aquatic organisms. In A.P.M. Cockwood (ed.), *Effects of Pollutants on Aquatic Organisms*. Cambridge: Cambridge University Press, 7–34.

Butcher, R.W. (1927). A preliminary account of the vegetation of the River Itchen. *J. Ecol.* **15**, 55–65.

Butcher, R.W. (1933). Studies on the ecology of rivers. I — On the distribution of macrophytic vegetation in the rivers of Britain. *J. Ecol.* **21**, 58–91.

Butcher, R.W. (1946). The biological detection of pollution. *J. Inst. Sewage Purif.* **2**, 92–7.

Butcher, R.W. (1947). Studies on the ecology of rivers VII. The algae of organically enriched waters. *J. Ecol.* **35**, 186–91.

Caffrey, J.M. (1986a). The impact of peat siltation on macrophyte communities in the River Suck: An Irish coarse fishery. *Proc. EWRS/AAB 7th Symposium on Aquatic Weeds*, 1986, 53–60.

Caffrey, J.M. (1986b). Macrophytes as biological indicators of organic pollution in Irish rivers. In D.H.S. Richardson (ed.), *Biological Indicators of Pollution*. Dublin: Royal Irish Academy, 77–87.

Cairns, J., Albaugh, D.W., Busey, F. and Chancy, M.D. (1968). The sequential comparison index — a simplified method for non-biologists to estimate relative differences in biological diversity in steam pollution studies. *J. Wat. Pollut. Control Fed.* **40**, 1607–13.

Cairns, J., Lanza, G.R. and Parker, B.C. (1972). Pollution-related structural and functional changes in aquatic communities with emphasis on freshwater algae and protozoa. *Proc. Nat. Acad. Sci. USA* **124**, 79–127.

Campbell, L.A. (1988). The impact of river engineering on water birds on an English lowland river. *Bird Study* **35**, 91–6.

Carbiener, R. and Ortscheit, A. (1987). Wasserpflanzengesellschaften als Hilfe zur Qualitätsüberwachung eines der grössten Grundwasser-Vorkommens Europas (Oberrheinebene). In A. Miyawaki (ed.), *Vegetation Ecology and Creation of New Environments*, Proc. Internal. Sym. Tokyo. Tokyo: Tokai University Press, 283–312.

Carpenter, K.E. (1924). A study of the fauna of rivers polluted by lead mining in the Aberystwyth region of Cardiganshire. *Ann. Appl. Biol.* **11**, 1–23.

Carpenter, K.E. (1926). the lead mine as an active agent in river pollution. *Ann. Appl. Biol.* **13**, 395–401.

Carpenter, K.E. (1928). *Life in Inland Waters*. London: Sidgwick & Jackson.

Casey, H. (1981). Discharge and chemical changes in a chalk stream headwater affected by the overflow of a commercial watercress bed. *Environ. Pollut. Ser. B* **22**, 373–85.

Casey, H. and Newton, P.V.R. (1973). The chemical composition and flow of the River Frome and its main tributaries. *Freshwat. Biol.* **3**, 337–53.

Casey, H. and Westlake, D.F. (1974). Growth and nutrient relationships of macrophytes in Sydling Water, a small unpolluted chalk stream. *Proc. Eur. Weed Res. Counc. 4th Int. Symposium on Aquatic Weeds, 1974*, 67–76.

Castella, E. (1987). Apport des macroinvertébrés aquatiques au diagnostic écologique des écosystèmes abandonnés par les fleuves. Recherches methodologiques sur le Haut-Rhône français. Thesis, Université Claude-Bernard, Lyon, France.

Castenholz, R.W. and Wickstrom, C.E. (1975). Thermal streams. In B.A. Whitton (ed.), *River Ecology*. Oxford: Blackwell Scientific Publications, 264–85.

Cave, T.G. (1981). Current weed control problems in land drainage channels. *Proc. Aquatic Weeds and their Control, 1981*, 5–14.

Chandler, J.R. (1970). A biological approach to water quality management. *Wat. Pollut. Control Lond.* **69**, 415–22.

Chanin, P. (1985). *The Natural History of Otters*. London: Croom Helm.

Chiaudani, G. and Marchetti, R. (1984). Po. In B.A. Whitton (ed.), *Ecology of European Rivers*. Oxford: Blackwell Scientific Publications, 401–36.

Choudry, G.C. (1983). Humic substances: interaction with environmental chemicals (excluding sorptive interactions). *Toxicol. Envir. Chem.* **6**, 231–57.

Clark, J.A. (1854–5. On trunk drainage. *J. Roy. Agric. Soc.* (1st series) **15**, 1–73.

Court, W.H.B. (1938). *The Rise of the Midland Industries 1600–1838*. Oxford: Oxford University Press.

Crabtree, K.T. (1972). Nitrate and nitrite variation in ground water. *Tech. Bull.* **58**, Department of Natural Resources, Madison, Wisconsin.

Cresser, M. and Edwards, A. (1987). *Acidification of Freshwaters*. Cambridge: Cambridge University Press.

Crisp, D.T. (1970). Input and output of minerals for a small watercress bed fed by chalk water. *J. Appl. Ecol.* **7**, 117–40.

Crisp, D.T. (1989). Some impacts of human activity on trout populations, *Salmon trutta, Freshwat. Biol.* **21**, 21–34.

Cummins, K.W. ,(1975). Macroinvertebrates. In B.A. Whitton (ed.), *River Ecology*. Oxford: Blackwell Scientific Publications, 170–98.

Curtis, E.J.C. and Harrington, D.W. (1970) Effects of organic wastes on rivers *Process Biochemistry* **5**, 44–6.

Czekanowski, J. (1913). *Zarys metod statystycznych*. Warsaw.

Darby, H.C. (1983). *The Changing Fenland*. Cambridge: Cambridge University Press.

Davies, L.J. and Hawkes, H.A. (1981). Some effects of organic pollution on the distribution and seasonal incidence of *Chironomidae* in riffles in the River Cole. *Freshwater Biol.* **11**, 549–59.

Dawson, F.H. (1976). Organic contribution of stream edge forest litter fall to the chalk stream ecosystem. *Oikos*, **27**, 13–18.

Dawson, F.H. (1978). Aquatic plant management in semi-natural streams: the role of marginal vegetation. *J. Envir. Mgmt.* **6**, 213–21.

Dawson, F.H. (1980). The origin, composition and downstream transport of plant material in a small chalk stream. *Freshwat. Biol.* **10**, 419–35.

Dawson, F.H. (1989). Ecology and management of water plants in lowland streams. *Freshw. Biol. Assoc. Ann. Rep.* **57**, 43–60.

Dawson, F.H. and Hallows, J.B. (1983). Practical applications of a shading material for macrophyte control in watercourses. *Aquat. Bot.* **17**, 301–8.

Dawson, F.H. and Haslam, S.M. (1983). The management of river vegetation with particular reference to shading effects of marginal vegetation. *Landscape Planning*, **10**, 147–69.

Dawson, F.H. and Kern-Hansen, U. (1979). The effect of natural and artificial shade on the macrophytes of lowland streams and the use of shade as a management technique. *Int. Revue Ges. Hydrobiol.* **69**, 437–55.

Dearder, P. (1982). Comparative risk assessment associated with the growth of Eurasian water milfoil (*Myriophyllum spicatum* L.) in the Okanagan Valley, British Columbia, Canada. *Proc. EWRS 6th Symposium on Aquatic Weeds, 1982*, 113–21.

Delvosalle, L. and Vanhecks, L. (1982). Essai de notation quantitative de la raréfaction d'espèces aquatiques et palustres en Belgique entre 1960 et 1980. In J.J. Symoens, S.S. Hooper and P. Compère (eds), *Studies on Aquatic Vascular Plants*. Brussels: Royal Botanical Society of Belgium, 403–9.

Descy, J.-P. (1976a). La végétation algale benthique de la Somme (France) et ses relations avec la qualité des eaux. *Mem. Soc. Roy. Bot. Belg.* **7**, 101–28.

Descy, J.-P. (1976b). Un appareillage pratique pour l'échantillonnage quantitatif du périphyton épilithique. *Bull. Soc. Roy. Belg.* **109**, 453–7.

Descy, J.-P. (1976c). Etude quantitative de peuplement algale benthique en vue de l'établissement d'une méthodologie d'estimation biologique de la qualité des eaux courantes. *Recherche et Technique au service de l'Environment (1976)*. Liège: Cebedoc, 159–206.

Descy, J.-P. (1976d). Utilisation des algaes benthiques comme indicateurs biologiques de la qualité des eaux courantes. In P. Pesson (ed.), *La pollution des eaux continentales. Incidence sur les biocénoses aquatiques*. Paris: Gauthier-Villars, 149–72.

Descy, J.-P. (1976e). Value of aquatic plants in the characterization of water quality and principles of methods used. In R. Amavis and J. Smeets (eds), *Principles and Methods for Determining Ecological Criteria on Hydrobiocenoses*, Proc. Eur. Sci. Colloquium, Luxembourg. Nov. 1975. Oxford: Pergamon Press for Commission of the European Communities, 157–83.

Descy, J.-P. (1979). A new approach to water quality estimation using diatoms. *Nova Hedwigia* **64**, 305–23.

Descy, J.-P. and Empain, A. (1984). Meuse. In B.A. Whitton (ed.), *Ecology of European Rivers*. Oxford: Blackwell Scientific Publications, 1–24.

Descy, J.-P., Empain, A. and Lambinon, J. (1981). *La Qualité des Eaux Courantes en Wallonie, Bassin de la Meuse*. Brussels: Secretariat d'Etat à l'Environnement, à l'Aménagement du Territoire et à l'Eau pour la Wallonie.

Descy, J.-P., Empain, A. and Lambinon, J. (1982). Un inventaire de la qualité des eaux du bassin wallon de la Meuse (1976–1980). *Trib. Cebedeau* **35**, 267–78.

Doarks, C. (1980). *Botanical Survey of Marsh Dykes in Broadland*. Norwich: BARS, Broads Authority.

Driscoll, R.J. (1983). Broadland dykes: the loss of an important wildlife habitat. *Trans. Norf. Norw. Nat. Soc.* **26**, 170–2.

Driscoll, R.J. (1986). Changes in land management in the Thurne Catchment area, Norfolk, between 1973 and 1983 and their effects on the dyke flora and fauna. *Proc. EWRS/AAB 7th Symposium on Aquatic Weeds, 1986*, 87–92.

Dubos, R. (1965). *Man Adapting*. New Haven, CT: Yale University Press.

Dunham, K.C. (1981). Mineralisation and mining in the Dinantian and Namurian rocks of the northern Pennines. In P.J. Say and B.A. Whitton (eds), *Heavy metals in Northern England: Environmental and Biological Aspects*. Department of Botany, University of Durham, 7–17.

Dykyjova, D. (1979). Selective uptake of mineral ions and their concentration factors in aquatic higher plants. *Folia Geobot. Phytotax. Praha* **14**, 267–325.

Economou-Amilli, A. (1980). Periphyton analysis for the evaluation of water quality in running waters of Greece. *Hydrobiol.* **74**, 39–48.

Edmunds, W.M. and Kinniburgh, D.G. (1986). The susceptibility of U.K. groundwater to acidic deposition. *J. Geol. Soc. Lond.* **143**, 707–20.

Edwards, A., Martin, D. and Mitchell, G. (eds), (1987). *Colour in Upland Waters*. Yorkshire Water, Leeds, and WRC, Medmenham.

Edwards, R.W., Williams, P.F. and Williams, R. (1984). Ebbw. In B.A. Whitton (ed.), *Ecology of European Rivers*. Oxford: Blackwell Scientific Publications, 83–112.

Egli, E. and Miller, H.R. (1959). *Europe from the Air*. London: George G. Harrap.

Eichenberge, E. and Weilenmann, H.U. (1982). The growth of *Ranunculus Fluitans* Lim. in artificial canals. In J.J. Symoens, S.S. Hooper and P. Compère (eds), *Studies on Aquatic Vascular Plants*. Brussels: Royal Botanical Society of Belgium, 324–32.

Eijk, M. van der (1978). Notes on the experimental introduction of grass carp (*Cetnopharyngodon idella*) into the Netherlands. *Proc. EWRS 5th Symposium on Aquatic Weeds, 1978*, 245–51.

Ekwall, E. (1984). *The Concise Oxford Dictionary of English Place-Names*. Oxford: Clarendon Press.

Eliot, G. (1859). *Adam Bede*, Bernhard Tauchnitz, Leipzig.

Eliot, G. (1967). *The Mill on the Floss*. London: Zodiac Press.

Ellenberg, E. (1971). Zeigerwerte der Gefässplanzen Mitteleuropas, 2nd edn. *Scripta Geobotanica* **9**.

Ellenberg, H. (1973). Chemical data and aquatic vascular plants as indicators for pollution in the Moosach river system near Munich. *Arch. Hydrobiol.* **72**, 533–49.

Elliot, J.M. (ed.) (1989a) Wild Brown Trout: the scientific basis for their conservation and management. *Freshwat. Biol.* **21**, 1.

Elliot, J.M. (1989b) Wild Brown Trout *Salmo trutta*: an important national and international resource. *Freshwater Biol.* **21**, 1–6.

Empain, A. (1976a). Les Bryophytes aquatiques utilisés comme traceurs de la contamination en métaux lourds des eaux douces. *Mem. Soc. Roy. Boat. Belg.* **7**, 141–56.

Empain, A. (1976b). Estimation de la pollution par métaux lourdes dans la Somme par l'analyse des bryophytes aquatiques. *Bull. Fr. Pisiculture* **48**, 138–42.

Empain, A., Lambinon, J., Mouvet, C. and Kirchmann, R. (1980). Utilisation des bryophytes aquatiques et subaquatiques comme indicateurs biologiques de la qualité des eaux courantes. In P. Rosson (ed.), *La Pollution des eaux continentales*, 2nd edn. Paris: Gauther-Villers, 195–223.

Engelen, G.B. and Roebert, A.J. (1974). Chemical water types and their distribution in space and time in the Amsterdam dunewater catchment area with artificial recharge. *J. Hydrol.* **21**, 339–56.

Fayed, S.E. and Abd-El-Shafay, H.I. (1985). Accumulation of Cu, Zn, Cd and Pb by aquatic macrophytes. *Environ. Int.* **11**, 77–87.

Fielding, M., Gibson, T.M., James, H.A., McLoughlin, K. and Steel, C.P. (1981). *Organic micropollutants in drinking water. Tech. Report* **TR 159**, Water Research Centre, Medmenham, Stevenage.

Fisher, R.A., Corbett, A.S. and Williams, G.B. (1943). The relation between the number of species and the number of individuals in a random sample of an animal population. *J. Anim. Ecol.* **12**, 42–58.

Fjerdingstad, E. (1975). Bacteria and Fungi. In B.A. Whitton (ed.), *River Ecology*. Oxford: Blackwell Scientific Publications, 129–40.

Flahault, C. and Schröter, C. (1910). Rapport sur la nomenclature phytogéographique. *Proc. 3rd Int. Bot. Congr. Brussels, 1910* **1**, 131–64.

Flanagan, P.J. (1974). *The National Survey of Irish Rivers. A Second Report on Water Quality*. Water Resources Division. Ireland: An Foras Forbatha.

Florczyk, H. and Gokowin, S. (1979). Self-purification of rivers polluted with heavy metals. *Pol. Arch. Hydrobiol.* **26**, 457–73.

Fowler, D. (1984) Transfer to terrestrial surfaces. *Phil. Trans. R. Soc. Lond.* B **305**, 281–97.

Fox, A.M., Murphy, K.J. and Westlake, D.F. (1986). Effects of diquat alginate and cutting on the submerged macrophyte community of a *Ranunculus* stream in northern England. *Proc. EWRS/AAB 7th Symposium on Aquatic Weeds 1986*, 105–12.

Friedrich, G. and Müller, D. (1984). Rhine. In B.A. Whitton (ed.), *Ecology of European Rivers*. Oxford: Blackwell Scientific Publications, 265–316.

Gessner, F. (1959). Die ökologische Bedeutung der Strömungsgeschwindigkeet fliessender Gewässer und ihre Messung auf kleinstem Raum. *Arch. Hydrobiol.* **43**, 159–65.

Giles, N. (1989). Assessing the status of British wild brown trout, *Salmo trutta*: a pilot study utilising data from game fisheries. *Freshwat. Biol.* **21**, 125–34.

Glading, P.R. (1986). A Botanical survey of ditches on the Pevensey Levels. England Field Unit, Project No. 25. Nature Conservancy Council.

Glanzer, V., Haber, W. and Kohler, A. (1977). Experimentelle Untersuchungen zur Belastbarkeit submerser Fliessgewässer-Makrophyten. *Arch. Hydrobiol.* **79**, 193–232.

Golterman, H.L. (1982). Preliminary observations on nutrient cycles in *Scirpus* and rice fields in the Camargue. In J.J. Symoens, S.S. Hooper and P. Compère (eds), *Studies in Aquatic Vascular Plants*. Brussels: Royal Botanical Society of Belgium, 200–1.

Golterman, H.L. (1984). *The impact of the Rhône on the Camargue*. In D. Kotzias (ed.) *et al* Mediterranean Scientific Association Environmental Protection (MESAEP). Munich, pp. 79–84.

Gorham, E. and Gordon, A.C. (1963). Some effects of smelter pollution upon aquatic vegetation near Sudbury, Ontario. *Can. J. Bot.* **46**, 371–8.

Grahame, K. (1908). *The Wind in the Willows*. London: Methuen.

Grant, I. and Hawkes, H.A. (1982). The effects of die 1 oxygen fluctuations on the survival of the freshwater shrimp (*Gammarus pulex*). Environ. Pollut. (Series A) **28**, 53–66.

Greenhaugh, J.G. (1980). The present use of the River Cam in relation to its historical perspective. M.Litt. thesis, University of Cambridge.

Gregory, K.J. (ed.) (1977). *River Channel Changes*. Chichester: Wiley.

Grose, M.P.B. and Allen, D.S. (1978). *Ouse Washes Water Plants*. Report, Royal Society for the Protection of Birds and Anglia Water Authority, Sandy.

Guilizzoni, P. (1979). Heavy metal distribution and their effects on photosynthesic rates of two submersed macrophytes of Lake Mezzola (N. Italy). Unpublished report. Freshwater Contact Group, Commission of the European Communities, Brussels.

Guilizzoni, P., Adams, M.S. and MacGaffey, N. (1984). The effect of chromium on growth and photosynthesis of a submersed macrophyte *Myriophyllum spicatum*. *Ecol. Bul.* **36**, 90–6.

Haan, F.A.M. de (1976). *Interaction mechanisms in soil as related to soil pollution and groundwater quality*. Institute for Land and Water Management Research, Wageningen. *Tech. Bull.* **95**.

Haggett, P. and Chorley, R.J. (1972). *Network Analysis in Geography*. London: Arnold.

Harbott, B.J. and Rey, C.J. (1981). The implications of long-term aquatic herbicide application: problems associated with environment impact assessment. *Proc. Aquatic Weeds and their Control, 1981*, 219–31.

Harper. D.B., Smith, R.V. and Gotto, D.M. (1977). BHC residues of domestic origin: a significant factor in pollution of freshwater in Northern Ireland. *Environ. Pollut.* **12**, 223–33.

Haslam, S.M. (1978). *River Plants*. Cambridge: Cambridge University Press.

Haslam, S.M. (1981). Changing rivers and changing vegetation in the past half century. *Proc. Aquatic Weeds and their Control, 1981*, 49–57.

Haslam, S.M. (1982a). A proposed method for monitoring river pollution using macrophytes. *Envir. Technol. Letters* **3**, 19–34.

Haslam, S.M. (1982b). *Vegetation in British Rivers*, 2 vols. London: Nature Conservancy Council.

Haslam, S.M. (1982c). Indices for dyke vegetation. *Nature Cambridgeshire* **25**, 34–40.

Haslam, S.M. (1982d). Major factors determining the distribution of macrophytic vegetation in the watercourses of the European Economic Community. *Proc. EWRS 6th Symposium on Aquatic Weeds, 1982*, 104–11.

Haslam, S.M. (1984). Stream vegetation and pollution in Italy and Mediterranean France. *Second International Symposium on Environmental Pollution and its impact on life in the Mediterranean Region.* Second International Symposium of the Mediterranean Scientific Association of Environmental Protection MESAEP). Munich, 19–91.

Haslam, S.M. (1986). Causes of changes in river vegetation giving rise to complaints. *Proc. EWRS/AAB 7th Symposium on Aquatic Weeds, 1986*, 151–6.

Haslam, S.M. (1987a). *River Plants of Western Europe*. Cambridge: Cambridge University Press.

Haslam, S.M. (1987b). Sources of watercourses pollution in Italy and Mediterranean France. *Chemosphere* **16**, 331–7.

Haslam, S.M. (1989). The influence of climate and man on the watercourses of Malta. *Toxic. Envir. Chem.* **20–21**, 85–92.

Haslam, S.M., Harding, J.P.C. and Spence, D.H.N. Authors not listed on Title Page, (1987). Methods for the Use of Aquatic Macrophytes for Assessing Water Quality, 1985–86. In *Methods for the Examination of Waters and Associated Materials*. London: Her Majesty's Stationery Office.

Haslam, S.M. with Molitor, A.M.M. (1988). The macrophytic vegetation of the major rivers of Luxembourg. *Bull. Soc. Nat. Luxemb.* **88**, 3–54.

Haslam, S.M., Sinker, C.S. and Wolseley, P.A. (1982). *British Water Plants* (reprint with corrections). Taunton: Field Studies Council.

Haslam, S.M. and Wolseley, P.A. (1981). *River Vegetation: Its Identification, Assessment and Management.* Cambridge: Cambridge University Press.

Hawkes, H.A. (1975). River zonation and classification. In B.A. Whitton (ed.), *River Ecology.* Oxford: Blackwell Scientific Publications.

Hawkes, H.A. (1978a). Conceptual basis, for the biological surveillance of river water quality. In H.A. Hawkes and J.G. Hughes (eds), *Biological Surveillance of River Water Quality.* University of Aston, Birmingham, 1–14.

Hawkes, H.A. (1978b). Invertebrates as indicators of river water quality. In L.M. Evison and A. James (eds), *Biological Indicators of Water Quality.* Chichester: Wiley, 2.1–2.43.

Hawkes, H.A. and Davies, L.J. (1971). Some effects of organic enrichment on benthic invertebrate communities in stream riffles. In E. Duffey and A.S. Watt (eds), *The Scientific Management of Animal and Plant Communities for Conservation.* Oxford: Blackwell Scientific Publications, 271–93.

Hayes, M.H.B. (1987). Concepts of the composition and structure of humic substances relevant to the discolouration of waters. In A. Edwards, D. Martin and G. Mitchell (eds), *Colour in Upland Waters.* Yorkshire Water Authority (Leeds) and Water Research Centre (Medmenham), 35–42.

Heath, S. (1971). *Pilgrim Life in the Middle Ages.* London: Kennikat Press.

Heise, P. (1984). Gudenå. In B.A. Whitton (ed.), *Ecology of European Rivers.* Oxford: Blackwell Scientific Publications, 25–50.

Hellawell, J.M. (1978). *Biological Surveillance of Rivers.* Water Research Centre, Stevenage.

Hellawell, J.M. (1986). *Biological Indicators of Freshwater Pollution and Environmental Management.* London: Elsevier Applied Science Publishers.

Hermens, L.C.M. (1975). Levendgroen, II. Groene beken in Limburg. *Tijdschr. K. ned. Heidemaatsch.* **86**, 476–81.

Heurteaux, F. (1979). Aquatic pollution in the Camargue. Ecology of rice fields and the role of rice cultivation in the contamination of the aquatic environment. Unpublished Report. Freshwater Contact Group, Commission of the European Communities, Brussels.

Higler, H.W.G. (1975). *Proposal for a Classification of International Water-courses.* Presentation to: European Convention for the Protection of International Watercourses Against Pollution. Strasbourg: Council of Europe. EXP/Eau (75) **58**.

Hills, J.W. (1924). *A Summer on Test.* London: Hodder & Stoughton.

Hodges, F.O.C. (1972). Report on all statutory main rivers giving details of schemes and maintenance requirements. Internal Report, Avon and Dorset River Authority, Poole.

Hoeks, J. (1976). *Pollution of soil and groundwater from land disposal of solid wastes.* Institute for Land and Water Management Research, Wageningen. *Tech. Bull.* **96**.

Hoeks, J. (1977). *Mobility of pollutants in soil and groundwater near waste disposal sites.* Institute for Land and Water Management Research, Wageningen. *Tech. Bull.* **105**.

Holdgate, M.W. (1971). The need for environmental monitoring. In *International symposium on identification and measurement of environmental pollutants.* Ottawa, Ontario, Canada, 1–80.

Holdgate, M.W. (1984). Concluding remarks. *Phil. Trans. R. Soc. Lond. B.* **305**, 569–77.

Holland, D.G. and Harding, J.P.C. (1984). Mersey. In B.A. Whitton (ed.), *Ecology of European Rivers.* Oxford: Blackwell Scientific Publications, 113–44.

Holmes, N.T.H. (1978). *Macrophyte Surveys of Rivers.* London: Nature Conservancy Council. Chief Scientists Team No. **7**.

Holmes, N.T.H. (1983). *Focus on Nature Conservation. 4. Typing British Rivers according to their Flora.* Shrewsbury: Nature Conservancy Council.

Holmes, P. (1974). *That Alarming Malady.* Ely: Cambridge Education Authority.

Hoogerkamp, M. and Rozenboom, G. (1978). The management of vegetation on the slopes of waterways in the Netherlands. *Proc. EWRS 5th Symposium on Aquatic Weeds, 1978*, 203–11.

Horkan, J.P.K. (1980). *Interim report on eutrophication and related studies of the Thurles area of the River Suir (May–July 1978).* An Foras Forbatha. Water Resources Division.

Horkan, K. (1984). Suir. In B.A. Whitton (ed.), *Ecology of European Rivers.* Oxford: Blackwell Scientific Publications, 385–406.

Hoskins, W.G. (1953). *The Making of the English Landscape.* London: Penguin.

Hoskins, W.G. (1973). *English Landscapes. How to Read the Man-made Scenery of England.* London: British Broadcasting Corporation Publications.

Howells, G. (1976). Introduction. Effects of pollutants on aquatic organisms. In A.P.M. Lockwood (ed.), *Effects of Pollutants on Aquatic Organisms.* Cambridge: Cambridge University Press, 1–16.

Howells, G.D. (1989). Fishery decline, mechanisms and predictions. *Phil. Trans. R. Soc. Lond. B* **305**, 529–48.

Hudson, K. (1984). *Industrial History from the Air,* Cambridge: Cambridge University Press.

Huet, M. (1951). Nocivité des boisements en Epicéas (*Picea excelsa* Link) pour certains cours d'eau de l'Ardenne Belge. *Verh. Int. Verein. Limnol.* **11**, 189–200.

Huet, M. (1954). Biologie, profils en long et en travers des eaux courantes. *Bull. Franc. Piscic.* **175**, 41–53.

Huet, M. (1959). Profiles and biology of Western European streams as related to fish management. *Trans. Am. Fish. Soc.* **88**, 155–63.

Huet, M. (1962). Influence de courant sur la distribution des poissons dans les eaux courants. *Schweiz. Z. Hydrobiol.* **24**, 413–32.

Hughes, G.M. (1976). Polluted fish respiratory physiology. In A.P.M. Lockwood (ed.), *Effects of Pollutants on Aquatic Organisms*. Cambridge: Cambridge University Press, 163–83.

Hunt, B. (1982a). The ancient history of the Valley. In F. Myatt (ed.), *The Deverill Valley*. The Deverill Valley History Group, Wellington Cottage, Longbridge Deverill, 21–30.

Hunt, B. (1982b). The Roman period. In F. Myatt (ed.), *The Deverill Valley*. The Deverill Valley History Group, Wellington Cottage, Longbridge Deverill, 31–8.

Huttermann, A. and Ulrich, B. (1984). Solid phase-solution–root interactions in soils subjected to acid deposition. *Phil. Trans. R. Soc. Lond.* B **305**, 369–82.

Hynes, H.B.M. (1960). *The Biology of Polluted Waters*. Liverpool: Liverpool University Press.

Hynes, H.B.M. (1970). *The Ecology of Running Waters*. Liverpool: Liverpool University Press.

Illies, J. and Botosaneau, L. (1963). Problèmes et méthodes de la classification et de la zonation écologique des eaux courantes, considerées surtout du point du vue faunistique. *Mitt. Int. Verein. theor. angew. Limnol.* **12**.

Ineson, P. (ed.) (1986). *Pollution in Cumbria*. Abbots Ripton: Institute of Terrestrial Ecology.

Jaccard, P. (1912). The distribution of the flora in the alpine zone. *New Phytol.* **11**, 37–50.

Johnson, G.A.L. (1981). An outline of the geology of North East England. In P.J. Say and B.A Whitton (eds), *Heavy Metals in Northern England: Environmental and Biological Aspects*. Department of Botany, University of Durham, 1–6.

Jones, H.R. and Peters, T.C. (1976). *Physical and biological typings of unpolluted rivers*. Water Research Centre and Central Water Planning Unit, Reading: CWPU Report.

Jones, H.R. and Peters, T.C. (1977). *Physical and biological typing of unpolluted rivers*. Water Research Centre, Technical Report No. 41.

Jones, J.G. (1975). Heterotrophic micro-organisms and their activity. In B.A. Whitton (ed.), *River Ecology*. Oxford: Blackwell Scientific Publications, 141–54.

Jones, J.R.E. (1940). A study of the zinc-polluted river Ystwyth in North Cardiganshire, Wales. *Ann. Appl. Biol.* **27**, 368–78.

Jong, J. de (1976). The purification of wastewater with the aid of rush or reed ponds. In J. Tourbier and R.W. Pierson (eds), *Biological Control of Water Pollution*, Philadelphia: University of Pennsylvania Press, 133–9.

Jong, J. de and Kok, T. (1978). The purification of wastewater and effluents using marsh vegetation and soils. *Proc. EWRS 5th Symposium on Aquatic Weeds, 1978*, 135–42.

Jong, J. de, Kok, T. and Koridon, A.H. (1977). The purification of sewage with the aid of ponds containing bulrushes or reeds in the Netherlands. *Rijkskienst voor de Ijsselmeerpolders*. Lelystad: Smedinghuis.

Jorgo, W. and Weise, G. (1977). Biomasseentwicklung submerser Macrophyten in langsam fliessenden Gewässern in Beziehung zum Sauerstoffhaushalt. *Revue Ges. Hydrobiol.* **62**, 209–34.

Junk, G.A. and Stanley, S.E. (1975). *Organics in Drinking Water. Part 1. Listing of identified chemicals*. US Energy Research and Development Administration.

Keller, T. (1984). Direct effects of sulphur dioxide on trees. *Phil. Trans. R. Soc. Lond.* **305**, 317–26.

Kelly, M. (1989). Monitoring water pollution: the role of plants. *Plants today*, 96–100.

Kendall, M.G. (1962). *Rank Correlation Methods*. London: Griffin & Co.

Kern-Hansen, U. (1978). The drift of *Gammarus pulex* L. in relation to macrophyte cutting in four small Danish lowland streams. *Verh. Internat. Verein. Limnol.* **20**, 1440–5.

Kern-Hansen, U. and Dawson, F.H. (1978). The standing crop of aquatic plants of lowland streams. *Proc. EWRS 5th Symposium on Aquatic Weeds, 1978*, 143–50.

Kern-Hansen, U. and Holm, T.F. (1982). Aquatic plant management in Danish streams. *Proc. EWRS 6th Symposium on Aquatic Weeds, 1982*, 122–31.

Kickuth, R. (1976). Degradation and incorporation of nutrients from rural waste waters by plant rhizosphere under limnic conditions. In *Utilisation of Manure by Land Spreading*. Commission of the European Communities, 335–47.

Kickuth, R. (1984). Das Wurzelraumverfahren in der Praxis. *Landsch. Stadt.* **16**, 145–54.

Kickuth, R. and Tittizer, Th. (1974a). Makrophyten limnischer Standorte und ihr Verhalten gegenüber p-Toluoslsulfonsä-re im Substrat. I. Über die Toleranz von Sumpf- und Wasserpflanzen gegenüber p-Toluolsulfonsäure. *Angew. Bot.* **48**, 185–94.

Kickuth, R. and Tittizer, Th. (1974b). Makrophyten limnischer Standorte und ihr Verhalten gegenüber p-Toluolsulfonsäure im Substrat. II. Elimination von p-Toluolsulfonsäure im Modellsystem mit Sumpf- und Wasserpflanzen. *Angew. Bot.* **48**, 195–207.

Kickuth, R. and Tittizer, Th. (1974c). Makrophyten limnischer Standorte und ihr Verhalten gegenüber p-Toluolsulfonsäure im Substrat. III. Abbau von p-Toluolsulfonsäure durch Pflanzen limnischer Standorte und ihre Rhizosphärengesellschaft. *Angew. Bot.* **48**, 355–70.

Kinsman, D.J.J. (1984). Ecological effects of deposited S and N compounds: effects on aquatic biota. *Phil. Trans. R. Soc. Lond.* B **305**, 479–85.

Kloeden, J.L. (1976). An ecological assessment of alluvial grassland in the Beult River Valley, Kent. *Discussion papers in Conservation II*. University College, London.

Knöpp, H. (1961). Der A–Z Test, ein neues Verfahren zur toxologischen Prüfung von Abwässern. *Dt.*

Gewässerkundl. Mitt. **5**, 66–73.

Kobayashi, J. (1971). Relation between the 'itai-itai' disease and the pollution of river water by cadmium from a mine. *Proc. 5th Internat. Wat. Poll. Res. Conf.* **1–25**. Oxford: Pergamon Press, 1–7.

Kohler, A. (1975a). Macrophytische Wasserpflanzen als Bioindikatoren für Belastungen von Fliessgewässerökosystemen. *Verh. Okologie, Wien* **3**, 255–76.

Kohler, A. (1975b). Veränderung natürlicher submerser Fliessgewässervegetation durch organische Belastung. *Daten und Dokumente zum Umweltschulz* **14**, 59–66.

Kohler, A. (1978). Methoden der Kartierung von Flora und Vegetation von Süsswasserbiotopen. *Landschaft und Stadt* **13**, 73–85.

Kohler, A. (1980). Gewässerbiotope in Agrarlandschaften. *Landwirtsch. Forsch.* **37**, 46–60.

Kohler, A. (1981). Die Vegetation bayerischer Fliessgewässer und einige Aspekte ihrer Veränderung. In *Fliess gewässern in Bayern.* Tagungsbericht 5/81. Akademie für Naturschutz und Landschaftspflege, 6–18.

Kohler, A. (1982). Wasserpflanzen als Belastungsindikatoren. *Decheniana* **26**, 31–42.

Kohler, A., Brinkmeier, R. and Vollrath, H. (1974). Verbreitung und Indikatorwert der submersen Makrophyten in den Fliessgewässern der Friedberger Au. *Ber. Bayer. Bot. Ges.* **45**, 5–36.

Kohler, A. and Labus, B.C. (1983). Eutrophication processes and pollution of freshwater ecosystems including waste heat. In O.L. Lange, P.S. Nobel, C.B. Osmond and H. Ziegler (eds), *Encyclopedia of Plant Physiology N.S. 12D, Physiological Plant Ecology IV.* Berlin: Springer Verlag, 413–64.

Kohler, A., Pensel, T. and Zeltner, G.H. (1980). Veränderungen von Flora und Vegetation in den Fliessgewässern der Friedberger Au (bei Augsburg) zwischen 1972 und 1978. *Ver. Ges. Ökol.* **8**, 343–50.

Kohler, A. and Schiele, S. (1985). Vernänderungen von Flora und Vegetation der Fliessgewässer der Friedberger Au (bei Augsburg) von 1972 bis 1982 nach veränderten Belastungsbedingungen. *Ver. Ges. Ökol.* **13**, 207–9.

Kohler, A. and Schoen, R. (1984). Versauerungsresisternz submerser Makrophyten in Gewässerversauerung in der Bundesrepublik Deutschland. *Matierialien* **1/84**. Erich Schmidt, Verlag, Berlin, 353–69.

Kohler, A., Vollrath, H. and Beisl. E. (1971). Zur Verbreitung, Vergesellschaftung und Ökologie der Gefass-Makrophyten im Fliesswässersystem Moosach (Münchener Ebene). *Arch. Hydrobiol.* **69**, 333–65.

Kohler, A., Wonneberger, R. and Zeltner, G. (1973). Die Bedeutung chemischer und pflanzlicher 'Verschmutzungsindikatoren' im Fliessgewässersystem Moosach (Münchener Ebene). *Arch. Hydrobiol.* **72**, 533–49.

Kohler, A. and Zeltner, G.H. (1974). Verbreitung und Ökologie von Makrophyten in Weichwasserflüssen des Oberpfälzer Waldes. *Denkschr. Regensb. Bot. Ges.* **33**, 171–232.

Kohler, A. and Zeltner, G.H. (1981). Der Einfluss von Be- und Entlastung auf die Vegetation von Fliessgewässern. *Daten und Dokumente zum Umweltschutz Sonderreihe Umweltagung,* Universität Hohenheim, **31**, 127–39.

Kolwitz, R. and Marsson, M. (1902). Grundsätze für die Biologische Beurteilung des Wassers nach seiner Flora und Fauna. *Mitt. a.d. Kgl Prüfungsanst. f. Wasserversorg. u. Abwsserbeseitingung zu Berlin* **1**, 33–72.

Kolkwitz, R. and Marsson, M. (1908). Oekologie der Pflanzen Saprobien. *Int. Rev. Ges. Hydrobiol.* **2**, 126–52.

Kolkwitz, R. and Marsson, M. (1909). Oekologie der tierischen Saprobien. *Ber. d. Deut. Bot. Gesell.* **26**, 505–19.

Kothé, P. (1962). Der 'Artenfehlbetrag', ein einfaches Gütekriterium und seine Anwendung bei biologischen Vorflutersuntersuchungen. *Dt. Gewässerkundl. Mitt.* **6**, 60–5.

Kothé, P., Otto, A. and Braukmann, U. (1980). Typology of running waters in rural areas. Unpublished Report, Freshwater Contact Group, Commission of the European Communities, Brussels.

Krause, A. (1977). On the effect of marginal tree rows with respect to the management of small lowland streams. *Aquat. Bot.* **3**, 185–92.

Krause, W., Hügin, G. and Bundesforschungsanstalt für Naturschutz und Landschaftsökologie (1987). Ökologische Auswierkungen von Altarmverbundsystemen am Beispiel des Altrheinausbaus. *Natur und Landschaft* **62**.

Kulczynski, S. (1928). Die Pflanzenassozionationen der Pienimen. *Bull. Int. Acad. Pol. Sci. Lett. B.,* Suppl. **2**, 57–203.

Labus, B.C. (1979). Der Einfluss des Waschrohstoffs Marlon A (Anionenaktives Tensid) auf das Wachstum und die Nettophotosynthese verschiedener submerser makrophytischer Wasserpflanzen unter besonderer Berücksichtigung primärer Standortsfaktoren. Degree dissertation, Institut fur Landeskultur und Pflanzenokologie, Universität Hohenheim, Stuttgart.

Labus, B.C. and Kohler, A. (1976). Die Wirkung anionenaktiver Tenside auf submerse Wasserpflanzen. *Dokumentationensstelle der Universität Hohenheim* **19**, 141–52.

Labus, B.C. and Kohler, A. (1981). Die Rolle einiger primärer ökologischer Faktoren bei Nettophotosynthese-Messungen zur Prüfung der Wirkung von Umweltchemikalien auf submerse Makrophyten. *Limnol. (Berlin)* **13**, 373–98.

Labus, B.C., Nobel, W., Smetana, R. and Kohler, A. (1976). Der Einfluss der Abwassersubstanzen Marlon A (Anionenaktives Tensid) und Bor auf die Photosyntheserate einiger submerser Makrophyten: *Verh.*

Ges. Ökol. Göttingen, 1976, 325–33.

Labus, B., Schuster, H., Nobel, N. and Kohler, A. (1977). Wirkung von taxischen Abwaserkomponenten auf submerse Makrophyten. *Angew. Bot.* **51**, 17–36.

Lange, L. de and Zon, J.C.G. van (1978). Evaluation of the botanical response to different methods of aquatic weed control, based on the structure and floristic composition of the macrophytic vegetation. *Proc. EWRS SH Symp. on Aquatic Weeds* 1978, 279–96.

Lange, L. de and Zon, J.C.J. van (1983). A system for the evaluation of aquatic biotypes based on the composition of the aquatic vegetation. *Biol. Cons.* **25**, 273–84.

Langford, T.E. (1983). *Electricity Generation and the Ecology of Natural Waters.* Liverpool: Liverpool University Press.

Lawson, T. (1985). Cultivating reeds for root zone treatment of sewage. Institute of Terrestrial Ecology, Grange-over-Sands, Cumbria.

Léglize, L. and Crochard, C. (1987). Vérification experimentale du choix de *Dreissena polymorpha* Pallas (Lamellibranchiate) comme bioindicateur de contaminatin métallique. *Naturaliste. Can. (Rev. Ecol. Syst.)* **114**, 315–23.

Lester, W.F. (1975). Polluted river: River Trent. In B.A. Whitton (ed.), *River Ecology.* Oxford: Blackwell Scientific Publications, 489–513.

Lewis, J. (ed.) (1981). *British Rivers.* London: George Allen & Unwin.

Limbrey, S. (1983). Archaeology and palaeohydrology 8. In K.J. Gregory (ed.), *Background to Palaeohydrology.* Chichester: Wiley.

Litav, M. and Lehrer, Y. (1978). The effects of ammonium in water on *Potamogeton lucens. Aquat. Bot.* **5**, 127–38.

Lloyd, R. and Swift, D.J. (1976). Some physical response by freshwater fish to low dissolved oxygen, high carbon dioxide, ammonia and phenol with particular reference to water balance. In A.R.M. Lockwood (ed.), *Effects of Pollutants on Aquatic Organisms.* Cambridge: Cambridge University Press, 46–69.

Lockwood, A.P.M. (ed.) (1976). *Effects of Pollutants on Aquatic Organisms.* Cambridge: Cambridge University Press.

Lohmeyer, W. and Krause, A. (1974). Über den Gehölzbewuchs und kleinen Fliessgewässern Nordwestdeutschlands und seine Bedeutung für den Uferschutz. *Natur. Landsch.* **49**, 323–30.

Lohmeyer, W. and Krause, A. (1975). Über die Auswirkungen des Gehölzbewuchses an kleinen Wassertaufen des Münsterlandes auf die Vegetation im Wasser und an den Böschungen im Hinblick auf die Unterhaltung der Gewässer. *Schriftenr. Vegetationsk.* **9**, 1–105.

Looman, J. and Campbell, J.B. (1960). Adaptation of Sorensen's K (1948) for estimating affinities in prairie vegetation. *Ecology* **41**, 409–16.

Lovegrove, R. and Snow, P. (1984) *River Birds,* London: Columbus Books.

Maas, D. and Kohler, A. (1983). Die Makrophytenbestände der Donau im Raum Tuttlingen. *Landschaft und Stadt* **15**, 49–60.

Macdonald, S.M. and Mason, C.F. (1983). Some factors affecting the distribution of otters. *Mamm. Rev.* **13**, 1–10.

Maesener, J. de, Paelinck, H., Verheven, R. and Hulle, D. van (1982). Use of artificial reed marshes for treatment of industrial waste-waters and sludge. In J.J. Symoens, S.S. Hooper and P. Compère (eds), *Studies on Aquatic Vascular Plants.* Brussels: Royal Botanical Society of Belgium, 363–9.

Magnuson, J.J., Baker, J.P. and Rohen, E.J. (1989). A critical assessment of effects of acidification on fisheries in N. America. *Phil. Trans. R. Soc. Lond.* B **305**, 501–16.

Margalef, R. (1951). Diversidad de especies en las comunidades naturales. *Publnes. Inst. Biol. Apl. (Barcelona)* **6**, 59–72.

Margalef, R. (1968). *Perspectives in Ecological Theory.* London: University of Chicago press.

Markmann, P.N. (1988). Marginal jorder — besky Helses braemmer og-zoner. *Vand and Miljo* 2/1988, 77–81.

Mason, C.F. and Macdonald, S.M. (1982). The input of terrestrial invertebrates from tree canopies to a stream. *Freshw. Biol.* **12**, 305–11.

Mason, C.F., Macdonald, S.M. and Aspen, V.J. (1982). *Metals in Freshwater Fishes in the United Kingdom 1980–81.* London: The Vincent Wildlife Trust.

McCarthy, D.T. (1975). The effects of drainage on the flora and fauna of a tributary of the River Boyne. Fishery Leaflet **68**, Department of Agriculture and Fisheries, Fisheries Division, Dublin.

McCarthy, D.T. (1977). The effects of drainage on the Trimblestown River. I. Benthic invertebrates and Flora. Irish Fisheries Investigations. Series A. **16**.

McIntosh, R.P. (1967). An index of diversity and the relation of certain concepts to diversity. *Ecology* **48**, 392–404.

Mellanky, K. (1979). A water pollution survey, mainly by British Schoolchildren. *Environ. Pollut.* **6**, 161–73.

Menhinick, E.F. (1964). A comparison of some species-individuals' diversity indices applied to samples of field insects. *Ecology* **45**, 859–61.

Mériaux, J.L. (1982a). Inventaire et distribution des espèces des genres *Callitriche, Elodea* et *Ranunculus* (sous genre *Batrachium*) dans le nord de la France. In J.J. Symoens, S.S. Hooper and P. Compère (eds), *Studies on Aquatic Vascular Plants.* Brussels, Royal Botanical Society of Belgium, 311–12.

Mériaux, J.L. (1982b). Espèces rares ou menacées des biotopes lacustres et fluviatiles du nord de la France. In J.J. Symoens, S.S. Hooper and P. Compère (eds), *Studies on Aquatic Vascular Plants.* Brussels, Royal

Botanical Society of Belgium, 398–402.

Mériaux, J.L. (1982c). L'Utilisation des macrophytes et des phytocoenoses aquatiques comme indicateurs de la qualité des eaux. *Les Naturalistes Belges* **63**, 12–28.

Miles, W.D. (1976). Land drainage and weed control. In *Aquatic Herbicides*. Oxford: British Crop Protection Council, 7–13.

Miller, S.N. and Skertchley, S.B.J. (1878). *The Fenland, Past and Present*. Wisbech: Leach.

Morgan, M.D. and Philipp, K.R. (1986). The effect of agricultural and residential development on aquatic macrophytes in the New Jersey Pine Barrens. *Biol. Cons.* **35**, 143–58.

Morisawa, M. (1968). *Streams, their Dynamics and Morphology*. Philadelphia: McGraw-Hill Book Co.

Morton, H.V. (1936). *In the Steps of St. Paul*. London: Rich & Cowan.

Mountford, M.D. (1962). An index of similarity and its application to classificatory problems. In P.W. Murphy (ed.), *Progress in Soil Zoology*. London: Butterworth.

Muniz, I.P. (1984). The effects of acidification on Scandinavian freshwater fish fauna. *Phil. Trans. R. Soc. Lond. B*, **305**, 517–28.

Murphy, K.J. and Eaton, J.W. (1981a). Ecological effects of four herbicides and two mechanical clearance methods used for aquatic weed control in canals. *Proc. Aquatic Weeds and their Control, 1981,* 201–17.

Murphy, K.J. and Eaton, J.W. (1981b). Water plants, boat traffic and angling in canals. *Proc. 2nd Brit. Freshw. Fish Conf., 1981,* 173–87.

Murphy, K.J. and Eaton, J.W. (1982). The management of aquatic plants in a navigable canal system used for amenity and recreation. *Proc. EWRS 6th Symposium on Aquatic Weeds, 1982,* 141–50.

Murphy, K.J. and Eaton, J.W. (1983). The effects of pleasure boat traffic on macrophyte growth in canals. *J. Appl. Ecol.* **20**, 713–29.

Murphy, K.J., Eaton, J.W. and Hyde, T.M. (1980). A survey of aquatic weed growth and control in the canals and river navigations of the British Waterways Board. *Proc. 1980 Brit. Crop Protection Conf. — Weeds* **70**, 7–14.

Murphy, K.J., Eaton, J.W. and Hyde, T.M. (1982). The management of aquatic plants in a navigable canal system used for amenity and recreation. *Proc. 6th EWRS Symposium on Aquatic Weeds, 1982,* 141–51.

Murphy, K.J., Hanbury, T.G. and Eaton, J.W. (1981). The ecological effects of 2-methyl-thiotriazine herbicides used for aquatic weed control in navigable canals. I. Effects on aquatic flora and water chemistry. *Arch. Hydrobiol.* **91**, 294–331.

Myatt, F. (ed.) (1982). *The Deverill Valley*. The Deverill Valley History Group, Wellington Cottage, Longbridge Deverill, Warminster, Wiltshire.

Newbold, C. (1975). Herbicides in aquatic systems. *Biol. Conserv.* **7**, 97–118.

Newbold, C. (1976). Environmental effects of aquatic herbicides. In *Aquatic herbicides*. Oxford: British Crop Protection Council, 78–90.

Newbold, C. (1981). The decline of aquatic plants and associated wetland wildlife in Britain — Causes and perspectives on management technique. *Proc. Aquatic Weeds and their control, 1981,* 241–53.

Newbold, C., Furseglove, J. and Holmes, N. (1983). *Nature Conservation and River Engineering*. London: Nature Conservancy Council.

Newbold, C. and Holmes, N.T.H. (1987). Nature Conservation: Water quality criteria and plants as water quality monitors. *Water Pollut. Control* **86**, 345–64.

Newton, C. (1944). Pollution of the rivers of West Wales by lead and zinc mine effluent. *Ann. Appl. Biol.* **31**, 1–12.

Nielsen, M.B. (1985). Grødesæringsforsøg i Surback. *Vand og Miljø* 3/1985. 120–4.

Nielsen, M.B. (1986). Sediment transport og vandløbsvedlinge-holdeskse — behav for projekter til belysning af økologiske konsek ven. Nordisk Hydrologick Program. *NHP-Rapport* **14**, 341–59. Geography Institute, University of Copenhagen.

Nobel, W. (1980). *Der Einfluss der Belastungsstoffe Chlorid, Borat und Phosphat auf die Photosyntheseleistung submerser Weichwassermakrophyten*. Institut für Landeskultur und Pflanzenökologie der Universität Hohenheim.

Nobel, W., Mayer, T. and Kohler, A. (1983). Submerse Wasserpflanzen als Testorganismen für Belastungsstoffe. *Z. Wasser Abwasser Forsch.* **16**, 87–90.

Noirfalise, A. and Dethioux, M. (1977). Synopsis des vegetations aquatiques d'eaux douce en Belgique. *Communs. Cent. ecol. forest. rurale (IRSIA)* **14 (N.S.)**.

Nuttall, P.M. (1972). The effects of sand deposition upon the macroinvertebrate fauna of the River Camel, Cornwall. *Freshwat. Biol.* **2**, 181–6.

Nyberg, P. (1984). Effects of liming on fisheries. *Phil. Trans. R. Soc. Lond. B* **305**, 549–60.

Ortscheit, A., Jaeger, P., Carbinier, R. and Kapp, E. (1982). Les Modifications des eaux et de la végétation aquatique du Waldhrein consecutives à la mise en place de l'ouvrage hydroelectrique de Gambsheim au nord de Strasbourg. In J.J. Symoens, S.S. Hooper and P. Compère (eds), *Studies on Aquatic Vascular Plants*. Brussels, Royal Botanical Society of Belgium, 277–83.

Otto, A. (1979). Typology of running waters in rural areas. Unpublished Report, Freshwater Contact Group, Commission of the European Communities, Brussels.

Otto, A. (1980). Gewässertypologie im landlichen Raum. Bundestalt für Gewässerkunde, *Zwischenbericht zum Forschungsauftrag,* **77**, Koblenz.

Palmer, M.A. (1984). A comparison of the flora and invertebrate fauna of watercourses in old pasture and arable land in the Pevensey Levels, Sussex, in 1983. Chief Scientist's Team, Nature Conservancy Council.

Palmer, M. (1986). The impact of a change from permanent pasture to cereal farming on the flora and invertebrate fauna of watercourses in the Pevensey Levels, Sussex. *Proc. EWRS/AAB 7th Symposium on Aquatic Weeds, 1986*, 233–8.

Pannett, D.J. (1981). Fish weirs of the River Severn. In *Evolution of Marshland Landscapes*. Oxford: University Department for External Studies, 144–57.

Pantle, R. and Buck, H. (1955). Die biologische Überwachung der Gewässer und die Darstellung der Ergebnisse. *Gas- u. Wasserfach* **96**, 604.

Parker, R. (1975). *The Village Stream*. London: Paladin.

Parkin, L. (1973). Last throw on the Mimram. *Association of River Authorities Year Book and Directory, 1973*. London: National Water Council, 199–205.

Patterson, G. and B.A. Whitton (1981). Chemistry of water, sediments and algal filaments in groundwater draining in old lead–zinc mines. In P.J. Say and B.A. Whitton (eds), *Heavy Metals in Northern England. Environmental and Biological Aspects*. Department of Botany, University of Durham, 60–72.

Pennington, W. (1984). 52nd Annual Report. Freshwater Biological Association, 28–46.

Persoone, G. (1978). *A Proposal for a Biotypological Classification of Watercourses in the European Communities*. Brussels: Commission of the European Communities.

Pinder, L.C.V. and Far, I.S. (1987). Biological surveillance of water quality — 3. The influence of organic enrichment on the macroinvertebrate fauna of small chalk streams. *Arch. Hydrobiol.* **109**, 619–37.

Pinder, L.C.V., Ladle, M., Gledhill, T., Bass, J.A.B. and Matthews, A.M. (1987). Biological Surveillance of water quality — 1. A comparison of macroinvertebrate surveillance methods in relation to assessment of water quality in a chalk stream. *Arch. Hydrobiol.* **109**, 207–26.

Pinter, I. and Backhaus, D. (1984). Neckar. In B.A. Whitton (ed.), *Ecology of European Rivers*. Oxford: Blackwell Scientific Publications, 317–44.

Pittwell, L.R. (1976). Biological monitoring of rivers in the community. In R. Amavis and J. Smeets (eds), *Principles and methods of determining ecological criteria on hydrobiocenoses*. Oxford: Pergamon Press, for the Commission of the European Communities, 223–61.

Potter, T.W. (1981). Mediaeval drainage in the classical world. In *Evolution of Marshland Landscapes*. Department of External Studies, Oxford University, 1–19.

Prat, N.F., Puig, M.A., Gonzalez, G., Tort, M.F. and Estrada, M. (1984). Llobregat. In B.A. Whitton (ed.), *Ecology of European Rivers*. Oxford: Blackwell Scientific Publications, 527–52.

Preston, F.W. (1948). The commonness and rarity of species. *Ecology*, **29**, 254–83.

Raabe, E.W. (1952). Über den 'Affinitätswert' in der Pflanzensoziologie. *Vegetatio, Haag* **4**, 53–68.

Rabe, R., Nobel, W. and Kohler, A. (1982). Effects of sodium chloride on photosynthesis and some enzyme activities of *Potamogeton alpinus*. *Aquat. Bot.* **14**, 159–65.

Rabe, R., Schuster, H. and Kohler, A. (1982). Effects of copper chelate on photosynthesis and some enzyme activities of *Elodea canadensis*. *Aquat Bot.* **14**, 167–75.

Rahtz, P.A. (1981). Medieval Milling. In D.W. Crossley (ed.), *Medieval Industry*. CBA Research Report **40**, Council for British Archaeology, 1–15.

Rasmussen, K. (1980). Undersøgelse of Rens Dambrugs ind flyd else på plants — og dyrelivet i Sønder. Unpubl. Report. Sønderamtskommune.

Rasmussen, L. and Sand-Jensen, K. (1979). Heavy metals in acid streams from lignite mining areas. *The Science of the Total Environment* **12**, 61–74.

Raven, P. (1986). Changes in the breeding population of a small clay river following flood alleviation works. *Bird Study*, **33**, 24–35.

Robson, T.O. (1973). *The Control of Aquatic Weeds*. Ministry of Agriculture, Fisheries and Food, *Bull.* **194**. London: HMSO.

Robson, T.O. (1974). Mechanical control. In D.S. Mitchell (ed.), *Aquatic Vegetation and its Use and Control*. Paris: Unesco, 72–84.

Robson, T.O. (1978). The present state of chemical aquatic weed control. *Proc. EWRS 5th Symposium on Aquatic Weeds, 1978*, 17–25.

Robson, T.O. and Fillenham, L.F. (1976). *Weed control in land drainage channels in Britain*. FAO. European Commission on Agriculture. Report ECA: WR/76/3(h).

Rodgers, J. (1947–8). *English Rivers*, London: B.T. Batsford.

Sand-Jensen, K. and Rasmussen, L. (1978). Macrophytes and chemistry of acidic streams from lignite mining areas. *Bot. Tidsskr.* **72**, 105–12.

Sanders, P.F. (1982). *An investigation into the growth of 'sewage fungus' in the lower River Don, Aberdeenshire*. Scottish Development Agency.

Sawyer, F. (1952). *Keeper of the Stream*. London: Adam & Charles Black.

Say, P.J. and Whitton, B.A. (1981a). Chemistry and plant ecology of zinc-rich streams in the Northern Pennines. In P.J. Say and B.A. Whitton (eds), *Heavy Metals in Northern England: Environmental and Biological Aspects*. Department of Botany, University of Durham, 55–63.

Say, P.J. and Whitton, B.A. (eds) (1981b). *Heavy Metals in Northern England: Environmental and Biological Aspects*. Department of Botany, University of Durham.

Say, P.J. and Whitton, B.A. (1981c). Aquatic mosses as monitors of heavy metals contamination in the River Etherow, Great Britain. *Environ. Pollut. Ser. B*, 295–307.

Schloesser, D.W. and Manny, B.A. (1989). Potential effects of Shipping on submersed macrophytes in the St Clair and Detroit Rivers of the Great Lakes. *Michigan Academician* **21**, 101–8.

Schmedtje, U. and Kohmann, F. (1987). Bioindication

by macrophytes — can macrophytes indicate saprobity? *Arch. Hydrobiol.* **109**, 455–69.

Schmitz, A., Gommes, R., Vander Borght, P. and Vets, A. (1982). L'Epuration en marais natural: Cussigny. In J.J. Symoens, S.S. Hooper and P. Compère (eds), *Studies on aquatic vascular plants.* Brussels: Royal Botanical Society of Belgium, 353–87.

Schuster, H., Kohler, A. and Krub, K. (1976). Eine neue Methode zur Beurteilung des Belastbarkeit von submersen Makrophyten. *Ver. Ges. Okol., Göttingen, 1976*, 335–45.

Seidel, K. (1956–7) Unsere Flechtbinden. *Hydrobiol. Anstalt der Max- Planck-Gesellschaft.* Druck, Eril Patzschke, Neustadt b. Coburg.

Seidel, K. and Kickuth, R. (1967). Biologische Behandlung phenolhaltiger Abwässer mit Hilge der Flechtbinse (*Scirpus lacustris* L.). *Wass. Wirt. Wass. Tech.* **G**, 209–10.

Seidel, K. Scheffer, F., Kickuth, R. and Schlemme, E. (1967). Auf natume und Umwandlung Organischer Stoffe durch die Flechtbinse. *Gas u Wassfach* **18**, 6, 138–9.

Shackleford, W.M. and Keith, L.H. (1976). Frequency of organic compounds identified in water. *Environmental Monitoring Series Report.* EPA–600/4–76–062.

Shannon, C.E. (1948). A mathematical theory of communication. *Bell Systems Tech. J.* **27**, 379–423, 623–56.

Shannon, C.E. and Weaver, W. (1949). *The Mathematical Theory of Communication.* Urbana: University of Illinois Press.

Sharpe, V. and Denny, P. (1976). Electron microscope studies on the absorption of localisation of lead in the leaf tissue of *Potamogeton pectinatus* L. *J. Exp. Bot.* **27**, 1155–62.

Sheail, J. (1971). The formation and maintenance of water-meadows in Hampshire, England. *Biol. Conserv.* **3**, 101–6.

Shimway, D.C., Warren, C.E. and Daudoroff, P. (1963). Influence of oxygen centration and water movement on the growth of steelhead trout and coho salmon embryos. *Trans. Amer. Fish. Soc.* **93**, 342–56.

Shimwell, D.W. (1971). *Description and Classification of Vegetation.* London: Sidgwick & Jackson.

Simpson, E.H. (1949). Measurement of diversity. *Nature* **163**, 688.

Sládaček, V. (1973a). The reality of three British biotic indices. *Wat. Res.* **7**, 995–1002.

Sládaček, V. (1973b). System of water quality from the biological point of view. *Arch. Hydrobiol.* (Ergbn. Limnol.) **7**, 1–218.

Slater, F.M., Curry, P. and Chadwell, C. (1987). A practical approach to the evaluation of the conservation status of vegetation in river corridors in Wales. *Biol. Cons.* **40**, 53–68.

Smith, I.R. (1975). *Turbulence in Lakes and rivers.*

Freshwater Biological Association. Scientific Publication, **29**.

Smith, I.R. (1979). The physics of freshwater systems. Model development and ecological applications. Unpublished Report. Institute of Terrestrial Ecology, Edinburgh.

Sokal, R.R. (1961). Distance as a measure of taxonomic similarity. *Syst. Zool.* **10**, 71–9.

Spearman, C. (1913). Correlation of sums and differences. *Brit. J. Psychol.* **5**, 417–26.

Steenvoorden, J.H.A.M. (1976). *Nitrogen, phosphate and biocides in groundwater as influenced by soil factors and agriculture.* Institute for Land and Water Management Research. *Tech. Bull.* **97**.

Stoke, E. (1967). Higher vegetation and nitrogen in a rivulet in Central Sweden. *Hydrobiologia* **29**, 107–23.

Stoke, E. (1968). Higher vegetation and phosphorous in a small stream in Central Sweden. *Hydrobiologia* **30**, 353–73.

Stott, B. (1981). Progress towards the practical use of grass carp for water weed control. *Proc. Aquatic Weeds and their Control, 1981,* 117–23.

Stott, B. and Robson, T.O. (1970). Efficiency of grass carp (*Ctenopharyngodon idella* Val.) in controlling submerged water weeds. *Nature* **226**, 870.

Stratton, R. (1982). Farming in the upper Deverills. In F. Myatt (ed.), *The Deverill Valley.* The Deverill Valley History Group, Wellington Cottage, Longbridge Deverill, Warminster, Wiltshire, 83–98.

Sutcliffe, D.W. and Carrick, T.R. (1986). Effects of acid rain on waterbodies in Cumbria. In P. Ineson (ed.), *Pollution in Cumbria.* Abbotts Ripton: Institute of Terrestrial Ecology, 16–25.

Sørensen, T. (1948). A method of establishing groups of equal amplitude in plant sociology based on similarity of species content and its application to analyses of the vegetation on Danish commons. *Biol. Skr.* (K. Danske. Vidensk. Selsk. N.S.) **5**, 1–34.

Tazik, P.F. (1986). Aquatic macrophyte production and mortality: demonstrating mortality due to herbivory. *Proc. EWRS/AAB 7th Symposium on Aquatic Weeds, 1986,* 345–50.

Tendron, G. and Ravera, O. (1976). Conclusions. In R. Amavis and J. Smeets (eds), *Principles and Methods for Determining Ecological Criteria on Hydrobiocenoses.* Oxford: Pergamon Press for Commission of the European Communities, 355–89.

Thieneman, A. (1954). Ein drittes biozöatisches Grundprinzip. *Arch. Hydrobiol* **49**, 421–2.

Urk, G. van (1984). Lower Rhine-Meuse. In B.A. Whitton (ed.), *Ecology of European Rivers.* Oxford: Blackwell Scientific Publications, 437–68.

Verneaux, J. (1976). Basic principles for the determination of ecological criteria for hydrobiocenoses of running water. In R. Amavis and J. Smeets (eds), *Principles and Methods for Determining Ecological Criteria on Hydrobiocenoses.* Oxford: Pergamon Press for Commission of the European Communities, 191–200.

Vrhovšek, D., Martinčič, A. Kralj, M. and Sremfeli, M. (1985). Pollution degree of two alpine rivers evaluated with Bryophyta species. *Biol. Vestn.* **33**, 95–106.

Voulgarapoulos, A., Fytianos, K., Apostolopoulou, A. and Gounaridou, X. (1987). Correlation of some organic pollution factors in water systems in northern Greece. *Chemosphere* **16**, 395–8.

Wade, P.M. (1981). The long-term effects of aquatic herbicides on the macrophyte flora of freshwater habitats — a review. *Proc. Aquatic Weeds and their Control, 1981*, 233–40.

Wade, P.M. (1982). The long-term effects of herbicide treatment on aquatic weed communities. *Proc. EWRS 6th Symposium on Aquatic Weeds, 1982*, 278–86.

Warnek, L. and Kohler, A. (1988). Veränderungen der Fliessgewässervegetation in der Fiedberger Au von 1972–1987. In A. Kohler and H. Rahmann (eds), *Gefährdung und Schutz von Gewässern*. Hohenheimer Arbeiten. Stuttgart, Ulmer, 159–68.

Warren, C.E. (1971). *Biology and Water Pollution Control*. Philadelphia: W.B. Saunders.

Watton, A.J. and Hawkes, H.A. (1984). Studies on the effects of sewage effluent on gastropod populations in experimental streams. *Water Res.* **18**, 1235–47.

Wehr, J.D., P.J. Say and B.A. Whitton (1981). Heavy metals in an industrially polluted river, the Team. In P.J. Say and B.A. Whitton (eds), *Heavy Metals in Northern England: Environmental and Biological Aspects*. Department of Botany, University of Durham, 99–107.

Weinberger, P., Greenhalgh, R., Moody, R.P. and Boulton, B. (1978). The fate of fenitrion in model aquatic ecosystems. *Abstract Intecol., 1978*.

Westlake, D.F. (1975). Macrophytes. In B.A. Whitton (ed.), *River Ecology*. Oxford: Blackwell Scientific Publications, 106–28.

Westlake, D.F., Casey, H., Dawson, F.H., Ladle, M., Mann, R.H.K. and Marker, A.F.H. (1972). The chalk-stream ecosystem. In Z. Kajak and A. Hillbricht-Ilowska (eds), *Productivity problems of freshwaters*. Warsaw and Cracow: IBP and Unesco, 615–37.

Westlake, D.F. and Dawson, F.H. (1986). The management of *Rununculus calcareus* by pre-emptive cutting in southern England. *Proc. EWRS/AAB 7th symposium on Aquatic Weeds, 1986*, 395–400.

Whitton, B.A. (ed.) (1975a). *River Ecology*. Oxford: Blackwell Scientific Publications.

Whitton, B.A. (1975b). Algae. In B.A. Whitton (ed.) *River Ecology*. Oxford: Blackwell Scientific Publications, 81–105.

Whitton, B.A. and Say, P.J. (1975). Heavy Metals. In B.A. Whitton (ed.), *River Ecology*. Oxford: Blackwell Scientific Publications, 286–381.

Whitton, B.A., Say, P.J. and Wehr, J.D. (1981). Use of plants to monitor heavy metals in rivers. In P.J. Say and B.A. Whitton (ed.), *Heavy Metals in Northern England: Environmental and Biological Aspects*. Depart-

ment of Botany, University of Durham, 135–45.

Whyle, L. (1962). *Medieval Technology and Social Change*. Oxford: Oxford University Press.

Wilhm, J.F. (1975). Biological indicators of pollution. In B.A. Whitton (ed.), *River Ecology*. Oxford: Blackwell Scientific Publications, 375–402.

Williams, J.H., Kingham, H.G., Cooper, D.J. and Regle, S.J. (1977). Growth regulator injury to tomatoes in Essex, England. *Environ. Pollut.* **12**, 145–57.

Williams, M. (1970). *The draining of the Somerset Levels*. Cambridge: Cambridge University Press.

Wilson, R.S. (1980). Classifying rivers using Chironomid pupal exaviae. In D.A. Murray (ed.), *Chironomidae*. Oxford: Pergamon Press, 209–16.

Wilson, R.S. and McGill, J.D. (1977). A new method of monitoring water quality in a stream receiving sewage effluent, using chironomid pupal exuviae. *Water Research* **11**, 959–62.

Winner, J.M. (1975). Zooplankton. In B.A. Whitton (ed.), *River Ecology*. Oxford: Blackwell Scientific Publications, 155–69.

Wisdom, A.S. (1976). *The Law of Rivers and Watercourses*. London: Shaw.

Wium-Anderson, S., Jorgensen, K.H., Christophersen, C. and Anthoni, U, (1987). Algal growth inhibitors in *Sium erectum* Herds. *Arch. Hydrobiol* **111**, 317–20.

Wolff, C. (1985). Analyse de la vegetation aquatique et de la vegetation riveraine de la Haute-Sure en fonction des perterbations du milieu. Unpublished thesis, Luxembourg.

Wolseley, P.A. (1986). The aquatic macrophyte communities of the ditches and dykes of the Somerset Levels and their relation to management. *Proc. EWRS/AAB 7th Symposium on Aquatic Weeds, 1986*, 407–19.

Wolseley, P.A., Palmer, M.A. and Williams, R. (1984). *The Aquatic Flora of the Somerset Levels and Moors*. Chief Scientist's Team, S.W. Region: Nature Conservancy Council.

Woodiwiss, F.S. (1964). The biological system of stream classification used by the Trent River Board. *Chemy. Indust.* **11**, 443–7.

Woodiwiss, F.S. (1981). *Biological monitoring of surface water quality*. Rep. Commission of the European Communities. Environment and Consumer Protection Service. Contract No. ENV/223/74–EN, 45 pp.

Zelinka, M. and Marvan, P. (1966). Bemerkung zu neuen Methoden der saprobiologischen Wasserbeurteilung. *Verh. Int. Verein. Theor. angew. Limnol.* **16**, 817–22.

Avon and Dorset River Authority (1973). Upper Wylye Investigation, Poole.

Centraal Bureau voor de Statistiek (1976a). *Statistiek van de Scheepvaartbeweging*. Voorburg.

Centraal Bureau voor de Statistiek (1976b). *Onderhoud Watergangen, 1976*. Voorburg.

Methods for the examination of waters and associated

materials (1987). *Methods for the Use of Aquatic Macrophytes for Assessing Water Quality, 1985.* London: HMSO.

Ministry for Transport and Public Works (1975). *The Combat against Surface Water Pollution in the Netherlands, 1975–79.* Prospective multiannual programme. The Netherlands.

Ministry for Transport and Public Works (1980). *Water Action Programme: 1980–4. The Netherlands. The principles and main outline of national policies to maintain the quality of Dutch surface waters.* The Netherlands.

Central Bureau voor de Statistiek: environmental statistics (1987). *Environmental Statistics of The Netherlands, 1987.* The Hague.

North East River Purification Board (1982). *Annual Report.* Aberdeen.

North East River Purification Board (1984). *Annual Report.* Aberdeen.

Partikulaert bundet Stoftransport i vand og jordevos jan (1986). *Nordisk Hydrologick Program NHP, Rapport* **14**, ed. B. Hasholt. København Universitet.

Regional Biology Unit (1976). *Ecological Investigation of the Hampshire Avon.* First Report. Wessex Water Authority, England.

RSPB and RSNC (Royal Society for the Protection of Birds and Royal Society for Nature Conservation) (1977). *Rivers and Wildlife.*

RSPB and RSNC (Royal Society for the Protection of Birds and Royal Society for Nature Conservation) (1980). *Rivers and Wildlife.*

Severn-Trent Water (1988). *Water Quality 1987/88.* Solihull.

Sønderjyllands Amstkommune (1982a). *Undersøgelse of Vidå med hensyn til okkerforurening og fiskebiologisk tilstand, 1979–80.* Hovedrapport, Bilag A.

Sønderjyllands Amstkommune (1982b). *Undersøgelse of Vidå med hensyn til okkerforurening og fiskebiologisk tilstand, 1979–80.* Vandekemiscke analyse resultater og fiskearternes ud bredlelse. Bilag B.

Sønderjyllands Amstkommune (1982c). *Undersøgelse of Vidå med hensyn til okkerforurening og fiskebiologisk tilstand, 1979–80.* Elfiskeblanketter.

Sønderjyllands Amstkommune (1982d). *Undersøgelse of okkerindholdets inflydelse pa invertebratfaunaen i vidå system et 1979–80.* Teknisk Forvaltning, Miljøafdelingen.

Sønderjyllands Amstkommune (1984). *Grødeskaerings-forsøg i surbaek 1982.* Miljøafdelingen og natur-forvaltnings-projektet.

Sønderjyllands Amstkommune (1986). *Fiskebestandene i vandløbene i Sønderjyllands amstkommune 1975–85.* Miljø-og Vandløbsvaesenet.

Sønderjyllands Amstkommune (1987). *Teknisk Forvaltning. Vandløbenes Forureningstilstand 1985–86.* Miljø-og Vandløbsvaesenet.

Sønderjyllands Amstkommune (1988a). *Teknisk Forvaltning. Vandløbenes Forureningstilstand 1987–88.* Miljø-og Vandløbsvaesenet.

Sønderjyllands Amstkommune (1988b). *Teknisk Forvaltning. Grødeundersøgelser i sønderjyske vandlobe, 1986.*

Sønderjyllands Amstråd (1978). *Vandløbenes forureningstilstand i Sønderjyllands Amstkommune, 1978. Teknisk forvaltning.*

Staatsministerium des Innern (1987). *Gewässergüte-karte Bayern, 1986.* Oberste Baubehörde im Bayer, RB/03B/87/12, Munchen.

U.E.R. d'Ecologie-Institut Européen d'Ecologie (1979). Cloitre des Recollets, 57 000 Metz France. Avenant au programme de recherche no. 212.77,1. ENVF Rapport préliminaire présenté lors de la réunion de la Commission Eau des Communatés européennes. **I.** Etude Ecologique et biologique des populations benthiques de Mollusques, ainsi que de leur interét du point des vue écotoxiologique 'in situ'. **II.** Ecotoxiologie ii. 1. Ecohistopathologie; ii. 2. Ecotoxicologie in vitro. Unpublished Report. Commission of the European Communities.

Watch on Stream (1987). Watch Trust for Environmental education, Lincoln.

Water Research Centre (1977). *Water Purification in the EEC. A State-of-the-art Review.* Oxford: Pergamon Press for the Commission of the European Communities.

Water Research Centre (1984) with 1987 Addendum. *An inventory of polluting substances which have been identified in various freshwater effluent discharges, aquatic animals and plants, and bottom sediments.* Concerted action analysis of organic micropollutants in water (COST 64b bis). Compiled by Water Research Centre, Stevenage. Commission of the European Communities.

Yorkshire Water and WRC (1987). Colour in Upland Waters. *Proc. of a Workshop held at Yorkshire Water, Leeds, 29 September, 1987.*

General index

Species names index

Note: This index lists species cited in the text and tables. Species cited within figures are not included here. For macrophyte patterns see river maps.

DATE DUE

FEB 1 0 1997		
DEC		
MAY 0 5 1998		
DEC 0 8 1998		
GAYLORD		

GAYLORD